D1600253

Introduction to Survey Quality

Introduction to Survey Quality

PAUL P. BIEMER

RTI International and the Odum Institute
for Research in Social Sciences at the
University of North Carolina at Chapel Hill

LARS E. LYBERG

Statistics Sweden

WILEY-INTERSCIENCE

A JOHN WILEY & SONS PUBLICATION

For general information on our other products and services please contact our Customer Care Department within the U.S. at 877-762-2974, outside the U.S. at 317-572-3993 or fax 317-572-4002.

Wiley also publishes its books in a variety of electronic formats. Some content that appears in print, however, may not be available in electronic format.

Library of Congress Cataloging-in-Publication Data Is Available

ISBN 0-471-19375-5

Printed in the United States of America

10 9 8 7 6 5 4 3 2

To Judy and Lilli

Contents

Preface

Survey research is a thriving industry worldwide. Rossi et al. (1983) estimated the gross income of the industry in the United States alone to be roughly $5 billion, employing 60,000 persons. There are no current estimates, but the market for survey work has only continued to grow in the last two decades. Accompanying this growth are revolutionary breakthroughs in survey methodology. The field of cognitive psychology has dramatically changed how survey researchers approach the public to request participation in surveys, and how they design questionnaires and interpret survey findings. There have also been breakthroughs in computer technology, and these have transformed the way data are collected. In addition, the use of new survey data quality evaluation techniques have provided more information regarding the validity and reliability of survey results than was previously thought possible. Today more than ever, the collection of survey data is both an art and a well-developed science.

Simultaneously, the industry has become increasingly competitive. Data users and survey sponsors are more and more demanding of survey organizations to produce higher-quality data for lower survey costs. In response to these demands, survey organizations have developed sophisticated data collection and data processing procedures which are complex and highly optimized. This high level of technical sophistication and complexity has created a demand for survey workers at all levels who are knowledgeable of the best survey approaches and can implement these approaches in actual practice. Because very few survey workers are academically trained in survey research, survey organizations are seeking postgraduate training in state of the art survey methodology for many of their employees. Unfortunately, the evolution of academic training in survey methods is lagging behind the growth of the industry. In the United States and elsewhere, there are few degree-granting programs in survey methodology, and the course work in survey methods is sparse and inaccessible to many survey workers (see Lyberg, 2002). Further, there are few alternative sources of training in the practical methods of survey research.

One resource that is available to almost everyone is the survey methods literature. A number of professional journals that report on the latest findings in survey methodology can be found at university and most corporate libraries. Unfortunately, much of the literature is considered incomprehensible by many survey workers who have no formal training in survey research or statistics. The terminology used and knowledge of survey methods assumed in the literature can present major obstacles for the average worker wishing to advance his or her knowledge and career by self-study.

Noting this knowledge gap in our own organizations and the paucity of resources to fill it, we decided that an introductory textbook in survey methodology is needed. The book should expose the beginning student to a wide range of terms, concepts, and methods often encountered when reading the survey methods literature. In addition, it was our intention that the book treat a number of advanced topics, such as nonsampling error, mean squared error, bias, reliability, validity, interviewer variance, confidence intervals, and error modeling in nontechnical terms to be accessible to the survey worker with little formal training in survey methodology or statistics.

Thus, the goal of the book is to address the need for a nontechnical, comprehensive introduction to the concepts, terminology, notation, and models that one encounters in reading the survey methods literature. The specific objectives of the book are:

1. To provide an overview of the basic principles and concepts of survey measurement quality, with particular emphasis on sampling and non-sampling error

2. To develop the background for continued study of survey measurement quality through readings in the literature on survey methodology

3. To identify issues related to the improvement of survey measurement quality that are encountered in survey work and to provide a basic foundation for resolving them

The target audience for the book is persons who perform tasks associated with surveys and may work with survey data but are not necessarily trained survey researchers. These are survey project directors, data collection managers, survey specialists, statisticians, data processors, interviewers, and other operations personnel who would benefit from a better understanding of the concepts of survey data quality, including sampling error and confidence intervals, validity, reliability, mean squared error, cost–error trade-offs in survey design, nonresponse error, frame error, measurement error, specification error, data processing error, methods for evaluating survey data, and how to reduce these errors by the best use of survey resources.

Another audience for the book is students of survey research. The book is designed to serve as a course text for students in all disciplines who may be involved in survey data collection, say as part of a master's or Ph.D. thesis, or later in their careers as researchers. The content of the book, appropriately supplemented with readings from the list of references, provides ample material for a two- or three-credit-hour course at either the undergraduate or graduate level.

The book is not designed to provide an in-depth study of any single topic, but rather, to provide an introduction to the field of survey measurement quality. It includes reviews of well-established as well as recently developed principles and concepts in the field and examines important issues that are still unresolved and which are being actively pursued in the current survey methods literature.

The book spans a range of topics dealing with the quality of data collected through the survey process. *Total survey error*, as measured by the *mean squared error* and its component parts, is the primary criterion for assessing the quality of the survey data. Chapter 1 traces the origins of survey research and introduces the concept of survey quality and data quality. Chapter 2 provides a nontechnical discussion of how data quality is measured and the criteria for optimizing survey design subject to the constraints of costs and timeliness. This chapter provides the essential concepts for data quality that are used throughout the book.

Then the major sources of survey error are discussed in some detail. In particular, we examine (1) the origins of each error source (i.e., its root causes), (2) the most successful methods that have been proposed for reducing the errors emanating from these error sources, and (3) methods that are most often used in practice for evaluating the effects of the source on total survey error. Chapter 3 deals with coverage and nonresponse error, Chapter 4 with measurement error in general, Chapter 5 with interviewer error, Chapter 6 with data collection mode, and Chapter 7 with data processing error. In Chapter 8 we summarize the basic approaches for evaluating data quality. Chapter 9 is devoted to the fundamentals of sampling error. Finally, in Chapter 10 we integrate the many concepts used throughout the book into lessons for practical survey design.

The book covers many concepts and ideas for understanding the nature of survey error, techniques for improving survey quality and, where possible, their cost implications, and methods for evaluating data quality in ongoing survey programs. A major theme of the book is to introduce readers to the *language* or terminology of survey errors so that they can continue this study of survey methodology through self-study and other readings of the literature.

Work on the book spanned a four-year period; however, the content was developed over a decade as part of a short course one of us (P.P.B.) has taught in various venues, including the University of Michigan Survey Research Center and the University of Maryland–University of Michigan/Joint Program in Survey Methodology. During these years, many people have contributed to the book and the course. Here we would like to acknowledge their contributions and to offer our sincere thanks for their efforts.

We would like to acknowledge the support of our home institutions (RTI International and Statistics Sweden) for their understanding and encouragement throughout the entire duration of the project. Certainly, working on this book on nights and weekends for four years was a distraction from our day jobs. Particularly toward the end of the project, our availability for work outside normal working hours was quite limited as we raced to finalize the draft chapters. We would also like to thank RTI International and the U.S. National Agricultural Statistics Service for their financial support, which made it possible for one of us (P.P.B.) to take some time away from the office to work on the book. They also paid partially for a number of trips to Europe and the United States, as well as for living expenses for the trips for both of us on both continents.

A number of people reviewed various chapters of the book and provided excellent comments and suggestions for improvement: Fritz Scheuren, Lynne Stokes, Roger Tourangeau, David Cantor, Nancy Mathiowetz, Clyde Tucker, Dan Kasprzyk, Jim Lepkowski, David Morganstein, Walt Mudryk, Peter Xiao, Bob Bougie, and Peter Lynn. Certainly, their contributions improved the book substantially. In addition, Rachel Caspar, Mike Weeks, Dick Kulka, and Don Camburn provided support in various capacities. We also thank the many students who offered suggestions on how to improve the course, which also affected the content of the book substantially.

Finally, we thank our families for their sacrifices during this period. There were many occasions when we were not available or able to join them for leisuretime activities and family events because work needed to progress on the book. Many thanks for putting up with us for these long years and for their encouragement and stoic acceptance of the situation, even though it was not as short-lived as we thought initially.

Research Triangle Park, NC   PAUL P. BIEMER
Stockholm, Sweden   LARS E. LYBERG
June 2002

Introduction to Survey Quality

CHAPTER 1

The Evolution of Survey Process Quality

Statistics is a science consisting of a collection of methods for obtaining knowledge and making sound decisions under uncertainty. Statistics come into play during all stages of scientific inquiry, such as observation, formulation of hypotheses, prediction, and verification. This collection of methods includes descriptive statistics, design of experiments, correlation and regression, multivariate and multilevel analysis, analysis of variance and covariance, probability and probability models, chance variability and chance models, and tests of significance, to mention just a few of the more common statistical methods.

In this book we treat the branch of statistics called *survey methodology* and, more specifically, *survey quality*. To provide a framework for the book, we define both a survey and survey quality in this chapter. We begin with the definition of a survey and in Section 1.2 describe some types of surveys typically encountered in practice today. Our treatment of surveys concludes with a short history of the evolution of survey methodology in social–economic research (Section 1.3). The next three sections of this chapter deal with the very difficult to define concept of quality; in particular, survey quality. We describe briefly what quality means in the context of survey work and how it has co-evolved with surveys, especially in recent years. What has been called a *quality revolution* is treated in Section 1.4. Quality in statistical organizations is discussed in Section 1.5. The measurement and improvement of process quality in a survey context are covered in Sections 1.6 and 1.7, respectively. Finally, we summarize the key concepts of this chapter in Section 1.8.

1.1 THE CONCEPT OF A SURVEY

The American Statistical Association's Section on Survey Research Methods has produced a series of 10 short pamphlets under the rubric "What Is a Survey?" (Scheuren, 1999). That series covers the major survey steps and high-

1

lights specific issues for conducting surveys. It is written for the general public and its overall goal is to improve survey literacy among people who participate in surveys, use survey results, or are simply interested in knowing what the field is all about.

Dalenius (1985) provides a definition of survey comprising a number of study prerequisites that must be in place. According to Dalenius, a research project is a survey only if the following list of prerequisites is satisfied:

1. *A survey concerns a set of objects comprising a population.* Populations can be of various kinds. One class of populations concerns a finite set of objects such as individuals, businesses, or farms. Another class of populations concerns a process that is studied over time, such as events occurring at specified time intervals (e.g., criminal victimizations and accidents). A third class of populations concerns processes taking place in the environment, such as land use or the occurrence of wildlife species in an area. The population of interest (referred to as the *target population*) must always be specified. Sometimes it is necessary to restrict the study for practical or financial reasons. For instance, one might have to eliminate certain remote areas from the population under study or confine the study to age groups that can be interviewed without obvious problems. A common restriction for the study of household populations is to include only these who are noninstitutionalized (i.e., persons who are not in prison, a hospital, or any other institution, except those in military service), of age 15 to 74, and who live in the country on a specific calendar day.

2. *The population under study has one or more measurable properties.* A person's occupation at a specific time is an example of a measurable property of a population of individuals. The extent of specified types of crime during a certain period of time is an example of a measurable property of a population of events. The proportion of an area of land that is densely populated is an example of a measurable property of a population concerning plane processes that take place in the environment.

3. *The goal of the project is to describe the population by one or more parameters defined in terms of the measurable properties. This requires observing (a sample of) the population.* Examples of parameters are the proportion of unemployed persons in a population at a given time, the total revenue of businesses in a specific industry sector during a given period, and the number of wildlife species in an area at a given time.

4. *To get observational access to the population, a frame is needed (i.e., an operational representation of the population units, such as a list of all objects in the population under study or a map of a geographical area).* Examples of frames are business and population registers, maps where land has been divided into areas with strictly defined boundaries, or all *n*-digit numbers which can be used to link telephone numbers to individuals. Sometimes no frame is readily accessible, and therefore it has to be constructed via a listing

procedure. For general populations this can be a tedious task, and to select a sample that is affordable, a multistage sampling procedure is combined with the listing by first selecting a number of areas using a map and then for sampled areas having field staff listing all objects in the areas sampled. For special populations, for instance the population of professional baseball players in the United States, one would have to combine all club rosters into one huge roster. This list then constitutes the frame that will be used to draw the sample. In some applications there are a number of incomplete listings or frames that cover the population to varying degrees. The job then is to combine these into one frame. Hartley (1974) developed a theory for this situation referred to as *multiple frame theory*.

5. *A sample of objects is selected from the frame in accordance with a sampling design that specifies a probability mechanism and a sample size*. The sampling literature describes an abundance of sampling designs recommended for various situations. There are basically two design situations to consider. The first involves designs that make it easier to deal with the necessity of sampling in more than one stage and measuring only objects identified in the last stage. Such designs ensure that listing and interviewer travel is reduced while still making it possible to estimate population parameters. The second type of design is one where we take the distribution of characteristics in the population into account. Examples of such situations are skewed populations that lend themselves to stratified sampling, or *cutoff sampling*, where measurements are restricted to the largest objects and ordered populations that are sampled efficiently by systematic sampling of every nth object. Every sampling design must specify selection probabilities and a sample size. If selection probabilities are not known, the design is not statistically valid.

6. *Observations are made on the sample in accordance with a measurement process (i.e., a measurement method and a prescription as to its use)*. Observations are collected by a mechanism referred to as the *data collection mode*. Data collection can be administered in many different ways. The unit of observation is, for instance, an individual, a business, or a geographic area. The observations can be made by means of some mechanical device (e.g., electronic monitors or meters that record TV viewing behavior), by direct observation (e.g., counting the number of wildlife species on aerial photos), or by a questionnaire (observing facts and behaviors via questions that reflect conceptualizations of research objectives) administered by special staff such as interviewers or by the units themselves.

7. *Based on the measurements, an estimation process is applied to compute estimates of the parameters when making inference from the sample to the population*. The observations generate data. Associated with each sampling design are one or more estimators that are computed on the data. The estimators may be based solely on the data collected, but sometimes the estimator might include other information as well. All estimators are such that they include sample weights, which are numerical quantities that are used to correct the

Table 1.1 Dalenius's Prerequisites for a Survey

Criterion	Remark
1. A survey concerns a set of objects comprising a population.	Defining the target population is critical both for inferential purposes and to establish the sampling frame.
2. The population under study has one or more measurable properties.	Those properties that best achieve the specific goal of the project should be selected.
3. The goal of the project is to describe the population by one or more parameters defined in terms of the measurable properties.	Given a set of properties, different parameters are possible, such as averages, percentiles, and totals, often broken down for population subgroups.
4. To gain observational access to the population a frame is needed.	It is often difficult to develop a frame that covers the target population completely.
5. A sample of units is selected from the frame in accordance with a sampling design specifying a probability mechanism and a sample size.	The sampling design always depends on the actual circumstances associated with the survey.
6. Observations are made on the sample in accordance with a measurement process.	Data collection can be administered in many different ways. Often, more than one mode must be used.
7. Based on the measurements an estimation process is applied to compute estimates of the parameters with the purpose of making inferences from the sample to the population.	The error caused by a sample being observed instead of the entire population can be calculated by means of variance estimators. The resulting estimates can be used to calculate confidence intervals.

Source: Dalenius (1985).

sample data for its potential lack of representation of the population. The error in the estimates due to the fact that a sample has been observed instead of the entire population can be calculated directly from the data observed using variance estimators. Variance estimators make it possible to calculate standard errors and confidence intervals; however, not all the errors in the survey data are reflected in the variances.

In Table 1.1 we have condensed Dalenius's seven prerequisites or criteria. Associated with each criterion is a short remark. These seven criteria define the concept of a survey. If one or more of them are not fulfilled, the study cannot be classified as a survey, and consequently, sound inference to the target population cannot be made from the sample selected. It is not uncommon, however, to find studies that are labeled as surveys but which have serious shortcomings and whose inferential value should be questioned.

Typical study shortcomings that can jeopardize the inference include the following:

- The target population is redefined during the study, due to problems in finding or accessing the units. For instance, the logistical problems or costs of data collection are such that it is infeasible to observe objects in certain areas or in certain age groups. Therefore, these objects are in practice excluded from the study, but no change is made regarding the survey goals.
- The selection probabilities are not known for all units selected. For instance, a study might use a sampling scheme in which interviewers are instructed to select respondents according to a quota sampling scheme, such that the final sample comprises units according to prespecified quantities. Such sampling schemes are common when studying mall visitors and travelers at airports. Self-selection is a very common consequence of some study designs.

For example, in a hotel service study a questionnaire is placed in the hotel room and the guest is asked to fill it out and leave the questionnaire at the front desk. Relatively few guests (perhaps only 10% or less) will do this; nevertheless, statements such as "studies show that 85% of our guests are satisfied with our services" are made by the hotel management. The percentage is calculated as the number of satisfied guests (according to the results of the questionnaire) divided by the number of questionnaires left at the front desk. No provision is made for the vast majority of guests who do not complete the questionnaire.

Obviously, such estimates are potentially biased because there is no control over who completes the survey and who does not. Other examples include the daily Web or e-mail questions that appear in newspapers and TV shows. Readers and viewers are urged to get on the Internet and express their opinions. The results are almost always published without any disclaimers and the public might believe that the results reflect the actual characteristics in the population. In the case of Internet surveys publicized by newspapers or on TV, self-selection of the sample occurs in at least four ways: (1) the respondent must be a reader or a viewer even to have an opportunity to respond; (2) he or she must have access to the Internet; (3) he or she must be motivated to get on the Internet; and (4) he or she must usually have an opinion, since "don't know" and "no opinion" very seldom appear as response categories. Quite obviously, this kind of self-selection does not resemble any form of random selection.

- Correct estimation formulas are not used. The estimation formulas used in some surveys do not have the correct sample weights or there is no obvious correspondence between the design and the variance formulas. Often, survey practitioners apply "off-the-shelf" variance calculation packages that are not always appropriate for the sampling design. Others might use a relatively complex sampling design, but they calculate the variance as if the sampling design were not complex.

These are examples of violations of the basic criteria or prerequisites and should not be confused with survey errors that stem from imperfections in the design and execution of a well-planned scientific survey. This book deals with the latter (i.e., error sources, error structures, how to prevent errors, and how to estimate error sizes). The term *error* sounds quite negative to many people, especially producers of survey data. Errors suggest that mistakes were made. Some prefer a more positive terminology such as *uncertainties* or *imperfections* in the data, but these are really the same as our use of the term *errors*. During recent decades the term *quality* has become widely used because it encompasses all features of the survey product that users of the data believe to be important.

Surveys can suffer from a number of shortcomings that can jeopardize statistical inference, including:

- Changing the definition of the target population during the survey
- Unknown probabilities of selection
- Incorrect estimation formulas and inferences

1.2 TYPES OF SURVEYS

There are many types of surveys and survey populations (see Lyberg and Cassel, 2001). A large number of surveys are one-time surveys that aim at measuring population characteristics, behaviors, and attitudes. Some surveys are continuing, thereby allowing estimation of change over time. Often, a survey that was once planned to be a one-time endeavor is repeated and then turned gradually into a continuing survey because of an enhanced interest among users to find out what happens with the population over time.

Examples of continuing survey programs include official statistics produced by government agencies and covering populations of individuals, businesses, organizations, and agricultural entities. For instance, most countries have survey programs on the measurement of unemployment, population counts, retail trade, livestock, crop yields, and transportation. Almost every country in the world has one or more government agencies (usually national statistical institutes) that supply decision makers and other users with a continuing flow of information on these and other topics. This bulk of data is generally called *official statistics*.

There are also large organizations that have survey data collection or analysis of survey data as part of their duties, such as the International Monetary Fund (IMF); the United Nations (UN) and its numerous suborganizations, such as the Food and Agricultural Organization (FAO) and the International

Labour Office (ILO); and all central banks. Some organizations have as their job coordinating and supervising data collection efforts, such as Eurostat, the central office for all national statistical institutes within the European Union (EU), its counterpart in Africa, Afristat, and the Office for Management and Budget (OMB), overseeing and giving clearance for many data collection activities in the United States.

Other types of data collection are carried out by academic organizations and private firms. Sometimes, they take on the production of official statistics when government agencies see that as fitting. The situation varies among countries. In some countries no agency other than the national statistical institute is allowed to carry out the production of official statistics, whereas in others it is a feasible option to let some other survey organization do it. Private firms are usually contracted by private organizations to take on surveys covering topics such as market research, opinion polls, attitudes, and characteristics of special populations. The survey industry probably employs more than 130,000 people in the United States alone, and for the entire world, the figure is much larger. For example, in Europe, government statistical agencies may employ as few as a half-dozen or so (in Luxembourg) and several thousands of staff.

The facilities to conduct survey work vary considerably throughout the world. At the one extreme, there are countries with access to good sampling frames for population statistics, advanced technology in terms of computer-assisted methodology as well as a good supply of methodological expertise. However, developing countries and countries in transition face severe restrictions in terms of advanced methodology, access to technology such as computers and telephones, or sufficiently skilled staff and knowledgeable respondents. For instance, in most developing countries there are no adequate sampling frames, and telephone use is quite low, obviating the use of the telephone for survey contacts. Consequently, face-to-face interviewing is the only practical way to conduct surveys. The level of funding is also an obstacle to good survey work in many parts of the world, not only in developing countries.

There are a number of supporting organizations that help improve and promote survey work. There are large interest organizations such as the Section on Survey Research Methods (SRM) of the American Statistical Association (ASA), the International Association of Survey Statisticians (IASS) of the International Statistical Institute (ISI), and the American Association for Public Opinion Research (AAPOR). Many other countries have their own statistical societies with subsections on survey-related matters. Many universities worldwide conduct survey research. This research is by no means confined to statistical departments, but takes place in departments of sociology, psychology, education, communication, and business as well. Over the years, the field of survey research has witnessed an increased collaboration across disciplines that is due to a growing realization that survey methodology is truly a multidisciplinary science.

Since a critical role of the survey industry is to provide input to world leaders for decision making, it is imperative that the data generated be of such quality that they can serve as a basis for informed decisions. The methods available to assure good quality should be known and accessible to all serious survey organizations. Today, this is unfortunately not always the case, which is our primary motive and purpose for writing this book.

1.3   BRIEF HISTORY OF SURVEY METHODOLOGY

Surveys have roots that can be traced to biblical times. Madansky (1986) provides an account of censuses described in the Old Testament, which the author refers to as "biblical censuses." It was very important for a country to know approximately how many people it had for both war efforts and taxation purposes. Censuses were therefore carried out in ancient Egypt, Rome, Japan, Greece, and Persia. It was considered a great indication of status for a country to have a large population. For example, as late as around 1700, a Swedish census of population revealed that the Swedish population was much smaller than anticipated. This census result created such concern and embarrassment that the counts were declared confidential by the Swedish government. The government's main concern was a fear that disclosure of small population size might trigger attacks from other countries.

Although survey sampling had been used intuitively for centuries (Stephan, 1948), no specific theory of sampling started to develop until about 1900. For instance, estimating the size of a population when a total count in terms of a census was deemed impossible had occupied the minds of many scientists in Europe long before 1900. The method that was used in some European countries, called *political arithmetic*, was used successfully by Graunt and Eden in England between 1650 and 1800. The political arithmetic is based on ideas that resemble those of ratio estimation (see Chapter 9). By means of birthrates, family sizes, average number of people per house, and personal observations of the scientists in selected districts, it was possible to estimate population size. Some of these estimates were later confirmed by censuses as being highly accurate. Similar attempts were made in France and Belgium. See Fienberg and Tanur (2001) and Bellhouse (1998) for more detailed discussions of these early developments.

The scientific basis for survey methodology has its roots in mathematics, probability theory, and mathematical statistics. Problems involving calculation of number of permutations and number of combinations were solved as early as the tenth century. This work was a prerequisite for probability theory, and in 1540, Cardano defined *probability* in the classical way as "the number of successful outcomes divided by the number of possible outcomes," a definition that is still taught in many elementary statistics courses. In the seventeenth century, Galilei, Fermat, Pascal, Huygens, and Bernoulli developed probability theory. During the next 150 years, scientists such as de Moivre, Laplace, Gauss, and Poisson propelled mathematics, probability, and statistics forward. Limit theorems and distributional functions are among the great contributions during this era, and all those scientists have given their names to some of today's statistical concepts.

The prevailing view in the late nineteenth century and a few decades beyond was that a sample survey was seen as a substitute for a total enumeration or a census. In 1895, a Norwegian by the name of Kiear submitted a proposal to the ISI in which he advocated further investigation into what he called *representative investigations*. The reason that this development was at all interesting was the same faced by Graunt and others. Total enumeration was often impossible because of the elaborated nature of such endeavors in terms of costs but also that a need for detail could not be fulfilled. Kiear was joined by Bowley in his efforts to try to convince the ISI about the usefulness of the representative method. Kiear argued for sampling at three ISI meetings, in 1897, 1901, and 1903. A decade later, Bowley (1913) tried to connect statistical theory and survey design. In a number of papers he discussed random sampling and the need for frames and definitions of primary sampling units. He outlined a theory for purposive selection and provided guidelines for survey design. It should be noted that neither Kiear nor Bowley advocated randomization in all stages. They first advocated a mixture of random and purposive selection.

For instance, one recommendation was that units and small clusters should be chosen randomly or haphazardly, whereas large clusters should be chosen purposively. Independent of these efforts, a very similar development was taking place in Russia led by Tschuprow, who developed formulas for estimates under stratified random sampling. In the mid-1920s the ISI finally agreed to promote an extended investigation and use of these methods. Details on how to achieve representativeness and how to measure the uncertainty associated with using samples instead of total enumerations were not at all clear, though. It would take decades until sampling was fully accepted as a scientific method, at least in some countries.

Some of the results obtained by Tschuprow were developed by Neyman. It is not clear whether Neyman had access to Tschuprow's results when he outlined a theory for sampling from finite populations. The results are to some extent overlapping, but Neyman never referred to the Russian when presenting his early works in the 1920s.

In subsequent years, development of a sample survey theory picked up considerable speed (see Chapter 9). Neyman (1934) delivered a landmark paper "On the Two Different Aspects of the Representative Method: The Method of Stratified Sampling and the Method of Purposive Selection." In his paper Neyman stressed the importance of random sampling. He also dealt with optimum stratification, cluster sampling, the approximate normality of linear estimators for large samples, and a model for purposive selection. His writings constituted a major breakthrough, but it took awhile for his ideas to gain prominence. Neyman's work had its origin in agricultural statistics, and this was also true for the work on experimental design that was conducted by Fisher at Rothamsted. Fisher's work, and his ideas on random experiments were of great importance for survey sampling. Unfortunately, as a result of a major feud between Neyman and Fisher—two of the greatest contributors to statistical theory of all time—development of survey sampling as a scientific discipline was perhaps considerably impaired.

In the 1930s and 1940s most of the basic survey sampling methods used today were developed. Fisher's randomization principle was used and verified in agricultural sampling and subsampling studies. Neyman introduced the theory of confidence intervals, cluster sampling, ratio estimation, and two-phase sampling (see Chapter 9).

Neyman was able to show that the sampling error could be measured by calculating the variance of the estimator. Other error sources were not acknowledged particularly. The first scientist to formally introduce other error estimates was the Indian statistician Mahalanobis. He developed methods for the estimation of errors introduced by field-workers collecting agricultural data. He was able to estimate these errors by a method called *interpenetration*, which is used to this day to estimate errors generated by interviewers, coders, and supervisors who are supposed to have a more-or-less uniform effect on the cases they are involved with, an effect that typically is very individual.

The concepts of sampling theory were developed and refined further by these classical statisticians as well as those to follow, such as Cochran, Yates, Hansen, and others. It was widely known by the 1940s, that sampling error was not synonymous with total survey error. For example, we have already mentioned Mahalanobis's discovery about errors introduced by field-workers. In the 1940s, Hansen and his colleagues at the U.S. Bureau of the Census presented a model for total survey error. In the model, which is usually called the U.S. Census Bureau survey model, the total error of an estimate is measured as the mean squared error of that estimate. Their model provides a means for estimating variance and bias components of the mean squared error using various experimental designs and study schemes. This model showed explicitly that sampling variance is just one type of error and that survey error estimates based on the sampling error alone will lead to underestimates of the total error. The model is described in a paper by Hansen et al. (1964) and the study schemes in Bailar and Dalenius (1969).

Although mathematical statisticians are trained to measure and adjust for error in the data, generally speaking, they are not trained for controlling, reducing, and preventing nonsampling errors in survey work. A reduction in nonsampling errors requires thoughtful planning and careful survey design, incorporating the knowledge and theories of a number of disciplines, including statistics, sociology, psychology, and linguistics. Many error sources concern cognitive and communicative phenomena, and therefore it is not surprising that much research on explaining and preventing nonsampling errors takes place in disciplines other than statistics. [See O'Muircheartaigh (1997) for an overview of developments across these disciplines.]

In the early developments of sampling theory, bias was seldom a concern other than as a technical issue related to characteristics of the estimator itself, like the "technical bias" associated with a ratio estimator (Chapter 9). Early statisticians were not particularly interested in models of response effects, the interaction between the interviewer and the respondent, the complexity of the task to respond or measure, and the realism in variables (i.e., the extent to which measured variables relate to constructs they are meant to describe). Other disciplines assumed that responsibility. There are, for instance, some very early writings on the effects of question wording, such as Muscio (1917). Formal attitude scales were developed by Likert and others during the period 1920–1950. In the 1940s, extensive academic research was conducted on survey instruments when numerous experiments were carried out to identify the strengths and weaknesses of various questionnaire designs.

O'Muircheartaigh also gives an example of an interesting debate in the survey methods literature concerning the roles of interviewers and respondents. The early view of the interviewer held that information was either available or it was not. When it was, it could easily be collected from respondents. Thus, years ago, the primary technique for interviewing respondents was conversational in nature. In one form of conversational interviewing, the interviewer conversed informally with the respondent without necessarily taking notes at the time of the conversation and summarized the information from the interview later. Another form in use was more formal, with the interviewer equipped with a set of prespecified questions that were asked in order as the interviewer took notes.

Interviewer influences on the responses were usually not a concern. Interviewing was primarily a method used in social surveys that, in those days, were generally not held in high regard, due to the lack of control and standardization. Standardization eventually came into greater acceptance. In 1942, Williams provided a set of basic instructions for interviewers at the National Opinion Research Center (NORC) in the United States. In 1946 a discussant at a conference in the United States identified the ideal interviewer as "a married woman, 37 years old, neither adverse to nor steamed up about politics, and able to understand and follow instructions." To some extent, this image of the interviewer, at least as a woman, has prevailed in some interviewing organizations to this day.

There was very little said about respondents' role in the early writings. Issues that deal with interviewer–respondent interaction were not studied until 1968, when schemes for coding these interactions were presented by Cannell and Kahn. In fact, the respondent was often viewed as an obstacle in the data collection process, and this attitude can also be seen today in some survey programs, especially in some of those that are backed up by laws stipulating mandatory participation. A few historical papers, in addition to those already mentioned, include Fienberg and Tanur (1996), Converse (1986), Kish (1995), Hansen et al. (1985), and Zarkovich (1966).

Many new developments have influenced today's survey work. For instance, we now have a sampling theory using model-assisted methods. An example of a modern textbook reflecting these new sampling methods is that of Särndal et al. (1991). Also, response errors can now be incorporated directly into statistical models, and issues of cognition during the interview continue to be studied. There is a continued interest in trying to understand the response process, and new knowledge has increased our ability to improve data collection modes. The development of new technology include computer-assisted data collection, scanning of forms, and using software that makes it possible to convert complex verbal descriptions automatically into numerical codes.

However, to this day, many of the basic problems associated with survey work remain despite vigorous research efforts. These basic problems include the presence of survey errors, the lack of adequate measurement tools and resources to handle the errors, and the lack of understanding by some survey producers, survey users, and survey sponsors as to how errors affect survey estimates and survey analyses. There is need for improved quality in survey work.

LANDMARK EVENTS IN THE HISTORY OF SURVEYS

- The first guidelines for survey design were developed early in the twentieth century.
- Neyman's landmark paper on the representative method was published in 1934.
- In the 1940s, Mahalanobis developed the method of interpenetration of interviewer assignments to estimate errors made by survey field-workers.
- In the early 1960s, Hansen and others developed the first survey model.

1.4 THE QUALITY REVOLUTION

During the last couple of decades, society has witnessed what has been called by its advocates, a *quality revolution* in society. Deming, Juran, Taguchi, Crosby, Ishikawa, Joiner, and others have stressed the need for better quality and how to improve it. For instance, Deming (1986) presented his 14 points and the seven deadly diseases, Juran and Gryna (1980) had their spiral of progress in quality, Taguchi (1986) developed a type of designed experiment where variation is emphasized, Crosby advocated avoiding problems rather than solving them, Ishikawa (1982) listed the seven quality control tools (data collection, histogram, Pareto diagram, fishbone diagram, stratification, plotting, and control charts), and Joiner (Scholtes et al., 1994) emphasized the triangle (quality, scientific approach, and teamwork). Much earlier, Shewhart had invented the control chart, and Dodge and Romig (1944) had invented acceptance sampling, thereby starting the development of *statistical process control*.

Unquestionably, these "bellwether innovators" have started a quality industry that today takes many guises. Despite legitimate criticisms that these new and not so new ideas have been oversold (Brackstone, 1999; Scheuren, 2001), there have been some good results, including, for example, the movement away from mass inspection for quality control, the movement toward employee empowerment, greater customer orientation, and increased emphasis on teamwork as opposed to top-down management.

1.5 DEFINITIONS OF QUALITY AND QUALITY IN STATISTICAL ORGANIZATIONS

Quality improvement always implies change, and there is a process for change just as there are processes for car manufacturing and statistics production. Successful organizations know that to stay in business, continuous improvement is essential, and they have developed measures that help them improve. Typically, such organizations have adopted a number of strategies identified as the core values of the organization: values that will help them to change in positive ways.

A survey organization is no different from any other organization as regards the need for continuous improvement. There is need for good quality output, but there is also need for an organization to be nimble and to adjust its processes according to new demands from users. In that sense, how should quality be defined? Since it is a vague concept, there are a number of definitions of quality in use. Perhaps the most general and widely quoted is Juran and Gryna's (1980) definition as simply "fitness for use." However, this definition quickly becomes complex when we realize that whenever there are a variety of uses (as is the case of statistical products), fitness for use must have multiple quality characteristics, where the importance of different characteristics varies among users.

Quality can be defined simply as "fitness for use." In the context of a survey, this translates to a requirement for survey data to be as *accurate* as necessary to achieve their intended purposes, be available at the time it is needed (*timely*), and be *accessible* to those for whom the survey was conducted. Accuracy, timeliness, and accessibility, then, are three *dimensions* of survey quality.

Another definition distinguishes between quality of design and quality of conformance (Juran and Gryna, 1980). An example of *design quality* is how data are presented. A publication with multicolored charts presenting statistical data may be aesthetically superior to monochromatic charts or simple tables. Thus, design quality is said to be higher in the former case. In general, design quality tends to increase costs. *Quality conformance*, on the other hand, is the degree to which a product conforms to its intended use. For surveys, production conformance can be a predetermined margin of error of an estimate of a population parameter. Admittedly, the distinction between design quality and conformance quality is not always obvious.

The quality of a statistical product is a multidimensional concept. Data quality contains components for accuracy, timeliness, richness of detail, accessibility, level of confidentiality protection, and so on. Later in this chapter we will see examples of quality frameworks that are used in official statistics. Traditionally, there has been an emphasis on survey quality being a function of survey error (i.e., data accuracy). However, like other businesses, it has become necessary for survey organizations to work with a much broader definition of quality since users are not just interested in the accuracy of the estimates provided; to varying degrees, they also need data that are relevant, timely, coherent, accessible, and comparable.

Some have argued that accuracy must be foremost. Without accuracy, other quality features are irrelevant. However, the opposite may also be true. Very accurate data are useless if they are released too late or if they are not relevant. Developments during the last decade suggest that statistical organizations have started to change because there are a number of problems associated with a quality concept related solely to accuracy features.

1. Accuracy is difficult and expensive to measure, so much so that it is rarely done in most surveys, at least not on a regular basis. Accuracy is usually defined in terms of total survey error; however, some error sources are impossible to measure. Instead, one has to *assure* quality by using dependable processes, processes that lead to good product characteristics. The basic thought is that product quality is achieved through process quality.

2. The value of postsurvey measures of total survey error is relatively limited. Except for repeated surveys, accuracy estimates have relatively small effects on quality improvement.

3. The mechanical quality control of survey operations such as coding and keying does not easily lend itself to continuous improvement. Rather, it must be complemented with feedback and learning where the survey workers themselves are part of an improvement process.

4. A concentration on estimating accuracy usually leaves little room for developing design quality components.

Twenty to thirty years ago, the user was a somewhat obscure player in the survey process. In most statistical organizations, contacts were not well developed unless there was one distinct user of the survey results (e.g., the survey sponsor). Also, the technology at the time did not allow quick releases or nice presentations of the data. It was not uncommon that census data were released years after the collection had been terminated, and the general attitude among both users and producers was that these things simply took time to carry out. Somewhat ironically, there was often sufficient time to conduct evaluation studies. There are examples of evaluation studies where it was possible to provide an estimate of the accuracy close to the release of the survey data. Also, many organizations had relatively large budgets that allowed them to perform quality control and accuracy studies of specific survey operations. Thus if timeliness and good accessibility were considered almost impossible to achieve, it is no wonder that producers concentrated primarily on data accuracy.

Today, the situation is changed. The funding has been cut for many national statistical institutes, and there are many more actors on the survey market than before. At the same time, technological advances have made it possible to achieve good design quality components. Data processing is fast today, as are data collection and various value-adding activities, such as data analysis and making excerpts from databases. Statistical organizations either have to deliver the entire package in a timely and coherent fashion or risk that a competitor will.

As a consequence of this new situation, more and more statistical organizations throughout the world are now working with quality management models, business excellence models, user orientation, audits, and self-assessments as means to improve their work. The alternative is to risk going out of business. Even the national statistical institutes are at risk. For example, there has been a shift from mass inspection or verification of production and postsurvey evaluation to the use of a process control during production. This movement is fueled by the belief that product quality is achieved through process quality. This *process view* of survey work extends to almost all processes in a survey organization because many processes that support survey work have an effect on the quality of statistics products. Examples of such processes are training, user contacts, proposal writing, benchmarking, project work, contacts with data suppliers, and strategic planning.

A number of statistical organizations have produced documents on how they work with new demands on quality. Australian Bureau of Statistics, Statistics New Zealand, Statistics Netherlands, Statistics Denmark, Statistics Sweden, U.K. Office for National Statistics, U.S. Census Bureau, and U.S. Bureau of Labor Statistics are among national statistical agencies that have produced documents on business plans, strategic plans, or protocols. For instance, Statistics New Zealand (not dated) has produced a number of protocols as a code of practice for the production and release of official statistics. These principles are listed below.

There are also many similar documents in place. For instance, the UN has compiled 10 Fundamental Principles of Official Statistics (United Nations, 1994a), and Franchet (1999) discusses performance indicators for international statistical organizations. Statistics Sweden, in its quality policy, emphasizes objectivity, accessibility, and professionalism. Statistics Canada has a number of policy documents, one being quality guidelines, and another is the corporate planning and program monitoring system (Fellegi and Brackstone, 1999). Statistics Denmark (2000) has released its strategy for the years 2000–2005. The document defines the official status of the institute, key objectives and strategies, and the institute's relationships with the general public and its own staff. In its long-term strategic plan the U.S. Bureau of the Census (1996) defines key strategies to accomplish bureau goals. Also in this document, core business, core staff competencies, and target customers are defined. The U.S. Bureau of Labor Statistics, U.S. National Agricultural Statistics Service, and other U.S. federal statistical agencies have developed similar documents. Most of these are available at the agencies' web sites.

STATISTICS NEW ZEALAND'S CODE OF PRACTICE FOR THE PRODUCTION AND RELEASE OF OFFICIAL STATISTICS

1. The need for a survey must be justified and outweigh the costs and respondent burden for collecting the data.
2. A clear set of survey objectives and associated quality standards should be developed, along with a plan for conducting the many stages of a survey to a timetable, budget, and quality standards.
3. Legislative obligations governing the collection of data, confidentiality, privacy, and its release must be followed.
4. Sound statistical methodology should underpin the design of a survey.
5. Standard frameworks, questions, and classifications should be used to allow integration of the data with data from other sources and to minimize development costs.
6. Forms should be designed so that they are easy for respondents to complete accurately and are efficient to process.
7. The reporting load on respondents should be kept to the minimum practicable.
8. In analyzing and reporting the results of a collection, objectivity and professionalism must be maintained and the data presented impartially in ways that are easy to understand.
9. The main results of a collection should be easily accessible and equal opportunity of access should be available to all users.

A key point in this discussion is that the concept of quality in statistical organizations has changed during the last decade. It seems as if the dominating approach today is built on the ISO8402 norm from 1986, which states that quality is "the totality of features and characteristics of a product or service that bear on its ability to satisfy stated or implied needs." ISO, the International Organization for Standardization, is an organization that develops documented agreements containing technical specifications or other precise criteria to be used consistently as rules, guidelines, or definitions of characteristics, to ensure that materials, products, and processes are fit for their purposes. Quality is an area where ISO has contributed extensively. Thus, in statistical organizations, accuracy is no longer the sole measure of quality. Quality comprises a number of dimensions that reflect user needs. In such a setting, quality can be defined along these dimensions, where accuracy is but one dimension. As an example, Eurostat's quality concept has seven dimensions, as shown in Table 1.2.

Table 1.2   Eurostat's Quality Dimensions

Quality Dimension	Remark
1. Relevance of statistical concept	A statistical product is relevant if it meets user needs. Thus user needs have to be established at the outset.
2. Accuracy of estimates	Accuracy is the difference between the estimate and the true parameter value. Assessing the accuracy is not always possible, due to financial and methodological constraints.
3. Timeliness and punctuality in disseminating results	In our experience this is perhaps one of the most important user needs. Perhaps this is so because this dimension is so obviously linked to an efficient use of the results.
4. Accessibility and clarity of the information	Results are of high value when they are easily accessible and available in forms suitable to users. The data provider should also assist the users in interpreting the results.
5. Comparability	Reliable comparisons across space and time are often crucial. Recently, new demands on cross-national comparisons have become common. This in turn puts new demands on developing methods for adjusting for cultural differences.
6. Coherence	When originating from a single source, statistics are coherent, in that elementary concepts can be combined in more complex ways. When originating from different sources, and in particular from statistical studies of different periodicities, statistics are coherent insofar as they are based on common definitions, classifications, and methodological standards.
7. Completeness	Domains for which statistics are available should reflect the needs and priorities expressed by users as a collective.

Source: Eurostat (2000).

Other organizations use slightly different sets of quality dimensions. Statistics Canada (Brackstone, 1999) uses six dimensions—relevance, accuracy, timeliness, accessibility, interpretability, and coherence—and Statistics Sweden (Rosén and Elvers, 1999) uses five—content, accuracy, timeliness, comparability/coherence, and availability/clarity. Each dimension can be further divided into a number of subdimensions.

Another important question is how cost is related to quality. The survey cost is not a quality dimension per se, but it plays an important role when alternative design decisions are considered. One should choose the design that is the least expensive given the existing constraints regarding the quality dimensions (i.e., for a specified level of accuracy, schedule, degree of completeness, etc.). Alternatively, for a fixed survey budget, one should choose the best design where *best* is defined as some combination of the quality dimensions. Thus, cost is a component in any efficiency criterion related to survey design.

There is a literature on the characteristics of statistical systems. Examples of contributions include Fellegi (1996) and De Vries (2001). Recently, a leadership group on quality released a report on recommendations for improving the European Statistical System (Lyberg et al., 2001).

There are a number of frameworks for assessing data quality apart from those already mentioned. For instance, there is one developed by the International Monetary Fund (Carson, 2000).

- Quality can be defined along a number of dimensions, of which accuracy is one.
- Product quality is achieved through process quality.
- Process quality depends on systems and procedures that are in place in an organization.

1.6 MEASURING QUALITY

Once a framework that defines quality has been established, it is important to measure the quality. If we accept a definition of survey quality as a set of dimensions and subdimensions, then quality is really a multidimensional concept where some components are quantitative and others are qualitative. Accuracy is quantitative and the other components are, for the most part, qualitative. We have found no instance where a total survey quality measure has ever been calculated (i.e., a combined single measure of quality is computed taking all dimensions into account). Instead, quality reports or quality declarations have been used where information on each dimension is provided. Ideally, the quality report should give a description and an assessment of quality due to user perception and satisfaction, sampling and nonsampling errors, key production dates, forms of dissemination, availability and contents of documentation, changes in methodology or other circumstances, differences between preliminary results and final results, annual and short-term results, and annual statistics and censuses.

Work on standard quality reports is under way in several countries. Examples are the development of business survey reports for French official statistics, the ruling in Sweden stating that every survey in official statistics should be accompanied by a quality declaration, and that some surveys or survey systems in the United States have produced quality profiles (see Chapter 8). A *quality profile* is a collection of all that is known about the quality of a survey or a system of surveys. Such profiles have been developed for the Survey of Income and Program Participation, the Annual Housing Survey, and the Schools and Staffing Surveys, to mention a few examples. The problem with a quality profile is that it cannot be particularly timely, since it compiles the results from studies of the quality, and such postsurvey activities take time, as we have already stated. Quality profiles, quality declarations, and quality reports are discussed in more detail in Chapters 8 and 10.

Many survey organizations have now adopted a new approach to measuring quality. This approach is characterized by assessing organization performance to form a basis for improvement. There are a number of different methods for accomplishing this. One method is performance assessment using quality management approaches and business excellence models based on principles espoused in the general philosophy of *total quality management* (TQM). TQM is a method of working and developing business that is based on the explicit core values of an organization. A typical set of such values might include customer orientation, leadership and the participation of all staff, process orientation, measurement and understanding of process variation, and continuous improvement.

TQM offers no guidance per se on practical implementation, and therefore more concrete business excellence models have been developed. Examples of such models are the Swedish Quality Award Guidelines, the Malcolm Baldrige National Quality Award, and the European Foundation for Quality Management (EFQM) model. These models are all developed so that organizations can assess themselves against the criteria listed in the model guidelines. As an example, the Malcolm Baldrige Award lists the following criteria: leadership, strategic planning, customer and market focus, information and analysis, human resource focus, process management, and business results. Organizational assessment of adherence to criteria under this model is essentially self-assessment, although assistance from a professional, external examiner is preferable.

For the Baldrige Award, the organization has to respond to three basic questions for each criterion:

1. Specifically, what has the organization done to address the criterion?
2. To what extent have these approaches been used throughout the entire organization?
3. How are these approaches evaluated and continuously improved?

One might think that these are fairly innocuous questions, but that is not the case. The typical scenario is that all organizations have implemented some approaches to enhance quality, but they are not used uniformly throughout the organization and are seldom evaluated. In fact, many organizations use an ad hoc and local approach when it comes to improvements. Good procedures are not always transferred into the entire organization. A successful approach simply does not spread automatically. Therefore, there must be a process of change, as we have already mentioned. Like any other organization, a statistical organization can benefit from such an assessment, since good ratings on the aforementioned business model criteria will have a bearing on the quality of the statistical product.

There are also other assessment tools available. One is the ISO certification, for which an organization striving for certification is required to produce documents on its organization of quality work, segmentation of authorities, procedures, process instructions, specifications, and testing plans. Thousands of organizations worldwide, including a few statistical firms, have been certified. In some countries and business segments, certification is a requisite for organizations that want to stay in business.

The balanced scorecard is another tool that emphasizes a balance between four different dimensions of business: customers, learning, finances, and processes (Kaplan and Norton, 1996). As an example, Statistics Finland has started using this tool. One reason the scorecard was developed is that so many organizations put so much weight on financial outcome that the other three dimensions are frequently undervalued or ignored.

Business process reengineering is a totally different approach to process improvement (Hammer and Champy, 1995). It essentially means starting over and rebuilding a process from the ground up. Reengineering an organization means throwing out old systems and replacing them with new and, hopefully, improved ones. It requires a process of fundamentally rethinking and radically redesigning business processes with the goal of achieving dramatic improvements in key measures of performance, such as cost, quality, service, and speed. Notice the inclusion of the words *fundamental*, *radical*, and *dramatic*. This suggests that this method employs very different methodologies than those associated with continuous improvement and incremental changes.

Some statistical organizations have recently started using employee climate surveys, customer surveys, simple checklists, and internal quality audits. These methods recognize the importance of periodically assessing the motivation, morale, and professionalism of the staff. For example, in the U.K., the Office for National Statistics has developed an employee questionnaire to obtain information on staff perceptions and attitudes on issues concerning their jobs, their line managers, the organization as a whole, internal communication, and training and development. Statistics Sweden, Statistics Finland, and Eurostat are other agencies that are using employee climate surveys.

Customer surveys are important tools for providing an overview of customers' needs and reviews of past performances on part of the survey organization. They can be used to determine what product characteristics really matter to customers and their perceptions of the quality of the products and services provided by the organization. Another line of questioning might concern the image of the organization and how it compares to the images of other players in the marketplace. As pointed out by Morganstein and Marker (1997), many customer satisfaction surveys suffer from methodological short-comings. For instance, they may use limited 3- or 5-point scales, with the frequent result that many respondents continually select the same value (e.g., *very satisfied*). In many cases, the response categories are labeled only at the extremes (e.g., *very satisfied* and *very dissatisfied*). Consequently, the meaning of the intermediate categories is unclear to the respondent.

There are also frequent problems with the concept of satisfaction and how it should be translated into questions. It is often difficult to identify the best respondent in a user or client organization, with the result that responses are uninformed and misleading. The abundance of customer satisfaction surveys in society (hotels, airlines, etc.) developed by people with no formal training in survey methodology probably contributes to the large nonresponse rates and lukewarm receptions that are commonly associated with these kinds of surveys. This is an area where professional survey organizations should take the lead and develop some insightful new approaches.

Another type of self-assessment is the simple quality checklist that can be filled out by the survey manager. An example is one from Statistics New Zealand. The checklist consists of a number of indicators or assertions. The survey manager has to answer yes or no to each of the questions and is given the opportunity to elaborate on his or her answers. Examples of items on the checklist are shown in Figure 1.1. This type of checklist can be developed by adding follow-up questions containing such key words as *when* and *how*.

Finally, there is the method of self-assessment or audit, that can be either external or internal. In an external audit, experts are called in to evaluate a process, a survey, a set of surveys, or parts of or the entire organization. Typically, the auditors compare the actual survey with similar surveys of high quality with which they are familiar. If the audit targets organizational performance, the auditors can also use one of the business excellence models mentioned above. Usually, the audit will result in a number of recommendations for improvement. Examples of good procedures are conveyed to other parts of the organization.

If the audit is internal, it is performed by the organization's own staff. Any audit should be based on internal documentation of products and processes, organizational guidelines and policies, and on observations made by the auditors. Audits have become used increasingly in statistical organizations. For instance, Statistics Netherlands has a system for rolling audits led by a permanent staff of internal auditors. Statistics Sweden has recently started a five-year program during which all surveys will be audited at least once.

A Quality Checklist for Statistics New Zealand

- The program has a good understanding of who the key users are and emerging new stakeholders.
- Questionnaires, definitions, and classifications reflect contemporary needs and situations.
- Documentation of processes and products is complete and accessible.
- The program contributes to professional associations and international developments.
- Data definitions are consistent with those of other survey programs.
- The sample is regularly redesigned.
- Long-term time series are available.
- Seasonal adjustment analysis is performed.
- Sources and methods documentation is available.
- Release dates are advertised in advance.
- Information is available in formats and media as required by users.
- Standards for time taken to meet requests are met.
- Information to respondents on the purpose of data collection is provided.
- Releases are checked for confidentiality.
- Indicators of quality are regularly measured and monitored.
- Data are archived.
- Requirements of the Statistics Act (a legal framework for the production of official statistics) are met.

Figure 1.1   Examples of quality checklist items for a survey organization. *Source*: Adapted from Statistics New Zealand (not dated).

APPROACHES TO MEASURING AND REPORTING QUALITY

- Develop quality reports according to a standard framework.
- Develop and use quality profiles.
- Assess organizational performance according to an "excellence model."
- Conduct employee climate surveys.
- Conduct customer surveys.
- Conduct internal and external audits.

1.7   IMPROVING QUALITY

The approaches described in Section 1.6 are all examples of methods and measures that can identify areas where improvements are needed. Sometimes a quality problem can be solved easily. It is simply a matter of changing the process slightly so that a certain requirement is better met. But sometimes there is need for more far-reaching remedies that necessitate an organized improvement effort or project. The improvement project usually concerns some process that is not functioning properly. An idea to which many national statistical offices adhere is to set up a team that uses quality management tools: for instance, the Ishikawa (1982) tools mentioned earlier.

Quality Improvement Project Goals

- Increase the quality and efficiency of occupation coding.
- Streamline the process for land-use statistics.
- Improve the editing of energy statistics.
- Simplify the data capture of the Farm Register.
- Assure the quality of interview work.
- Improve the quality of user contacts.
- Improve the staff recruitment process.

Figure 1.2 Examples of quality improvement goals for improvement projects at Statistics Sweden.

A project team will typically have a quality facilitator who helps the team adhere to well-documented and approved work principles: for example, those found in Scholtes et al. (1994). The tools are deliberately simple to make it easy for all participants to contribute but there might, of course, also be a need for more complex statistical tools, such as designed experiments. Statistics Sweden has conducted over 100 such improvement projects since it started systematic work on quality improvement in 1993. Examples of goals in these projects are shown in Figure 1.2. From these examples of project goals it should be quite clear that there is a great value in having all staff levels represented on the team. Those who work on the processes should also be responsible for their improvements. Similarly, if one has been part of the improvement work, one is much more willing to help implement the changes leading to improvements.

Some processes are common to many different parts of an organization. Such processes include questionnaire development, coding, editing, nonresponse reduction and adjustment, hiring, staff performance evaluation, data collection, and so on. It is rather typical that such common processes are conducted in very different ways in an organization. Variation in approach will generally lead to variation in the characteristics of the final product, and not all approaches will be equally efficient. The best strategy is to eliminate unnecessary variation by standardizing the process. The current best method (CBM) approach is one way to do just that.

The process of developing a CBM is described in Morganstein and Marker (1997). It begins by assigning a team to conduct an internal review of some process to identify good practices. In addition, these practices are compared to those of other organizations, an approach called *benchmarking*. Then the team develops a draft CBM that is reviewed by a larger group of the staff. Comments and suggestions are collected and the CBM is revised. Once accepted, the CBM is implemented and data on its performance are collected. Typically, a CBM has to be revised every four years or so. At Statistics Sweden, CBMs have been developed for editing, nonresponse reduction, nonresponse adjustment, confidentiality protection, questionnaire design, and project work (see Chapter 10). Having a CBM in place makes further improvements of the process much easier since there documentation is available. Also, it helps the training of new staff in a consistent way so that new staff members can carry out their tasks more quickly.

There are a number of accompanying measures that aim at standardizing processes, including establishment of minimum standards, quality guidelines, and recommended practices. A minimum standard is supposed to ensure a basic decency level, quality guidelines are directed toward what to do rather than how to do it, and recommended practices provide a collection of procedures to choose from. A useful discussion of these instruments is found in Colledge and March (1997).

1.8   QUALITY IN A NUTSHELL

In this chapter we discussed the meaning of the quality concept and found that quality is a multidimensional concept. One dimension of it is accuracy measured by total survey error. The other dimensions are labeled differently depending on organization. Our book is about controlling the accuracy of survey data using quality-oriented methods.

To achieve error prevention and continuous quality improvement, a process perspective should be adopted. Accurate data can be achieved only if there are accurate processes generating the data (i.e., data quality is achieved through process quality). Inaccuracies stem from imperfections in the underlying processes, and it is therefore important to control key process variables that have the largest effect on characteristics of the survey output, such as data accuracy.

The chapters in this book provide many examples of how survey processes can be controlled and improved. It is not just a matter of using good survey methods; it is also a matter of letting all staff participate in improvement work and incorporating the best ideas of the collective. Some of the tools that we advocate are not feasible without a team approach. The development of CBMs is one example where practical knowledge and experience with a process are essential to producing a tool that will generate real improvements in that process.

Since there are really many dimensions to quality, why should we focus only on data accuracy? The answer is that accuracy is the cornerstone of quality, since without it, survey data are of little use. If the data are erroneous, it does not help much if relevance, timeliness, accessibility, comparability, coherence, and completeness are sufficient. Further, although all these other quality dimensions are important, we view them more as constraints on the process rather that attributes to be optimized. For example, we are seldom in a situation where time to complete the survey should be minimized. More often, we are given a date when the data should be available. In that sense, timeliness is a constraint just as cost is a constraint. The goal then is to provide data that are as accurate as possible subject to these cost and timeliness constraints. All quality dimensions other than accuracy can be viewed in the same way. Thus, this book is about designing surveys to maximize accuracy.

There is an abundance of literature pertaining to survey quality. Unfortunately, there are no textbooks on survey methodology that cover all the known survey error sources. Books that approach this ideal include Anderson et al. (1979), Groves (1989), and Lessler and Kalsbeek (1992). A vast majority of survey methodology textbooks cover sampling theory in detail and non-sampling errors in a chapter or two. There are also books that cover specific design aspects such as questionnaire design, survey interviewing, and non-response. During the last decade a series of edited monographs on survey methodology topics have been produced. One purpose of this endeavor has been to try to fill a void in the survey textbook literature. Monographs released so far cover panel surveys (Kasprzyk et al., 1989), telephone survey methodology (Groves et al., 1988), measurement errors in surveys (Biemer et al., 1991), survey measurement and process quality (Lyberg et al., 1997), computer-assisted survey information collection (Couper et al., 1998), and survey nonresponse (Groves et al., 2002). Recently, a discussion on survey theory advancement was initiated by Platek and Särndal (2001) in which many complex issues related to survey quality are penetrated.

CHAPTER 2

The Survey Process and Data Quality

In this chapter we review the survey process and describe the major sources of error associated with each stage of the process. Then our focus will shift to developing a means for quantifying the level of error in survey data using a measure referred to as the *mean squared error*. This measure of survey accuracy will guide all our efforts throughout this book to identify the major sources of survey error and to reduce them to the extent possible within the budgetary and scheduling constraints of the survey. The mean squared error will also serve as a device for comparing alternative methods in order to choose the best, most accurate method. Thus, the concept of mean squared error as a measure of survey error is fundamental to the study of data quality.

2.1 OVERVIEW OF THE SURVEY PROCESS

As mentioned in Chapter 1, there is an insatiable need today for timely and accurate information in government, business, education, science, and in our personal lives. To understand the present and to plan for the future, data are needed on the preferences, needs, and behaviors of people in society as well as other entities, such as business establishments and social institutions. For many researchers and planners, sample surveys and censuses are major sources of this information.

The word *survey* is often used to describe a method of gathering information from a *sample of units*, a fraction of the persons, households, agencies, and so on, in the population that is to be studied. For example, to measure the size of the workforce the government may ask a sample of people questions about their current employment. A business may use information from a survey to compare the costs of its production against the costs of other similar businesses (see Chapter 1). In this section we present an overview of the process for plan-

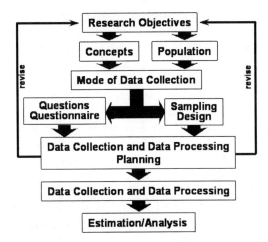

Figure 2.1 Survey process. The planning stages of the survey process are largely iterative. At each stage of planning, new information may be revealed regarding the feasibility of the design. The research objectives, questionnaire, target population, sampling design, and implementation strategy may be revised several times prior to implementing the survey.

ning and conducting sample surveys. Understanding the survey process is central to measuring and controlling survey quality.

As shown in Figure 2.1, the survey process is composed of a number of steps that are executed more or less sequentially, from determining the research objectives to analyzing the data. In what follows, we discuss each major step of the survey process in the context of a hypothetical study that might be commissioned by a government entity to draft new legislation or possibly to evaluate existing legislation. As a means of illustrating the concepts, let us assume that this agency is the U.S. Health Care Finance Administration (HCFA).

HCFA is responsible for administering the Medicare Health Insurance program, which provides health care benefits to U.S. citizens aged 65 and older and citizens with disabilities. Suppose that HCFA is interested in monitoring the health status over time of people who receive Medicare, referred to as *Medicare beneficiaries*. They are particularly interested in measuring the general health of new recipients of Medicare benefits (i.e., recipients who recently reached their sixty-fifth birthday) and how the health characteristics of this population change over time as the population ages and continues to receive Medicare benefits. This study is mandated by the U.S. Congress, which has specified a time frame for starting the study (two years) and a total budget that should not be exceeded for the first three years of the survey. Using this example, we consider the process for designing and conducting a survey to obtain information on the health of this population of older U.S. citizens.

Determining the Research Objectives

The first step in the survey process is to determine the research objectives (i.e., the primary estimates that will be produced from the survey results or the key data analyses that are to be conducted using the survey data). A well-specified set of research objectives is a critical component of the survey process and will facilitate many of the decisions involved in survey design. Defining the research objectives is often accomplished best by identifying a small set of key research questions to be answered by the survey. This is usually done in collaboration with the survey sponsor or researcher(s) commissioning the survey—in this case the survey sponsor is HCFA.

As an example, an important but very general question for Medicare analysts is whether and how the Medicare program contributes to the health and well-being of its beneficiaries. Before a questionnaire can be designed to obtain information on these abstract concepts, a series of steps must be taken. First, HCFA might convene a meeting of experts on the health and well-being of senior adults. The experts would determine the various dimensions of the concepts that should be measured to describe and evaluate the concepts adequately. For example, they may decide that data on food intake, exercise, medical diagnoses, quality of life, and so on, should be collected. They may also identify a number of existing measures, instruments, or scales to assess these concepts that have been validated in other studies and are therefore well understood. The experts may decide further that this information should be collected for all persons 65 years of age and older as they enter the Medicare system and then, following them over time, collected again at periodic intervals to determine how these characteristics change as the beneficiaries age.

The subject matter experts may also recommend that data be collected on visits to the doctor; medications received; current medical conditions; personal characteristics such as height, weight, blood pressure, and functional status of the respondent (i.e., sight, hearing, mobility, mental health); life satisfaction; frequency of depression; and other mental conditions. Further, they may decide that the survey should be repeated for the same sample of persons to determine how these characteristics change over time as the need for medical services increases.

Quite often, the time spent in the development of a comprehensive set of research questions is time saved in the questionnaire design step since eventually each question posed for the research can be linked to one or more data elements to be collected in the data collection phase of the process. These data elements or items are in turn linked to one or more questions on the survey questionnaire or form. In fact, it is good practice to ensure that every question on the questionnaire corresponds to at least one research question, to avoid the situation where questions that are superfluous and really not needed for the purposes of the survey somehow find their way onto the questionnaire. This process of linking research objectives and survey questions also ensures that all survey questions necessary to address the research objectives fully are included in the questionnaire (Table 2.1). As we will see later in this chapter, adherence to this approach will minimize the risk of *specification error* in the results. Specification errors are errors that arise when the survey questions fail

Table 2.1   Correspondence Between Research Questions and Survey Questions[a]

Research Questions	Survey Questions							
	SQ1	SQ2	SQ3	SQ4	SQ5	SQ6	SQ7	← SQ7 is an unnecessary question; could be deleted
RQ1	✓	✓						
RQ2	✓		✓					
RQ3				✓				
RQ4					✓	✓		
RQ5								← No questionnaire item to address RQ5

[a] A table such as this is useful for identifying redundant or unnecessary questions in the questionnaire or unaddressed research questions.

to ask respondents about what is essential to answer the research questions (i.e., the subject-matter problem).

Defining the Target Population

The next step in the survey process is to define the population to be studied or the target population for the survey. The target population is the group of persons or other units for whom the study results will apply and about which inferences will be made from the survey results. In the Medicare study, the target population is defined as "persons living in the United States who are aged 65 years or older and are enrolled in the Medicare system." Note that this definition does not include persons under the Medicare system living outside the United States or persons older than 65 years who do not receive Medicare benefits. However, it does include persons enrolled in Medicare whether or not they receive Medicare benefits. Decisions about whom to include and exclude in the target population are important. As we will see later, these decisions guide other important decisions about the survey design in virtually every subsequent stage of the survey process (see also Chapter 3).

Determining the Mode of Administration

Having specified the research objectives and defined the target population, the next step in the process is to determine the mode of administration for the survey. Here we consider whether to use mail questionnaires, telephone, or face-to-face interviewing or some other mode of collecting the data. These decisions must be made before designing the questionnaire, since different modes of data collection often require very different types of questionnaires. The mode of administration will also constrain the sampling design choices that can be used for the survey. Face-to-face interviewing will usually require a sample that is highly clustered (i.e., a sample composed of clusters of units such as persons living within the same neighborhoods). This is done to reduce interviewer travel costs. Telephone and mail survey samples are usually dispersed geographically or unclustered since interviewer travel costs are not incurred for these modes. However, a telephone survey requires either a fairly complete list of telephone numbers for the persons in the target population or a practical and cost-effective method to generate a random sample of telephone numbers that is representative of the target population. A mail survey requires a fairly complete list of addresses for the persons in the target population. If an adequate address list is not available, a mail survey may not be possible.

In deciding on the mode of administration, one of the first constraints one encounters is costs. Face-to-face interviewing, even with highly clustered samples, can be several times more costly than collecting data by telephone or by mail. The budget available for the survey often limits the choices regarding administration mode. Another important consideration relates to the topics to be surveyed. Interviewers can affect the responses to questions on sensitive topics, so if this is a concern, a more private mode of administration such as a mail self-administered questionnaire may be preferable.

One should also consider how important it is to have visual communication with the respondent during the interview. Is it important to use flash cards, for example, to identify the pills and other medications respondents may take? Or are there long lists of medical problems or procedures from which the respondent will be asked to choose? The timing of the survey is also an important consideration in deciding the mode of administration. How quickly are the data needed? If less than two months is available for data collection, a mail survey may not be the best choice.

After some discussion, the Medicare survey design team may determine that a self-administered mail survey is feasible and cost-effective since the questionnaire could be kept simple enough for a sample member to complete without the aid of an interviewer, and since mailing out questionnaires is less expensive than the other modes under consideration. Further, the current addresses of all target population members are available on the Medicare database, so mailing questionnaires to the appropriate addresses would not pose any difficulties.

Finally, to ensure an adequate response rate (about 75% is a typical minimum rate for U.S. government surveys), a telephone follow-up of the mail nonrespondents should also be included as part of the data collection design. Specifying one mode as a primary mode of administration and another mode as a secondary or follow-up mode is a common feature of data collection designs. Referred to as mixed-mode data collection, such strategies are often necessary to maximize response rates for the survey. Additional considerations for determining the best mode of administration are discussed in Chapter 6.

Developing the Questionnaire

The next step of the survey process is the development of the questionnaire or instrument. In this step, the research objectives developed previously are used to determine the data elements to be collected in the survey (i.e., the variables that will be used to address each research question). Each data element corresponds to a single response to a question on the questionnaire. For example, one data element may be the date of birth or a response to a question about medication.

A set of data elements may be used to create a new data element during the postsurvey processing stage. For example, suppose that a research question relates to the mental well-being of Medicare beneficiaries and how this changes over time. Specifically, the researchers may wish to know whether Medicare beneficiaries are generally depressed or contented and how these attributes vary as a person ages. The primary measure for this question is actually a *score*, which is a summary measure derived from a group of data elements on the questionnaire. The score then summarizes the information about a person's mental state into a single measure. The result is a mental health status score which is a single, continuous variable that increases as a person's level of happiness increases and decreases with the onset of depression. Note, however, that this measure requires not just one data element but multiple data elements, all of which provide some information on individual mental health. Thus, it is not uncommon that a number of data elements are needed to address a single research question (see Table 2.1).

As mentioned previously, the design of the questionnaire should also take into account the mode of administration and the capabilities of the target population members to provide information under the desired mode of administration. For example, if in our study a mail self-administered questionnaire is chosen, the design of the questionnaire may use a larger font and incorporate special features to help the oldest respondents complete the questionnaire. These and other considerations for instrument development are discussed further in Chapter 4.

Designing the Sampling Approach

Having defined the target population and the research objectives and determined the mode of administration, the next stage of the survey process can begin, that of specifying the sampling design. The sampling design specification describes the sampling frame (i.e., the list of population members) to be used for the survey, the methods used for randomly selecting the sample from the frame, and the sample sizes that are required. The *sampling frame* is simply the list of target population members from which the sample will be drawn. It may also be a combination of several lists, a map, or any other device that can be used to select the sample. As mentioned previously, the frame chosen for sampling depends to a large extent on the mode of administration for the survey. For our survey, a logical frame is the Medicare list of all persons who are registered in the Medicare program. The coverage of this frame (i.e., the proportion of target population members contained in it) is approximately 100%. This means that every member of the target population has a chance of being selected for the survey. Further, all the information needed to mail the questionnaires to the sample members is available on the Medicare frame.

Since the survey is to be conducted by mail with telephone follow-up of the nonrespondents, interviewer travel costs are not a consideration for the Medicare survey. Thus, the sample could be drawn completely at random without attempting to cluster the sample. However, a sampling plan that involves stratifying the frame into homogeneous groups (e.g., by age) might be used since such a design results in better precision in the survey estimates with no appreciable increase in survey costs.

Finally, after considering the required precision in the estimates for the most important population characteristics to be measured in the study, the sample size is determined. Determining the required sample size for the Medicare survey should take into consideration the loss of sample units that is inevitable as a result of refusals to respond, death, incorrect location information, and loss of sample members resulting from other types of nonresponse.

As shown in Figure 2.1, the process to this point is somewhat iterative. For example, quite often in the process of developing the questionnaire, it is necessary to rethink the survey objectives since to address them all would require a questionnaire or interview that is either longer than can be afforded with the available budget or too burdensome for the sample members, who are thus likely to refuse to participate in the survey. Further, it may be determined that some objectives cannot be addressed adequately with the chosen mode of administration. Consequently, it is necessary either to drop some research questions from the study or to reconsider the mode of data collection.

Similarly, during the sample design development step, it may be realized that an adequate sampling frame does not exist or is too expensive to develop. This could require the use of more than one sampling frame or modifying the definition of the target population to exclude those groups that are too difficult to reach. A common occurrence is that the sample size must be reduced as a result of cost considerations. Thus, several iterations of the foregoing steps of the design process may be necessary before the final design is determined. Additional aspects of the sample design specification are considered in Chapter 9.

Developing Data Collection and Data Processing Plans
Once the initial, basic design decisions are made, the data collection and data processing plans can be developed. These steps involve specifying the process of fielding the survey, collecting the data, converting the data to computer-readable format, and editing the data both manually and by computer. For the Medicare survey, the process would also involve developing procedures for controlling the flow of cases, checking in the mail returns, moving cases to the telephone follow-up operation, keying or scanning the data from paper questionnaires, and merging the data from the mail operation and the telephone operation. Plans are also developed for editing the survey data (i.e., for correcting stray or inappropriate marks on the questionnaire returns, errors that occur during keying or scanning the paper questionnaires, inconsistent

responses, and other problems with the data). The structure of the final data files should also be determined so that data analysis would be facilitated.

In the Medicare survey design process, there may be concerns about whether the elderly will complete the forms accurately; whether the response rates will be adequate using the mail mode; how to efficiently handle persons in institutions; whether to accept information from informants other than the sample persons on behalf of the sample persons; and so on. To address these questions and others, the initial design should be tested in a pretest of the survey procedures and questionnaire. The pretest can indicate whether certain aspects of the design do not function well so those aspects of the design can be modified for the main study. As an example, there may be problems in the design of the questionnaire or in the methods used for determining the telephone numbers of the mail nonrespondents for the telephone follow-up operation (see Chapter 10).

Collecting and Processing the Data

The next step of the survey process involves implementing the data collection and data processing plans developed in the previous steps. Interviewers must be recruited, trained, and sent into the field or to a telephone center to collect the data. If the survey is to be conducted by mail, the questionnaires must be mailed and plans for following up nonrespondents must be implemented. Even in a well-planned survey, unforeseen problems can develop which require deviations from the plans. Here it is important for the project staff to monitor carefully the progress of the data collection operations via measurements on key process variables to identify potential problems before they develop into real problems. Thus, an important aspect of the data collection plan is a process for routine monitoring of data collection and obtaining feedback from the supervisory staff. For the Medicare study, this would involve developing the procedures for mailing the questionnaires and checking in the returns, training the telephone interviewers who will contact sample members who do not return their questionnaires, scanning the mail questionnaires into the computer, and conducting quality control operations to ensure that these activities are conducted as planned.

Once the data are in computer-readable form, they can be edited, cleaned, and prepared for estimation and analysis. Editing the data involves correcting out-of-range or inconsistent responses, possibly recontacting respondents to obtain additional information, and generally, cleaning the data of many discernible errors. Information obtained from an open-ended question—that is, a question that elicits an unstructured response—is often converted into code numbers that summarize the verbal information provided by the respondent (see Chapter 7).

Estimation and Data Analysis

Finally, the data are *weighted* to compensate for unequal probabilities of selection, missing data, and frame problems, and the estimates are computed following the plans previously developed for estimation and analysis. Weighting the data essentially involves determining an appropriate multiplier for each observation so that the sample estimates better reflect the true population parameter. The estimation and analysis plan lists the major research questions that should be addressed in the analysis, the estimates that will be computed, and the statistical analyses that will be performed. The latter includes detailed specifications for weighting the data and compensating for nonresponse in the final estimates.

In remaining chapters of the book we discuss many of the decisions that must be made in the survey design process and provide a general background for understanding how these decisions are made. Unfortunately, there are no absolute criteria to dictate the best choice of mode, questionnaire design, data collection protocol, and so on, to use in each situation. Rather, survey design is guided more by past experience, theories, and good advice on the advantages and disadvantages of alternative design choices so that we can make intelligent decisions for each situation we encounter. As will become apparent, the emphasis will be on the general theory of good design rather than on specific guidelines to follow for each set of special circumstances. The aim of good design is to use practical and reliable processes whose outcomes are reasonably predictable. Thus, our guiding philosophy is that it is more useful to learn a few basic techniques for dealing with the underlying causes of survey error and the general theories leading to their development rather than to learn numerous ad hoc methods that essentially treat the same causes of survey error but under a variety of special circumstances.

2.2 DATA QUALITY AND TOTAL SURVEY ERROR

To many users of survey data, data quality is purely a function of the amount of error in the data. If the data are perfectly accurate, the data are of high quality. If the data contain a large amount of error, the data are of poor quality. For estimates of population parameters (such as means, totals, proportions, correlation coefficients, etc.), essentially the same criteria for data quality can be applied. Assuming that a proper estimator of the population parameter is used, an estimate of a population parameter is of high quality if the data on which the estimate is based are of high quality. Conversely, if the data themselves are of poor quality, the estimates will also be of poor quality. However, in the case of estimates, the sample size on which the estimates are based is also an important determinant of quality. Even if the data are of high quality, an estimate based on too few observations will be unreliable and potentially unusable. Thus, the quality of an estimator of a population parameter is a function of the *total survey error*, which includes components of error that arise solely as a result of drawing a sample rather than conducting a complete census called *sampling error components*, as well as other components that are related to the data collection and processing procedures called *nonsampling error components*.

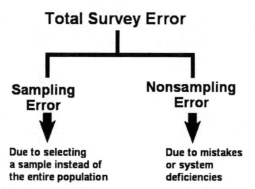

Figure 2.2   Total survey error. Total survey error can be partitioned into two types of components: sampling error and nonsampling error.

In what follows we use the term *estimator* to refer to the formula or rule by which estimates from a survey are produced. For example, an estimator of the population mean for some characteristic in the survey is the sum of the values of the characteristic for all sample members who responded divided by the number of sample members who responded. Suppose that for one particular implementation of the survey design, the value of the estimator is 22. Then 22 is called the *estimate* of the population mean.

Simply stated, the *total survey error* (Figure 2.2) of an estimate is the difference between the estimate and the true value of the population parameter. To illustrate the concept of total survey error, consider a very simple survey aimed at estimating the average annual income of all the workers in a small community of 5000 workers. Thus, the population parameter in this case is the average income over all 5000 workers. Suppose the survey designer determines that a sample of 400 employees drawn at random from the community population should be sufficient to provide an adequate estimate of the population average income. The designer also determines that the best estimator of average annual income is just the simple average of the incomes of the 400 workers in the sample. Thus, the sample is drawn, interviewers are hired and trained, the data are collected, and the sample average is computed from the survey data.

Suppose that average annual income for the persons in the sample is $32,981. Thus, $32,981 is the survey estimate of the population parameter. Finally, suppose that the actual population average income (i.e., the population parameter) for this community is $35,181. This value, of course, is not known since otherwise there would be no need for a survey to estimate it; however, for purposes of this illustration, assume it is known so that we can compute the error in the sample estimate. The difference between the survey estimate of annual income and the unknown true annual income for the community is the total survey error in the estimate of annual income. In this case, the total survey error in the estimate is $32,981 − $35,181 = −$2200.

> *Total survey error* is the difference between a population mean, total, or other population parameter and the estimate of the parameter based on the sample survey (or census).

As noted previously, the true value of the population parameter is not known, but sometimes it can be approximated using the methods discussed in Chapter 8. Therefore, the total survey error in an estimate is also not known but may be approximated using special methods for evaluating surveys. Next, we examine some of the reasons why survey error is unavoidable.

One major reason that a survey estimate will not be identical to the population parameter is *sampling error*, the difference between the estimate and the parameter as a result of only taking a sample of 400 workers in the community instead of the entire population of 5000 workers (i.e., a complete census). Another sample of 400 workers would very likely have different incomes and would therefore produce a different estimate from the first sample estimate. The only way to eliminate the sampling error from the estimation process is to take a complete census of the community. In that case, the average income for all 5000 workers in the "sample" should be the same as the population average income.

However, even if we could afford to observe the entire community in an attempt to measure the true annual income without sampling error, our estimate would not be exactly $35,181 because of another type of error, referred to as *nonsampling error*. Each step of the survey process is a potential source of nonsampling error. *Nonsampling error* encompasses all the various kinds of errors that can be made during data collection, data processing, and estimation except sampling error. The cumulative effect of these errors constitutes the nonsampling error component of the total survey error. In our example, nonsampling errors could arise from the following sources:

- *The respondent.* Respondents may not want to reveal their true income or may unintentionally exclude some sources of income in their response to the survey, such as tips, gifts, bonuses, winnings, and so on.
- *The interviewer.* Interviewers may make mistakes in entering the information on the survey form, or may cause the respondent to make an error, for example, by giving the respondent incorrect information about what to include as income.
- *Refusals to participate.* Some of the 400 persons contacted from the survey may refuse to reveal their incomes or even refuse to participate in the interview.
- *Data entry.* The income values entered on the survey questionnaire may be miskeyed during the data-entry process.

Any and all of these errors could result in the wrong income being recorded and thus cause the estimate of annual income to deviate from the true value of the population parameter. Thus, nonsampling errors can be viewed as mistakes or unintentional errors that can be made at any stage of the survey process. Despite our best efforts to avoid them, nonsampling errors are inevitable particularly in large-scale data collections. Sampling errors, on the other hand, are *intentional* errors in the sense that we can control their magnitude by adjusting the size of the sample. With a sample size of 1, sampling error is at its maximum, and as we increase the sample size to the population size (5000 in our example), sampling error becomes smaller and smaller. When the sample size is the same as the population size (as in a census), the sampling error is zero, and completely absent from the estimates. Thus, sampling error can be made as small as we wish (or can afford) to make it by manipulating the sample size. Later in this chapter we see further illustrations of the sampling error. A more thorough treatment of sampling error is left for Chapter 9.

Nonsampling error, on the other hand, is unpredictable and not so easily controlled. For example, the expected level of nonsampling error may actually increase with increases in the sample size. This may be the result of having to hire a larger staff of interviewers who may be less experienced, who are more prone to certain types of error, or who receive less adequate supervision. Alternatively, the scale of the survey operations may become such that the quality control systems become overloaded and less effective at preventing some types of error. Processing the survey data may be subject to similar control problems, thus resulting in larger data processing errors.

$$\text{total survey error} = \text{sampling error} + \text{nonsampling error}$$

Only in the last 50 years have survey researchers realized that, in many cases, nonsampling error can be much more damaging than sampling error to estimates from surveys. As stated previously, an important goal of this chapter, as well as this book, is to explain how this can happen and why we need to be just as concerned about controlling nonsampling errors in surveys as we are the sampling errors.

There is a considerable literature on nonsampling errors in surveys: the sources and causes of nonsampling error, the design of surveys to minimize them in the final results, statistical methods and models for assessing their effects on the survey results, methods for making postsurvey adjustments to reduce their effects on the estimates, and so on. In this book we try to cover all of these aspects to some extent since the key to survey data quality is understanding the root causes of nonsampling errors and how to minimize them. As mentioned in the preface, our goal is breadth of coverage of these topics, not depth of any specific topic. However, depending on the interests of the reader, more depth of coverage of each topic can be obtained through readings in the extensive literature on survey error, particularly the references that are provided throughout the book.

2.3 DECOMPOSING NONSAMPLING ERROR INTO ITS COMPONENT PARTS

The objective of any survey design is to minimize the total survey error in the estimates subject to the constraints imposed by the budget and other resources available for the survey. As we shall see later in this chapter, reducing non-sampling error while controlling survey costs sometime means increasing sampling error (by reducing the sample size) to reduce some important sources of nonsampling error. Optimizing a survey design means finding a balance between sampling errors and nonsampling errors so that the overall total survey error is as small as possible for the budget available for the survey. This entails allocating the survey resources to the various stages of the survey process so that the major sources of error are controlled to acceptable levels. It does not entail conducting every stage of the process as accurately as possible (without considering the costs involved) since this could result in exceeding the survey budget by a considerable margin.

To stay within the survey budget, training interviewers adequately may require eliminating or limiting the quality control activities conducted at the data processing stage. Increasing the response rate to the survey to an acceptable level may require substantial cuts in the sample size, and so on. How should these decisions be made? Making these trade-offs wisely requires an understanding of the sources of nonsampling error and how they can be controlled. To this end, in the next section, we consider each of the major sources of nonsampling error in surveys in some detail.

2.3.1 The Five Components of Nonsampling Error

Table 2.2 shows a decomposition of nonsampling error into five major sources: specification error, frame error, nonresponse error, measurement error, and processing error. All of the nonsampling errors that we consider in this book can be classified as originating from one of these five sources.

Specification Error
Specification error occurs when the concept implied by the survey question and the concept that should be measured in the survey differ. When this occurs, the wrong parameter is being estimated in the survey, and thus inferences based on the estimate may be erroneous. Specification error is often caused by poor communication between the researcher, data analyst, or survey sponsor and the questionnaire designer. For example, in an agricultural survey, the researcher or sponsor may be interested in the value of a parcel of land if it were sold at fair market value. That is, if the land were put up for sale today,

**Table 2.2 Five Major Sources of Nonsampling Error
and Their Potential Causes**

Sources of Error	Types of Error
Specification error	Concepts
	Objectives
	Data elements
Frame error	Omissions
	Erroneous inclusions
	Duplications
Nonresponse error	Whole unit
	Within unit
	Item
	Incomplete information
Measurement error	Information system
	Setting
	Mode of data collection
	Respondent
	Interview
	Instrument
Processing error	Editing
	Data entry
	Coding
	Weighting
	Tabulation

what would be a fair price for the land? However, the survey question may simply ask: "For what price *would you sell* this parcel of land?" Thus, instead of measuring the market value of the parcel, the question may instead be measuring how much the parcel is worth *to the farm operator*. There may be quite a difference in these two values. The farm operator may not be ready to sell the land unless offered a very high price for it, a price much higher than market value. Since the survey question does not match the concept (or construct) underlying the research question, we say that the question suffers from specification error.

To take this example a step further, suppose that the survey analyst is interested only in the value of the parcel without any of the capital improvements that may exist on it, such as fences, irrigation equipment, airfields, silos, outbuildings, and so on. However, the survey question may be mute on this point. For example, it may simply ask: "What do you think is the current market value of this parcel of land?" Note that this question does not explicitly exclude capital improvements made to the land, and thus the value of the land may be inflated by these improvements without the knowledge of the researcher. A more appropriate question might be: "What do you think is the current market value of this parcel of land? Do not include any capital improvements in your estimate, such as fences, silos, irrigation equipment, and so on."

The question, "What do you think is the current market value of this parcel of land?" is not necessarily a poorly worded question. Rather, it is the wrong question to ask considering the research objectives. A questionnaire designer who does not clearly understand the research objectives and how data on land values will be used by agricultural economists and other data users may not recognize this specification error. For that reason, identifying specification errors usually requires that the questions be reviewed thoroughly by the research analyst or someone with a good understanding of the concepts that need to be measured to address the research objectives properly. The research analyst should review each question relative to the original intent as it relates to the study objectives and determine whether the question reflects that intent adequately. For the land values example, the agricultural economist or other analyst who will use the data on land values would be the best person to check the survey questionnaire for specification errors. In general, detecting specification error usually requires a review of the survey questions by researchers who are responsible for analyzing the data to address the research objectives and who know best about what concepts should be measured in the survey.

Note that in some disciplines (e.g., econometrics), specification error means including the wrong variables in a model, such as a regression model, or leaving important variables out of the model. In our terminology, specification error does not refer to a model but to a question on the questionnaire.

Frame Error

The next source of nonsampling error is error that arises from construction of the sampling frame(s) for the survey. The sampling frame is usually a list of target population members that will be used to draw the sample. In the Medicare survey example above, the frame was the list of all persons receiving Medicare benefits. However, the frame may also be an area map, as in the agricultural land values example, where the sample for the survey is selected by a random selection of parcels of land delineated on the map.

A sampling frame may not even be a physical list, but rather, a conceptual list. For example, telephone survey samples are often selected using a method referred to as random-digit dialing (RDD). For RDD surveys conducted in the United States and Canada, the frame is a conceptual list of all 10-digit numbers that are potential telephone numbers. No physical lists may exist. Instead, telephone numbers are randomly generated as needed using an algorithm for generating random 10-digit numbers.

To ensure that samples represent the entire population, every person, farm operator, household, establishment, or other element in the population should be listed on the frame. Further, to weight the responses using the appropriate probabilities of selection, the number of times that each element is listed on the frame should also be known.

There are a number of errors that can occur when the frame is constructed. Population elements may be omitted or duplicated an unknown number of times. There may be elements on the frame that should not be included (e.g., businesses that are not farms in a farm survey). Erroneous omissions often occur when the cost of creating a complete frame is too high. Quite often, we must live with omissions due to the survey budget. Duplications on the frame are a common problem when the frame is a combination of a number of lists. For the same reason, erroneous inclusions on the frame usually occur because the available information about each frame member is not adequate to decide which entry is in the target population and which is not. In Chapter 3 we discuss how the problems of omissions, erroneous inclusions, and duplications affect the error in a survey estimate.

Nonresponse Error

Nonresponse error, the next error in Table 2.2, is a fairly general source of error encompassing unit nonresponse, item nonresponse, and incomplete response. A *unit nonresponse* occurs when a sampling unit (household, farm, establishment, etc.) does not respond to any part of the questionnaire. For example, a household refuses to participate in the survey, or a mail questionnaire is never returned from an establishment in the survey. *Item nonresponse* occurs when the questionnaire is only partially completed (i.e., some items are skipped or left blank that should have been answered). As an example, in a household survey, questions about household income are typically subject to a great deal of item nonresponse because respondents frequently refuse to reveal their incomes even though they may answer many other questions on the questionnaire. Finally, *incomplete responses* to open-ended questions are also a type of nonresponse error. Here, the respondent may provide some information, but the response is very short and inadequate. As an example, the open-ended question "What is your occupation?" that appears on all labor force surveys around the world is subject to this type of nonresponse. The respondent may provide some information about his or her occupation, but perhaps not enough information to allow an occupation and industry coder to assign an occupation code number later during the data processing stage. This type of error is discussed in detail in Chapter 7.

Measurement Error

Measurement error has been studied extensively and reported in the survey methods literature, perhaps more than any other source of nonsampling error. For many surveys, measurement error is also the most damaging source of error. The key components of measurement error are the respondent, the interviewer, and the survey questionnaire. Respondents may either deliberately or unintentionally provide incorrect information. Interviewers can cause errors in a number of ways. They may falsify data, inappropriately influence responses, record responses incorrectly, or otherwise fail to comply with the survey procedures. The questionnaire can be a major source of error if it is poorly designed. Ambiguous questions, confusing instructions, and easily misunderstood terms are examples of questionnaire problems that can lead to measurement error.

We also consider the errors that arise from the information systems that respondents may draw on to formulate their responses. For example, a farm operator or business owner may consult records that may be in error, and thus cause an error in the data reported. It is also well known that the mode of administration can have an effect on measurement error. For example, information collected by telephone interviewing is, in some cases, less accurate than the same information collected by face-to-face interviewing. Finally, the setting or environment within which the survey is conducted can also contribute to measurement error. For example, for collecting data on sensitive topics such as drug use, sexual behavior, fertility, and so on, a private setting for the interview is often more conducive to obtaining accurate responses than one in which other members of the household are present. In establishment surveys, topics such as land use, loss and profit, environmental waste treatment, and resource allocation can also be sensitive. In these cases, assurances of confidentiality may reduce measurement errors due to intentional misreporting.

These sources of nonsampling error can have a tremendous effect on the accuracy of a survey estimate. To illustrate, consider the previous example of a survey to estimate the income in a community where the unknown, true average income is $35,181. With a sample of 400 persons, we might expect the error in our estimate due to sampling error to be around $500. (See Chapter 9 for the details on how this sampling error prediction is constructed.) That is, the estimate from the survey could be as low as $34,681 and as high as $35,681. However, as a consequence of nonsampling errors from all the sources described above, the level of error in the survey estimate could be much higher. For example, it is not unreasonable to expect the error to be $1000—twice the size of the error for sampling alone! As a result, the survey estimate could be as low as $34,181 and as high as $36,181 when the true parameter value is $35,181 (see Figure 2.3).

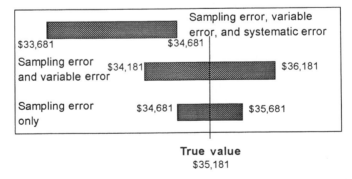

Figure 2.3 Range of estimates produced by a sample survey subject to sampling error, variable error, and systematic error. Shown is the range of possible estimates of average income for a sample of size 400. The range is much smaller with sampling error only, and when systematic nonsampling error is introduced, the range of possible estimates may not even cover the true value.

Even more damaging errors in the estimate can result when the errors of respondents who overreport their incomes do not balance against the errors of respondents underreporting their incomes; that is, if reporting errors tend to be in one direction, which tends to bias the estimate. For example, in the case of income, the negative errors may be the dominant errors since respondents, in general, may have a greater tendency to underreport their income than to overreport it. This type of situation leads to a *negative bias* in the estimates, which means that we expect that the survey estimate will always be lower than the true population parameter value by some unknown amount. In this case, the expected range for the income estimate might be more like $33,681 to $34,681 when the actual value is $35,181. The concepts of biasing *or systematic errors* and nonbiasing or *variable errors* are discussed further in the next section.

Processing Error

The fifth and final source of error in Table 2.2 is processing error, errors that arise during the data processing stage, including errors in the editing of data, data entry, coding, the assignment of survey weights, and the tabulation of survey data. As an example of editing error, suppose that a data editor is instructed to call back the respondent to verify the value of a budget item whenever the value of the item exceeds a specified limit. In some cases, the editor may fail to apply this rule correctly, thus causing an error in the data.

For open-ended items that are coded, coding error is another type of data processing error. The personnel coding the data may make mistakes or deviate from prescribed procedures. The system for assigning the code numbers—for variables such as place of work, occupation, industry in which the respondent is employed, field of study for college students, and so on—may itself be quite ambiguous and very prone to error. As a result, code numbers may be assigned inconsistently and inappropriately, resulting in significant levels of coding error.

The survey weights that compensate statistically for unequal selection probabilities, nonresponse error, and frame coverage errors may be calculated erroneously or there may be programming errors in the estimation software that computes the weights. Errors in the tabulation software may also affect the final data tables. For example, a spreadsheet used to compute the estimates may contain a cell-reference error that goes undetected. As a result, the weights are applied incorrectly and the survey estimates are in error. See Chapters 7 and 9 for details.

2.4 GAUGING THE MAGNITUDE OF TOTAL SURVEY ERROR

As we saw in Section 2.3, the development of a survey design involves many decisions that can affect the total error of a survey estimate. These are decisions regarding the sample size, mode of administration, interviewer training and supervision, design of the questionnaire, and so on, that ultimately will determine the quality of the survey data. Further, these decisions are often influenced by the costs of the various options and their effects on the duration of the survey. A mail survey may be less expensive than a survey conducted by personal visit, but the time allowed for data collection may be such that the mail survey option is not feasible. Face-to-face interviewing may not be affordable due to interviewer costs and other field costs, and less expensive options for collecting the data within the time limits available for the survey must be considered. Telephone interviewing may be both affordable and timely; however, the quality of the data for some items may not be adequate. For example, questions requiring the respondent to consider visual information such as pill cards or magazine covers he or she may have seen are not feasible by telephone. Thus, in determining the design of a survey, one must consider and balance several factors simultaneously to arrive at the design that is best in terms of data quality while meeting the schedule, budget, and other resource constraints for the survey. The resulting design is then a compromise which reflects the priorities attributed to the multiple users and uses of the data.

Making the correct design decisions requires the simultaneous consideration of many quality, cost, and timeliness factors and choosing the combination of design elements that minimizes the total survey error while meeting the budget and schedule constraints. An important aid in the design process is a means of quantifying the total error in a survey process. In this way, alternative survey designs can be compared not only on the basis of cost and timeliness, but also in terms of their total survey error.

As an example, consider two survey designs, design A and design B, and suppose that both designs meet the budget and schedule constraints for the survey. However, for the key characteristic to be measured in the study (e.g., the income question), the total error in the estimate for design A is + or − $3780, while the total error in estimate for design B is only + or − $1200. Since design B has a much smaller error, this is the design of choice, all other things being equal. In this way, having a way of summarizing and quantifying the total error in a survey process can provide a method for choosing between competing designs.

Such a measure would have other advantages as well. For example, suppose that we could establish that most of the error in the survey process under design B is due to nonresponse error. This indicates that nonresponse is the most important source of error for design B, and thus, efforts to improve the quality of the survey data further under design B should focus on the reduction of nonresponse error. To free up resources to reduce nonresponse error, the survey design could consider substituting less expensive procedures for more costly ones in other areas of the survey process. Even though other sources of error may increase by these modifications, the overall effect would be to reduce survey error by the reduction in nonresponse error. In this way, the total error associated with design B could be reduced without increasing the total costs of the survey.

As will become clear in this section, there are many ways of quantifying the total survey error for a survey estimate. However, one measure that is used most often in the survey literature is the total *mean squared error* (MSE). Each estimate that will be computed from the survey data has a corresponding MSE which reflects the effects on the estimate of all sources of error. The MSE gauges the magnitude of total survey error, or more precisely, the magnitude of the effect of total survey error on the particular estimate of interest. A small MSE indicates that total survey error is also small and under control. A large MSE indicates that one or more sources of error are adversely affecting the accuracy of the estimate. As we have said, this information is important since it can influence the way the data are used as well as the way the data are collected in the future should the survey ever be repeated.

One of the primary uses of the MSE is as a measure of the accuracy of the survey data. Unfortunately, it is usually not possible to compute the MSE directly from the survey data, particularly when the data are subject to large nonsampling errors. In most situations, special evaluation studies that are supplemental to the main survey are needed to measure the total MSE. Still, measures of data accuracy are important for the proper interpretation of survey results.

As an example, in the 2000 U.S. population census, a *postenumeration survey* (PES) was conducted following the census to estimate the number of persons missed by the census as well as potentially to use the estimates of number of persons missed for correcting the final census numbers. Special studies were conducted during the census and the PES to measure the MSE of the estimated census total with and without adjustment for the undercount. One important use of the census count is to determine how the 435 seats in the U.S. House of Representatives should be distributed among the 50 states, a process referred to as *congressional apportionment*. The amount of the improvement in the quality of the census counts as measured by the total MSE was an important consideration in the decision not to use the adjusted numbers for apportionment in 2000.

Thus, the concept of a total survey error measure is fundamentally important to the field of survey design and improvement. Indeed, the primary objective of survey design can be stated simply as minimizing the MSEs of the key survey estimates while staying within budget and on schedule. Therefore, the remainder of this chapter is devoted to developing and understanding these critical concepts. In the next section we discuss another way of classifying the nonsampling errors that arise from the survey process: errors that are *variable* and errors that are *systematic*. As we shall see, variable error and systematic error are the essential components of the total MSE since the former determines the variance of a survey estimate and the latter the bias. Later in the chapter we show that the MSE is essentially the sum of variance and bias components contributed by the many sources of error in the survey process.

> The *primary objective of survey design* is to minimize the MSEs of the key survey estimates while staying within budget and on schedule.

Variable Errors

In this discussion it will be useful to consider a specific item or question on the survey questionnaire: for example, the income question. For this item, the nonsampling errors that arise from all the various error sources in a survey have a cumulative effect on the survey responses, so that the value of an item for a particular person in a survey is either higher or lower than the true value for the person. In other words, the cumulative effect of the total error for a particular observation is either positive or negative. This is true for all observations: The cumulative effect of all errors will be positive for some persons and negative for others. Suppose that we wish to estimate the mean income for the population using the average of the sample observations (i.e., the sum of the observations in the sample divided by the number of observations). Further suppose that persons in the population are just as likely to make positive errors as they are to make negative errors in reporting their incomes. In this situation, the negative errors will, to some extent, offset the positive errors and the net effect of the errors on the average will be very small. That is, the negative errors in the observations tend to cancel the positive errors, so that nonsampling errors will have essentially no biasing effect on the estimate of the population mean.

Further, if the survey process for collecting income data were to be repeated for the population, a very similar result would occur (i.e., the negative errors would approximately cancel the positive errors). Error sources that produce these types of errors are called *variable error sources* and the errors arising from them are referred to as *variable errors*. When the frequency of variable errors in the data is high, the data are often referred to as *noisy*, since variable error limits our ability to understand what the data are telling us just as a noisy room limits our ability to hear a speaker.

Another concept that is closely related to variable error and often encountered in the survey literature is *data reliability*. *Reliability* refers to the ratio of two types of variation in the observations: the variation in the true values among the population members, and the total variation, which includes the true value variation as well as the additional variation due to variable error. The ratio of these two variances is referred to as the *reliability ratio*. Thus, the reliability ratio ranges from 0.0 to 1.0. *Perfect reliability* occurs when there are no variable errors in the data. In this case, the numerator of the reliability ratio is equal to the denominator and thus the reliability ratio is 1. As the amount of variable error increases, the denominator of the ratio increases and thus the reliability ratio decreases. For example, when the reliability ratio is 0.50 or 50% for a characteristic being measured, the variation in the true values of the characteristic in the population is equal to the variation in the values observed due to variable nonsampling error. This is considered by most standards to be very poor reliability.

In some cases, unreliable data can be recognized by a close examination of the variables that are related in the survey. For example, in an attitudinal survey, the attitudes that a person expresses toward similar issues should show very good agreement. If they do not, this may be a sign of poor reliability on the attitudinal measures. However, in most situations, determining whether the observed data are reliable requires special studies to evaluate the reliability. An example of one type of study to evaluate reliability is a reinterview study, in which the interview is repeated for a sample of households a few days after the original interview. Assuming that the first interview does not influence the responses to the second interview in any way, a comparison of the results of the two interviews will reveal whether the data are reliable. If the data are reliable, there will be good agreement between the first and second responses. Considerable disagreement is an indication of unreliable data. Thus, reliability is often referred to as *test–retest reliability* (referring to first and second measurement), a term that is rooted in the educational psychometric literature (see Lord and Novick, 1968).

Systematic Errors

In many situations, the negative and positive errors do not exactly cancel. For example, positive errors may be much more prevalent than negative errors, and consequently, when the observations are averaged together, the average may be much larger than the true population average. The sample average is then said to exhibit a positive bias. Conversely, the number of respondents in the sample who make negative errors may be considerably larger than the number who make positive errors, and thus the estimate of the mean is smaller than what it would have been without nonsampling error (i.e., it is negatively biased). Errors that do not sum to zero when the sample observations are averaged are referred to as *systematic errors* (Figure 2.4). When the systematic errors are such that the errors in the positive direction dominate (or outnumber) the errors in the negative direction, the sample average will tend to be too high or positively biased. Similarly, when the systematic errors are such that the negative errors dominate, the sample average will be negatively biased.

It is important to note that in our discussion of nonsampling error, the definitions of systematic error and variable error do not refer to what happens in the one *particular* sample that is selected. Rather, they refer to the collection of samples and outcomes of the same survey process over many repetitions under essentially the same survey conditions. This concept of the survey as a repeatable process is similar to the assumptions made in the literature on statistical process control. For example, consider a process designed for the manufacture of some product, say a computer chip. What is important to the designers of the process is the quality of the chips produced by the process over many repetitions of the process, not what the process yields for a parti-

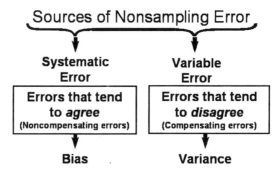

Figure 2.4 Two types of nonsampling error. All nonsampling error sources produce variable error, systematic error, or both. Systematic error leads to biased estimates; variable error affects the variance of an estimator.

cular chip. (Of course, that may be of primary interest to the consumer who purchases the chip!) Similarly, a survey is a process—one that produces data. Although we as consumers of the data are interested primarily in what happens in a particular implementation of the survey, the theory of survey data quality is more concerned about the process and what it yields over many repetitions.

Example 2.4.1 The survey question can be a source of either systematic or variable error in a survey. For example, consider a question that asks about a person's consumption of alcohol in the past week. Respondents may try to estimate their consumption rather than recall exactly the amount they consumed. However, many respondents may deliberately underreport their alcohol consumption to avoid embarrassment, an effect referred to in the literature as *social desirability bias*. As a result of this systematic underreporting, the average amount of alcohol consumed across all sample members will be biased downward, and the estimate of the average amount of alcohol consumed per person will be underestimated.

An example of variable errors occurs when respondents try to estimate events they wish to report accurately, such as the number of trips to the grocery store in the last six months. Rather than try to recall and count the number of trips in six months, many respondents might use some method to estimate the number. For example, some might say that they usually go to the grocery store about twice a week and multiply this number by 26, roughly the number of weeks in the six-month interval. Others may use some other method of estimation. The result is that some respondents may report a number that is slightly higher than the actual number and others may report a slightly lower number than actual. When the entire sample is considered, however, the average number of trips to the grocery store could still be very close to the actual average since the positive and negative errors cancel each other when the data are summed up.

Effects of Systematic and Variable Errors on Estimates

Although both systematic and variable error reduce accuracy, which type of error is more harmful to accuracy? The answer to this question depends on what is being estimated. As we have shown, for *linear estimates* such as estimates of population means, totals, and proportions—in other words, estimates which are sums of the observations in the sample—systematic errors will lead to biases in the estimates, whereas variable errors will tend to cancel one another out and are therefore often nonbiasing. Thus, for linear estimates, systematic errors may be more damaging than variable error, due to their biasing effects. In addition, as we see later, the effect of variable nonsampling errors on linear estimates is very similar to the effect of sampling error on linear estimates; that is, both sampling error and variable nonsampling error can be reduced by increasing the sample size. Therefore, one way to compensate for the effects of variable errors on linear estimates is by increasing the sample size for the survey. However, increasing the sample size will have no effect on systematic error. As mentioned previously, it is possible for the systematic errors to increase as the sample size increases, as a result of increasing the scope of work for the survey and potentially losing some control over the nonsampling error sources.

For *nonlinear estimates* such as estimates of correlation coefficients, regression estimates, standard error estimates, and so on, the answer to the question of what type of error is more damaging is not so simple. For these types of estimates, both systematic errors and variable errors can lead to bias. For example, it can be shown that estimates of regression coefficients are attenuated (i.e., biased toward zero) in the presence of variable error, while for systematic error, the direction of the bias is unpredictable. Therefore, for nonlinear estimates, there is little to choose between systematic error and variable error. Understanding exactly why this is true is not within the scope of this book; however, see Fuller (1987) and Biemer and Trewin (1997) for useful discussions for those interested in pursuing this topic.

Many times, the primary purpose of a sample survey is to report means, totals, and proportions for some target population (i.e., descriptive studies of the population). For this reason, in designing surveys for the reduction of total error, priority is usually given to the identification and elimination of the major sources of systematic error. Although the goal of survey design is the minimization of both types of error, the survey designer often must decide which types of errors are most damaging and control for those while other types of errors are either ignored or controlled much less.

For example, the designer may have to decide whether it is better to allocate more survey resources on interviewer training than on further refinement of the questionnaire. Another decision might be whether it is better to devote more survey resources to the follow-up of survey nonrespondents than to spend those resources on more extensive quality control checks for data-entry errors. In these situations, it is useful to have some idea as to whether a particular error source produces predominately systematic error or variable error.

In most situations, eliminating the source that produces systematic error should take priority over the error sources where the primary risk is variable error. However, there are no hard-and-fast rules about this. Nevertheless, in our subsequent discussion of the error sources in the survey process, some consideration of the risk of each systematic error from the error source will be useful.

Error Sources Can Produce Both Systematic and Variable Errors

Some error sources produce errors that are primarily systematic. For example, nonresponse error is primarily a systematic error rather than a variable error since the cumulative effect of the nonresponse error on a particular survey estimate is to bias the estimate. As we shall see in Chapter 3, the magnitude of the bias is a function of the nonresponse rate and the difference in the characteristics under study between the average respondent and the average non-respondent. Frame noncoverage error behaves in much the same way. If a particular type of population member is missing from the frame—for example, that small farms are missing in most agricultural survey frames—repeated implementations of the survey process using the incomplete frame will tend to err in the same direction.

Some error sources produce errors that are primarily variable error. For example, keying error is typically variable error, since errors data keyers make tend to be haphazard and omnidirectional. Rarely will a group of keyers make errors that tend to either increase or decrease the value of an estimate. Rather, for the most part, these types of errors will tend to cancel one another out. Another example used earlier is respondent estimation, as in estimating the number of trips to the grocery store. Some respondents may guess high while others may guess low, so that, on balance, the average of the guesses may be very near zero.

Other error sources produce both variable error and systematic error. For example, the errors committed by interviewers in carrying out their assignments may be a combination of variable and systematic errors. Consider the income question again. Some interviewers, by the mannerisms, dress, comments made earlier in the interview, and so on, may have a tendency to elicit responses that are higher than the true incomes, while other interviewers may tend to have just the opposite influence. However, particularly at lower income levels, the general tendency in the population may be to overstate actual income as a result of the respondents wanting to appear better off than they really are. Thus, although the overall tendency (over many repetitions of the survey process) is to overreport income, interviewers also provide a variable error component so that the income data are both biased and unreliable.

In the next section we discuss a method for summarizing the combined effects of variable errors and systematic errors on a survey estimate using the mean squared error of the estimate.

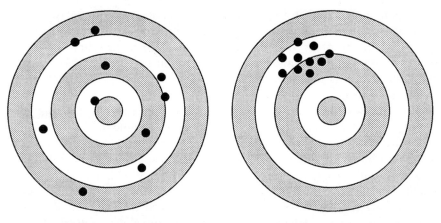

(*a*) **Large variance and small bias**   (*b*) **Large bias and small variance**

Figure 2.5 Systematic and variable error expressed as targets. If these targets represented the error in two survey designs, which survey design would you choose? The survey design in part (*a*) produces estimates having a large variance and a small bias, while the one in part (*b*) produces estimates having a small variance and a large bias.

2.5 MEAN SQUARED ERROR

Analogy of the Marksman and the Target

To better understand how total survey error can be decomposed into components for systematic error and variable error—or, equivalently, bias and variance—consider the picture in Figure 2.5. In this figure we use the process of shooting at a target to illustrate the process of using a survey to estimate some parameter of the population. The bull's-eye on the target represents the population parameter we wish to estimate with the survey data. Of course, in practice, there may be many parameters that we wish to estimate with a survey, but for now we concentrate on a particular parameter, such as the average income for the population. Conducting a survey to estimate the parameter is analogous to a marksman taking aim and shooting at the bull's-eye on the target. If the marksman's aim is accurate, he or she scores a bull's-eye; otherwise, he or she misses the bull's-eye by some distance. The distance between the point where the marksman hits the target and the bull's-eye is the total error in the marksman's aim. Similarly, if we conduct a survey, the goal is to estimate the population parameter exactly, but we miss the parameter because of survey error. The "distance" between the estimate from the survey and the population parameter is the total error in the estimate and is analogous to the total error in the marksman's aim.

Now suppose the marksman shoots repeatedly at the target, each time aiming at the bull's-eye. That is analogous to repeating the survey process a number of times under the same conditions, each time attempting to estimate the population parameter. In practice, we would implement the survey process only once to estimate the population parameter. However, if we could repeat the survey many times, the variation we observed in the survey estimates would tell us something about the total error in the survey process, just as the pattern of hits on the target tells us something about the error in the marksman's aim.

Now the error in a survey estimate, like the marksman's aim, consists of systematic and variable error components. For example, if the marksman's sights are not properly adjusted, he or she will probably miss the bull's-eye. Further, if his or her aim is consistent, the distance between each hit and the bull's-eye will be roughly the same, due to this sight misalignment. The marksman's sight misalignment is analogous to a biased survey process. The survey process may produce very consistent results each time it is implemented; however, the estimates all differ from the parameter value by roughly the same amount and in the same direction.

In addition to sight misalignment, the marksman may miss the target for a number of other reasons. For example, the marksman's aim may not be steady, and therefore each time he or she shoots, the bull's-eye will be missed by an unpredictable, random amount. Sometimes the hit is to the left of the bull's-eye, and other times the hit may veer to the right, and above or below the bull's-eye. Other factors, such as the wind or weather, the shape of the projectiles being fired, and the weapon itself may also have unpredictable, random effects on the accuracy of each shot. These factors are analogous to the variable errors in a survey. Each time the survey process is repeated, random variation due to a whole host of factors may affect the accuracy of the estimate and add to the total error (the distance between the hit and the bull's-eye).

The two targets in Figure 2.5 could correspond to two different marksmen with two different weapons. Note that the pattern of hits on the target on the left suggests that systematic error may be a problem for that marksman; that is, there is something inherently wrong with either the weapon or some other aspect of the shooting process that affects all the shots at the target in the same way. The pattern of hits for the left target suggests that the systematic error is smaller, but variable error is a problem. That is, the cumulative effect of many factors associated with shooting at the target causes the marksman to miss the target in seemingly random ways.

These targets can be used not only to help us understand systematic error (bias) and variable error (variance) in a survey process, but also how to measure them. Suppose that we were to repeat the same survey process many times under the very same conditions. That is, we use the same sampling procedure (but a different sample of respondents at each replication), the same questionnaire, the same process for hiring and training interviewers, the same data collection procedures, and so on. Each replication of the survey will yield one estimate of the population parameter represented by a hit on the target. The target on the right corresponds to one type of survey process for estimating the parameter, and the target on the left represents another survey process for estimating the same parameter. Thus, the two survey processes produce a different mixture of systematic and variable error components. The survey corresponding to the left target has a considerable systematic error; however, the variance (i.e., variable error) for that process is relatively small. The survey corresponding to the right target has a large variable error, but the systematic error is small.

Now suppose that both survey processes have approximately the same cost. Which survey process should be chosen for estimating the population parameter: the survey corresponding to the left target or the one corresponding to the right target? In making this decision, it would be very helpful if there were some way of quantifying the accuracy of the two surveys by combining the effects of systematic and variable error into a single dimension; in other words, a way of summarizing the total error into a single number. Then the survey process producing the smaller level of total survey error would be the preferred survey process.

Summarizing the Total Error of an Estimator Using the Mean Squared Error

There are many ways to summarize error in processes that produced the patterns or more generally, the total error in a survey process, depicted in Figure 2.1. However, one that is used in the statistical literature universally because of its favorable statistical properties is the *mean squared error*, a measure of the average closeness of the hits to the bull's-eye, where *closeness* is defined as the squared distance between a hit and the bull's-eye. To compute the mean squared error, we measure the distance between each hit and the bull's-eye, square that distance, and average these squared distances across all the hits.

As an example, consider the situation where each survey process, process A and process B, has been repeated 10 times, each time under identical conditions. For these 20 implementations, the error in the estimates for survey process A (distances for the left target) and survey process B (the right target) is shown in Figure 2.5. The average of the squared distances of the 10 hits for the right target is 0.15 and the average for the left target is 4.5. Therefore, by the mean squared error criterion, the survey process represented by the left target is preferred because the mean squared error of the estimate is smaller.

The *mean squared error* (MSE) of the marksman's aim is the average squared distance between the hits on the target and the bull's-eye. The MSE of a survey estimate is the average squared difference between the estimates produced by many hypothetical repetitions of the survey process (corresponding to the hits on the target) and the population parameter value (corresponding to the bull's-eye).

To put this illustration more concretely into a survey context, suppose that the right target represents a survey process where the data are collected from respondents by interviewers (i.e., interviewer-assisted mode) and the left target represents a survey process where the same data are collected by respondents recording their responses directly on the questionnaire (self-administered mode). Let us assume that both survey processes are based on the same sample size so that any differences in variance are due strictly to nonsampling variable error. Suppose that the interviewer-assisted mode has a larger bias, due to the influencing effect of the interviewer on the respondent; however, the variance is smaller as a result of reduction of respondent comprehension due to the assistance provided by the interviewer. Suppose further that the self-administered mode eliminates the bias resulting from interviewer influence but the process introduces larger variable error as a result of respondents interpreting the questions in different ways. Thus, the error in the interviewer-assisted survey resembles the target on the right in Figure 2.5, and the error in the self-administered mode resembles the left target. This example illustrates how two survey processes based on different modes but aimed at collecting the same information can have very different bias and variance characteristics.

A key aspect of the definition of the mean squared error given above is based on the idea that the same survey process is repeated many times for the same population and under the same survey conditions. The mean squared error is then the average squared difference between the estimate from each replication of the process and the population parameter value. There are two problems with this definition. First, it is usually impossible to repeat the survey process under identical conditions each time. Our world is dynamic and ever changing and the survey conditions may also change considerably from one implementation to the next. Many factors (the weather, politics, etc.) influence the outcomes of surveys, and these factors vary over time. The population parameter value itself may change over time. Further, it would not be cost-effective to repeat the survey process multiple times, or even twice, for the purpose of estimating the MSE. There are more efficient and effective ways of estimating the components of the MSE, and many of these are discussed in Chapter 8.

Another difficulty encountered in trying to assess the MSE of a survey estimate is determining the true value of a population parameter so that the error in the estimate can be quantified. Simply conducting another survey to estimate the parameter is not likely to be sufficient unless special procedures are put in place to ensure that the estimate is highly accurate. Other methods for assessing the true values of the characteristic for the sample members could involve the use of very accurate administrative records or the use of more expensive and elaborate measurement devices that eliminate most of the error inherent in the original survey process. Usually, these methods are conducted after the original survey data have been collected. In Chapter 8 we discuss these methods in some detail.

Estimation of the MSE is usually a complex and costly process, and therefore the total MSE is seldom estimated in practice. When it is estimated, the result is often only a rough approximation of the actual MSE. In addition, quite often only a few of the most important components of the MSE are estimated. For example, a typical approach to computing the MSE involves estimating several bias and/or variance components associated with the major sources of systematic and variable errors and then combining this error information to produce the mean squared error. The computational formula described above does not lend itself readily to computing the mean squared error in this manner. In the next section we describe an alternative method for computing the mean squared error that addresses this shortcoming.

Decomposing the Mean Squared Error into Bias and Variance Components

Let us revisit the targets in Figure 2.5 to discuss another way to view the total error in a survey estimate. Note that for each target, the hits form a cluster of points on the target. Consider the target on the right first and locate the point that is approximately the center (or centroid) of the hits or estimates for this target. The point in the center of the cluster represents the mean or average error of the survey process. The distance between the center of the cluster of hits and the bull's-eye is the systematic error or bias in the survey process. Further, the distances between the individual hits and the cluster center represent the variable errors in the survey process.

Thus, the bias in a survey estimate can be computed by computing the average of the estimates produced by repeating the survey process many times under the same conditions, averaging these estimates, and then subtracting from this average the value of the true population parameter. The variance of the estimator from the survey can be computed as the average squared distance between each survey estimate from the repetitions and the average survey estimate.

This discussion suggests that there are two ways to compute the mean squared error of a survey estimator:

METHOD 1

MSE = average squared distance between the hits and the bull's-eye, or, in survey terms,

= average squared difference between the survey estimates from repeating the survey many times and the true population parameter value

---

METHOD 2

MSE = squared distance between the center of the hits and the bull's-eye
          + average squared distance between the hits and the center of the
          hits, or, in survey terms,
      = squared distance between the average value of the estimator
          over replications of the survey process and the true population
          parameter value + average squared difference between the
          estimates from the replications and the average value of the
          replications

Both methods of computing the mean squared error will produce the same value; however, method 2 is often preferred because it decomposes the mean squared error into two components: squared bias and variance. The *squared bias* is the squared distance between the average value of the estimator over replications of the survey process and the true population parameter value. The *variance* is the average squared difference between the estimates from the replications and the average value of the replications. Therefore, another way of writing the formula for method 2 is

$$MSE = \text{Squared bias} + \text{Variance} \tag{2.1}$$

This formula says that the total MSE for an estimate is equal to the bias squared plus the variance. If we know that the bias of an estimate is 0, the MSE is simply the variance of the estimate. However, when the bias is not zero, the bias must be estimated in order to compute an estimate for the MSE. Computation of the bias requires knowledge of the true parameter value; however, the variance can be computed without knowing the true parameter value.

To see this, consider Figure 2.6, where we show the hits on the targets in Figure 2.5 without the targets. Notice that we can still compute the variance, which is the squared distance from each hit to the center or average of the hits. However, we cannot compute the bias since there is no bull's-eye or true parameter value. This figure also illustrates the fallacy of choosing between two survey processes on the basis of variance alone. Note that the survey process on the right would always be chosen over the survey process on the left, regardless of the bias in that design, since the variance is all that is known.

To illustrate that both methods for computing the MSE will yield the same results, consider the data in Tables 2.3 and 2.4. In Table 2.3 we compute the MSE for the two survey designs in Figure 2.5 using method 1. We measured the distance from each hit in Figure 2.5 to its corresponding bull's-eye and then squared these distances. The average squared distance for each target is the MSE for the corresponding process. For the survey process on the left, this

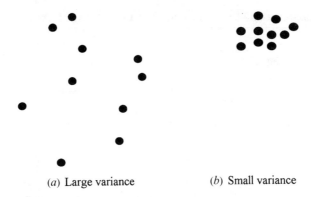

(a) Large variance   (b) Small variance

Figure 2.6 Truth is unknown. Determining the better survey process when the true value of the population parameter is unknown is like trying to judge the accuracy of two marksmen without having a bull's-eye. All that is known about the sets of survey estimates is that the survey design in part (a) produces a large variance and the survey design in part (b) produces a small variance.

Table 2.3 Computation of the Mean Squared Error Using Method 1[a]

	Left Target			Right Target	
Hit	Distance from Hit to Bull's-eye	Squared Distance to Bull's-eye	Hit	Distance from Hit to Bull's-eye	Squared Distance to Bull's-eye
1	2.2	4.8	1	3.1	9.6
2	−3.6	13.0	2	3.7	13.7
3	−4.5	20.3	3	5.3	28.1
4	6.8	46.2	4	4.9	24.0
5	5.1	26.0	5	6.1	37.2
6	−7.2	51.8	6	4.4	19.4
7	−3.9	15.2	7	2.8	7.8
8	5.3	28.1	8	6.1	37.2
9	−1.8	3.2	9	4.5	20.3
10	3.1	9.6	10	4.1	16.8
Avg or center	0.15	21.8 (= MSE)	Avg or center	4.5	21.4 (= MSE)

[a] The mean squared error is the average squared distance from the hit (or estimate) to the bull's-eye (or parameter value). For method 1 we compute the distance (or error) from the hit to the bull's-eye for each hit, square the result, and then average over all 10 hits.

average is 21.8, and for the process on the right it is 21.4 (see the last row in Table 2.3).

In Table 2.4 the MSE is computed using method 2. For this method we locate the center of the hits and measure the distance between each hit and the center of the hits and square the result. The average of these squared distances (i.e., 21.81 for the left target and 1.17 for the right target) is the

Table 2.4 Computation of the Mean Squared Error Using Method 2[a]

	Left Target			Right Target	
Hit	Distance from Hit to the Center of the Hits	Squared Distance from Center	Hit	Distance from Hit to the Center of the Hits	Squared Distance from Center
1	$(2.2 - 0.15) = 2.05$	4.20	1	$(3.1 - 4.5) = -1.40$	1.96
2	$(-3.6 - 0.15) = -3.75$	14.06	2	$(3.7 - 4.5) = -0.80$	0.64
3	-4.65	21.62	3	0.80	0.64
4	6.65	44.22	4	0.40	0.16
5	4.95	24.50	5	1.60	2.56
6	-7.35	54.02	6	-0.10	0.01
7	-4.05	16.40	7	-1.70	2.89
8	5.15	26.52	8	1.60	2.56
9	-1.95	3.80	9	0.00	0.00
10	2.95	8.70	10	-0.40	0.16
Avg	0.0	21.81 (= Variance)	Avg	0.0	1.17 (= Variance)
Bias = $(0.15 - 0.0) = 0.15$		Bias^2 = 0.023	Bias = $(4.5 - 0.0) = 4.5$		Bias^2 = 20.25
MSE = Bias^2 + Variance = 21.8			MSE = Bias^2 + Variance = 21.4		

[a] The mean squared error can be computed in two stages. First, measure the distance from each hit to the center of the hits and square the result. Average these square distances over all 10 hits. Finally, compute the distance between the center of the hits to the bull's-eye and square the result. The MSE is the sum of these two quantities. Note that these results agree with the results from Table 2.3, demonstrating that the two methods yield the same results.

variance for the estimate. Note that the variance for the left target is many times larger than the variance for the right target. Then the distance from the center of the hits and the bull's-eye is measured to produce 0.15 for the left target and 4.5 for the right target. As we have already observed, the right target shows a much larger bias than the left target. Putting the variance and bias components together, we have MSE (left target) = $(0.15)^2 + 21.81 = 21.83$ and MSE (right target) = $(4.5)^2 + 1.17 = 21.42$. This agrees with the MSEs we computed in Table 2.3. Thus, we have demonstrated that the two methods for computing the MSE are equivalent.

Major Components of the MSE

Each source of error in Table 2.2 can contribute both bias and variance components to the total MSE; however, some error sources pose a greater risk for bias, some for variance, and some error sources can contribute substantially to both bias and variance. Table 2.5 lists each major error source along with an assessment of the risk of variable error and systematic error for each. These risks will depend on the specifics of the survey design and the population to be surveyed, and there are no hard-and-fast rules regarding which sources of error are more problematic for systematic error or variable error. Nevertheless, Table 2.5 provides an indication of the risk for a typical survey.

Table 2.5 Risk of Variable Errors and Systematic Errors by Major Error Source

MSE Component	Risk of Variable Error	Risk of Systematic Error
Specification error	Low	High
Frame error	Low	High
Nonresponse error	Low	High
Measurement error	High	High
Data processing error	High	High
Sampling error	High	Low

For example, in a typical survey using acceptable random sampling methods, the risk of bias resulting from sampling error is quite small, and sampling variance is inevitable. Conversely, for specification error, the error in the estimate is primarily bias, since errors in the specification of the survey question is more likely to lead to systematic error than variable error. Nonresponse error also poses a greater risk to nonsampling bias than to variance; although some nonresponse adjustment methods can contribute substantially to the variance when the nonresponse rate is quite high. Frame error, particularly error due to population members missing from the frame, is viewed primarily as a biasing source of error. However, as we see later in this book, measurement error and data processing error can pose a risk for both bias and variance in the survey estimates.

Using Table 2.5 as a guide, we can write an expanded version of the MSE equation in (2.1). The squared-bias component can be expanded to include bias components for all the sources of error in the table that have a high risk of systematic error (i.e., specification bias, B_{SPEC}; nonresponse bias, B_{NR}; frame bias, B_{FR}; measurement bias, B_{MEAS}; and data processing bias, B_{DP}). Note that these components of bias sum to produce the total bias component, called *bias*. Similarly, the variance component is the sum of the components for the major sources of variance in the table (i.e., sampling variance, Var_{SAMP}; measurement variance, Var_{MEAS}; and data processing variance, Var_{DP}. Thus, the expanded version of the MSE formula showing components for all the major sources of bias and variance is as follows:

$$MSE = Bias^2 + Variance$$
$$= (B_{SPEC} + B_{NR} + B_{FR} + B_{MEAS} + B_{DP})^2$$
$$+ Var_{SAMP} + Var_{MEAS} + Var_{DP} \tag{2.2}$$

In practice, one method of assessing the MSE of an estimate is to estimate the eight components shown in (2.2). Similarly, one method of reducing the MSE is to develop a survey design that minimizes the contributions of each of the eight components in (2.2) to the total MSE. This approach to survey design is the basis of the fundamental principles of survey design discussed throughout this book. The illustration discussed in Section 2.6 shows how this approach can be used to identify the best survey design for a given survey budget among several alternative designs.

2.6 ILLUSTRATION OF THE CONCEPTS

To illustrate the usefulness of the mean squared error as a survey quality measure, consider a scenario in which a survey planner must choose among three alternative survey designs, labeled A, B, and C, for collecting data to meet the same set of research objectives. Design A specifies that the data will be collected by face-to-face interviewing, design B specifies telephone interviewing, and design C specifies data collection by mail, which is a self-administered mode. Since the costs associated with each mode of administration differ, the designs have been adjusted so that the total data collection costs are the same for each. For example, since face-to-face interviewing is more expensive than telephone or mail data collection, the sample size under design A must necessarily be smaller than that for the other two designs to meet the same total cost. Similarly, the per interview cost of design B is higher than the per interview cost of design C, and thus to cost the same, the sample size for design B must be smaller than that for design C.

In this illustration we assume that we have fairly complete information regarding the biases and variances associated with various sources of survey error for each design. In particular, we know roughly the biases associated with nonresponse, frame coverage, and measurement error. Since design A is expected to achieve the highest response rate, we estimate that the bias due to nonresponse will be the lowest for this design. Design C is expected to have the lowest response rate so its nonresponse bias will be highest.

Similarly, for design A, an area frame sampling approach will be used which implies that all housing units in a sample of areas will be listed and sampled. This intensive listing and sampling process will ensure complete coverage of the target population and therefore frame bias will be zero. By contrast, design B will use a random-digit-dial (RDD) telephone frame. For this frame, 10-digit telephone numbers are generated randomly so that all housing units that have a telephone have a chance of being selected. However, nontelephone housing units have no chance of being selected, and consequently, a small frame bias is expected as a result of the noncoverage of nontelephone housing units. Design C specifies the use of a telephone directory-type listing for obtaining the addresses of target population members, and thus both nontelephone and telephone households with unlisted numbers will be missed by this frame. Consequently, frame bias is expected to be the highest for this design.

The last source of bias for which information is available for all three designs is measurement error. As discussed previously, measurement bias arises from many sources, including the questionnaire, the respondent, the interviewer (designs A and B only), the setting, and the mode of administra-

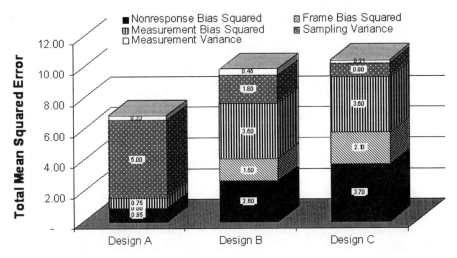

Figure 2.7 Comparison of the total mean squared error for three survey designs. Design A is preferred since it has the smallest total error, even though sampling error is largest for this design.

tion. This combination of systematic error sources is expected to be smaller for face-to-face interviewing for the primary objectives of the survey. Designs B and C are expected to cause larger measurement errors in these data; however, the net effect of measurement error bias will not differ appreciably for the two designs.

Next, with regard to measurement variance, interviewer error variance is expected to be particularly problematic for the primary contents of this survey. Interviewer variance is related to the influencing effects of interviewers on responses as a result of their expectations regarding respondent reactions to questions, their feedback to respondents, any inconsistent probing for "acceptable" responses, and many other biasing behaviors. The effect is expected to be much worse for face-to-face interviewing than for telephone interviewing and, of course, nonexistent for the mail survey. Thus, measurement variance is highest for design A and lowest for design C.

Finally, the sampling variance will be approximately proportional to the sample sizes for each design. Thus, design C, with the largest sample size, has a very small sampling variance; design A has the largest sampling variance, corresponding to its relatively small sample size.

Given this information, how does one proceed to choose the best or *optimal* design? Since the sample sizes for the designs are such that the total data collection costs for each are the same, and assuming that the time required to complete the data collection under each design is not an important criterion, the optimal design is the one that achieves the smallest mean squared error.

To compute the mean squared error, we simply sum up the squared bias and variance components as in Figure 2.7. As we see from this figure, the optimal design by this criterion is design A, the design with the smallest sample size. Note that had our criterion been to choose the design that minimizes the sampling error without regard to the nonsampling error components, design C would have been optimal. However, when nonsampling and sampling error components are considered jointly, this design ranks last.

Figure 2.7 leads to the following final lessons for this chapter:

- The mean squared error is the sum of the total bias squared plus the variance components for all the various sources of error in the survey design.
- Costs, timeliness, and other quality dimensions being equal for competing designs, the optimal design is the one achieving the smallest mean squared error.
- The contributions of nonsampling error components to total survey error can be many times larger than the sampling error contribution.
- Choosing a survey design on the basis of sampling error or variance alone may lead to a suboptimal design with respect to total data quality.

CHAPTER 3

Coverage and Nonresponse Error

Missing data, sometimes called *errors of nonobservation*, are encountered in almost all large data collection efforts and can be particularly problematic in survey work. Errors of nonobservation occur in two different ways:

- A failure to include all units of the target population on the survey frame (e.g., incomplete lists, out-of-date maps, etc.), which will result in *frame coverage errors.*
- A failure to obtain measurements on some units of the survey sample, resulting in *nonresponse error. Unit nonresponse* refers to the situation where no measurements (or insufficient data) are obtained on a unit, while *item nonresponse* refers to the situation where measurements are obtained on some (most) variables under study but not all variables.

As we will see, coverage error and nonresponse error have very much the same kind of effect on the mean squared error. One can even view population units that are missing in the target population frame as a type of nonresponse, since in both cases the data for these elements are missing. The methods available to reduce the two types of error are quite different, however.

In this chapter we examine coverage and nonresponse error. For each we discuss a simple model that can provide some useful insight regarding the effects of these errors. Quite often, we have no idea how large the coverage and nonresponse errors are. Most of the time we only have information on the extent of nonobservation (i.e., the noncoverage rates and the nonresponse rates); other times, even this information is not available. Based on assumptions, we can perform "what if" analyses to identify population characteristics that could cause error problems. Although it is important not to make any judgments based solely on the rates of nonobservation (i.e., nonresponse rates or noncoverage rates), unfortunately, this is a common practice for many survey data users. As we will see, for each type of nonobservation, the biasing effects of nonobservation errors is a function of the missing data rate and the

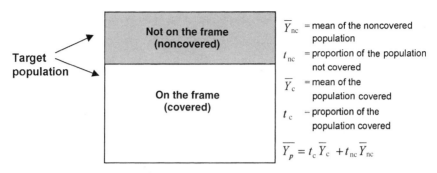

Figure 3.1 Basic coverage problem. Only the unshaded region of the box is covered by the sampling frame. The shaded region is not covered and is therefore missed by the sample. To the right of the box are the components that make up the bias due to noncoverage error.

difference in measurements between elements that are missing and those that are not. Also included is a review of some methods that prevent and reduce these errors.

3.1 COVERAGE ERROR

Prior to discussing the characteristics of coverage error, we review a few concepts covered in Chapters 1 and 2. Recall that the set of units about which data are sought and inferences are to be made is called the *target population.* The set of units from which the survey sample is actually selected is called the *frame population.* The *frame* consists of the materials used to define population members and could consist of a list of elements, a map, or a list of possible telephone numbers. The basic coverage problem is the extent to which the frame population corresponds to the target population. As soon as there is not a one-to-one correspondence between the two, we face a potential problem. This is illustrated in Figure 3.1, where the large square represents the entire target population. The unshaded area in the square represents the part of the total population listed on the frame; the shaded area then represents that part of the total population that is missing from the frame. For the moment, we ignore the mathematical symbols in this figure and will return to them later.

This figure illustrates several different situations related to the degree of correspondence between the target and the frame population. The *ideal* case, of course, is when there is a one-to-one correspondence between the frame and target population units (i.e., each target unit is included once and only once on the frame, and the frame does not contain units that are not in the target population). That case corresponds to no shading in the figure. Deviations from the ideal case can occur in several ways.

- First, some units of the target population may be missing from the frame population. Thus the frame is incomplete, due to omissions of various kinds. We call this case *frame undercoverage* or *noncoverage* and it is a major coverage problem. In effect, units that are missed have a zero probability of being included in the sample. Typically, the degree of non-coverage varies from region to region and between population groups. (A related concept is the undercount in censuses, which is the failure to count all members of a population of individuals, farms, businesses, etc.)

Example 3.1.1 One example of frame noncoverage occurs when the frame population is defined by all possible telephone numbers and the target population is defined as all households. In the United States, about 6% of the households do not have a telephone, so they are all missed when the survey is conducted solely by telephone. In Sweden the corresponding rate is just 1%, but in some other countries, the rate can be 50% or more. When the non-coverage rate increases, the telephone frame should be combined with some other frame that contains households without telephones. In developing countries the nontelephone rate is so large that telephone surveys are not feasible, due to the potential coverage bias.

- Another deviation from the ideal case occurs when some units in the frame population are not members of the target population (i.e., they are *ineligible* or *out of scope* for the survey). Since they are of no interest to the survey researcher, such units should be identified before the frame is sampled and deleted from the frame. However, this may not be possible without first contacting the unit to determine eligibility for the survey. In this case, the ineligible units are identified during the data collection and deleted with no biasing effects on the survey data. In the worst case, some units that are out of scope for the survey are mistaken for eligible units, and consequently, the estimates based on the survey results may be biased to the extent that this occurs and to the extent that the erroneously included units have characteristics that differ from the target population.

- The third problem with the frame occurs when more than one frame unit corresponds to the same target population unit (i.e., the existence of duplicated units on the frame). This case is sometimes called *overcoverage*. Of course, an obvious remedy for this problem is to identify the frame duplicates and remove them prior to sampling. If that is feasible, it may be possible to collect information during data collection to determine the number of times the units in the sample are listed on the frame. For example, in an RDD telephone survey, the interviewer can collect information on the number of telephone numbers that will reach the household and are used for voice communication. This information can be used in the estimation stage to adjust the probabilities of selection that are used to compute the sample weights and to correct the estimates for overcoverage.

- The final deviation from the ideal case occurs when one frame unit corresponds to several target population units. An example of this is when the frame consists of addresses for housing units and the target population is comprised of individuals. For each frame unit, the address corresponds to one or more persons. To determine which individuals in a housing unit should be sampled, it is necessary to list these people correctly without duplications or omissions. But this process can be quite problematic, due to privacy concerns of some household members or the definitions as to who in the household is eligible for the survey: for example, whether persons staying in the household or who are absent from the household at the time of the interview actually reside in the housing unit or may have their own place of residence. Depending on the response, a decision is made as to whether or not a person should be included in the household.

> *Frame noncoverage* results in zero selection probabilities for some target population units, and various forms of overcoverage (frame duplication) result in larger selection probabilities than intended. If not dealt with properly, both of these problems will bias survey estimates.

The problem of imperfect correspondence between frame and target populations is not the only problem associated with coverage. Frames can be more or less effective, depending on the amount and quality of frame data available, which could be used for identifying, classifying, contacting, and linking units.

Sampling frames can be based on numerous sources of information, which depend on administrative and financial resources available. In some countries there is a tradition of having registers for administrative purposes, and these registers could also be used for statistical purposes. Sometimes the registers of persons, businesses, cars, estates, financial transactions, and so on, are of such high quality that they can be used as sampling frames with only minimal coverage problems.

For example, the Swedish Register of the Total Population of Individuals is updated twice a month. Undercoverage for this register is essentially zero. The main coverage problem concerns people born abroad who move back to their country of origin without notifying authorities. Often, this coverage problem initially manifests itself as a nonresponse problem. A sampled person cannot be contacted, which indicates a nonresponse, but in some of these cases the person is simply no longer a member of the target population, since he or she has moved out of the country.

As a second example, twice a month Statistics Sweden receives information from the Swedish Tax Board about "births" and "deaths" of businesses in Sweden (i.e., changes associated with new businesses entering the population and other businesses exiting the population). All businesses are listed in a business register that can be used as a sampling frame for business statistics. The frame of businesses is much more complicated than the frame of the total population of individuals. Business frames usually cannot be updated con-tinuously because of coordination needs between surveys that use the same register as a base for their frames. Instead, the frames for the business sur-veys in Swedish official statistics are updated four times a year. However, not many countries are in a position to accurately update their business frames that regularly.

Often, a frame has to be developed by building it up from multiple sources of information about the members of the target population. A number of incomplete frames may be available that can be merged and unduplicated to create a single frame that can be used as a good starting point for creating a complete frame. But it still may not be possible to list all population members, and less than full coverage of the population should be accepted.

A widely used technique, particularly in agricultural and land-use surveys, is to use maps as frames. The map may cover an entire country or parts of a country, depending on the survey purpose. Areas with clearly defined bound-aries, called *primary sampling units*, are then defined. The entire country (or some other geographic entity of interest) is divided into primary sampling units. Then a sample of areas (primary sampling units) is selected, and in those areas new maps are created showing the second-stage areas (see Chapter 9 for a discussion of sampling issues). This process continues until we have reached the final sampling stage. The process results in a sample of ultimate area units. Within this ultimate sample, a special staff lists the final stage units. In this listing process, the listing staff may make the errors discussed previ-ously, such as omissions, erroneous inclusions, and duplications. Best practices for listing of units have been developed over the years and are often very similar to procedures used in censuses based on enumeration. An example of an area frame map is given in Figure 3.2.

Coverage problems should be distinguished from the case where we delib-erately and explicitly exclude sections of the population from the originally defined target population. Such exclusions are intentional and are made out of practical considerations due to, for example, the associated cost of obtain-ing an interview, the difficulty in accessing the units, and the unavailability of qualified data collectors in some areas, and so on. In some cases the target population can be redefined to exclude some of the units that are excluded from the frame either intentionally or unintentionally. In this way, the frame noncoverage problem is essentially defined away by declaring these excluded units as ineligible for the survey.

However, this practice may cause problems for the survey researcher in terms of the relevance of survey results for inference and his or her ability to achieve the primary objectives of the survey. If redefining the target popula-tion to better correspond with the sampling frame is done, it is critical that the new target population be described explicitly and clearly, to avoid misinter-pretations and erroneous inferences in using the survey results. In the end, however, only the user of the results can judge whether this procedure meets his or her needs.

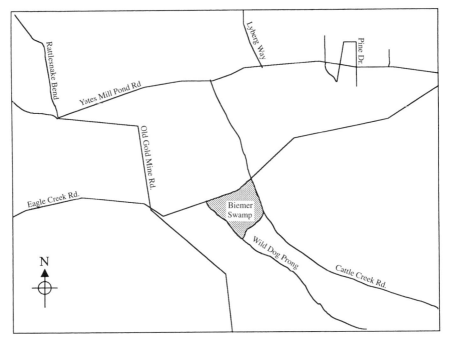

Figure 3.2   Area frame map.

As an example, in U.S. surveys of individuals, it is quite common to exclude people who are in hospitals, prisons, or in other institutions where access is restricted. Therefore, the target population is defined as the uninstitionalized population. Many nongovernment U.S. surveys of individuals that require face-to-face interviewing might exclude Alaska and Hawaii from the frame and define the target population as persons living in the 48 contiguous states. Although such restrictions may be justified for the reasons mentioned, they can have an effect on quality dimensions, such as relevance.

Useful overviews of the problems associated with coverage concepts and frame development are provided by Groves (1989) regarding household surveys, Colledge (1995) regarding business surveys, and the Food and Agriculture Organization of the United Nations (1996, 1998) regarding agricultural surveys.

3.2   MEASURES OF COVERAGE BIAS

In this section we consider in more detail the potential effects of noncoverage error on survey estimates. In connection with Figure 3.1 we defined a few relevant characteristics related to noncoverage bias. These quantities will be used to develop a simplified expression for the bias and relative bias resulting from coverage errors. Related to the figure are the following mathematical symbols and their definitions:

$\overline{Y}_c$ = mean characteristic in the population for persons covered by the frame

$\overline{Y}_{nc}$ = mean characteristic in the population for persons not covered by the frame

t_c = proportion of the target population that is covered by the frame (coverage rate)

t_{nc} = proportion of the target population that is not covered by the frame (noncoverage rate)

$\overline{Y}_p$ = mean characteristic for target population

In our discussion we consider the last quantity in this list, the target population mean, as the parameter we wish to estimate in the survey and derive an expression for the bias in survey estimates of this parameter due to noncoverage errors. We first note that the population mean can be expressed as a weighted combination of the means for the covered and noncovered populations. We denote the population mean as $\overline{Y}_p$ or, in words, "Y-bar sub p," and note that

$$\overline{Y}_p = t_c\overline{Y}_c + (1 - t_c)\overline{Y}_{nc} \tag{3.1}$$

or, in words,

population mean = (coverage rate)×(mean of the covered population)
+ (1 − coverage rate)
× (mean of the noncovered population)

From this relationship we can derive an expression for the bias (see Chapter 2) in estimates of the population mean $\overline{Y}_p$ due to coverage error. This expression can be helpful in trying to assess the effects of such an error. The mathematical details of this derivation are omitted but may be found in Groves (1989) and Lessler and Kalsbeek (1992). It is shown there that the bias in the estimate of the population mean due to noncoverage error is

$$B_{nc} = (1 - t_c)(\overline{Y}_c - \overline{Y}_{nc}) \tag{3.2}$$

which in words is

bias due to noncoverage = (noncoverage rate)
× (difference between the means of covered and
and noncovered populations)

This formula suggests that to determine the effect of noncoverage bias on the mean squared error, we need to know two quantities: the proportion of the target population missing from the frame, $t_{nc} = 1 - t_c$, and the difference between the means of the characteristic of interest for those who are represented on the frame and those who are missing from the frame, $\overline{Y}_c - \overline{Y}_{nc}$.

As an example, suppose that 20% of the population is missing from the frame and it is known that for some characteristic, say income, the difference between persons on the frame and those missing is approximately $10,000 annually. Then for estimating average annual income, the bias due to noncoverage is 0.20 × $10,000, or $2000. That is, samples drawn from this frame will tend to overestimate the annual income for the target population by an amount equal to $2000.

Another useful measure is the *relative bias* due to noncoverage error. This measure expresses the bias as a proportion or percentage of the total population parameter. It is easily computed by dividing the bias in (3.2) by $\overline{Y}_p$. For example, suppose it is known (perhaps from a previous survey) that the average annual income for the population considered above is $40,000. Then, using the estimate of the bias, $2000, the relative bias due to undercoverage is $2000/$40,000 = 0.05, or 5%. In other words, the estimate of annual income using this frame will be about 5% too high as a result of undercoverage bias. The computational formula for the relative bias is

$$\text{RB}_{nc} = (1 - t_c)\frac{\overline{Y}_c - \overline{Y}_{nc}}{\overline{Y}_p} \tag{3.3}$$

or, in words,

relative bias due to noncoverage = (noncoverage rate)
$\times$ (difference between the means of covered and noncovered populations)
$\div$ population mean

Several points can be made from these expressions for the coverage bias and relative bias. First, note that if the noncoverage rate is close to zero (i.e., if t_c is close to 1), the bias will be small no matter how large the difference is between the covered and noncovered populations. As the noncoverage rate increases, the size of the resulting bias also increases, but how fast it increases depends on the difference between the covered and noncovered populations. If the mean of the units covered by the frame is very close to the mean of the units not covered by the frame, there will not be much bias due to noncoverage.

Example 3.2.1   Consider the coverage bias associated with a random-digit-dial (RDD) household survey (see Chapter 1). An RDD survey only covers households that have telephones. In the United States, this currently means about 94% of all households. Then $1 - t_c$ (i.e., the proportion of nontelephone households) is 0.06. Consider a survey aimed at estimating the average income for the population households. Using data from Massey and Botman (1988), we estimate that for households that have a telephone, $\overline{Y}_c$ is \$18,700. For households that do not have telephones, $\overline{Y}_{nc}$ is \$11,500. Using formula (3.1) for $\overline{Y}_p$, the average for all households is therefore $0.94 \times \$18,700 + 0.06 \times \$11,500$, or \$18,268. Now we can apply the formula for the relative bias due to noncoverage in (3.3) to obtain the following:

$$RB_{nc} = \frac{0.06\,(18,700 - 11,500)}{18,268} = 0.024 \quad \text{or} \quad 2.4\%$$

Thus, the relative bias is 2.4%. Since the bias has a positive sign, the frame will produce overestimates. In other words, estimates derived from random samples drawn from the frame will tend to overestimate the true average annual income by 2.4%. A negative bias or relative bias would indicate that estimates will tend to underestimate the population parameter. Although it is quite simple, this coverage error model can provide some important insights regarding when coverage bias will be a problem and when it will not, for various types of populations.

Does a relative bias of 2.4% pose a problem for the survey? The answer depends on the objectives of the survey and how data will be used. If the relative bias of 2.4% drastically changes decisions or conclusions that analysts make from the data, 2.4% is an important bias. However, a bias of 2.4% could also be considered as inconsequential if analysts are really interested in differences of 10% or more in the estimates they are comparing in the survey. In that case, a 2.4% bias may be acceptable. But even if it is not acceptable, there may be few options for dealing with it. A few options will be discussed subsequently.

If a telephone survey is all that can be afforded, the analyst may have no choice but to live with the coverage bias due to the exclusion of nontelephone households from the frame. There might be an option to obtain data from nontelephone households through the mail if a good list of addresses exists for persons without telephones in the population. Another alternative is to do a parallel area sample using face-to-face interviewing. Such an approach is called a *dual frame survey*. In our example, this means that parallel to the RDD survey, an area frame survey is conducted where units are interviewed face-to-face. In each interview, respondents are asked whether their households have a (land-based) telephone. By doing this it is possible to check if the telephone frame covered the household as well. This could be a very expensive approach, however, since face-to-face interviewing usually costs much more than telephone interviewing.

A third option could be to attempt to adjust the estimator for the noncoverage bias. The survey sampling literature describes numerous techniques for making these types of adjustments. However, the literature for adjusting for nontelephone coverage bias suggests that these coverage adjustments are not very successful in most cases. Although postsurvey adjustments can eliminate some of the coverage bias, much of it remains, and the only option then is to collect data on nontelephone households. Postsurvey adjustments require data from some external source, such as a recent census or other major survey.

When the data collection mode is based on other types of frames, such as a listing of all households in the area (possibly from the postal service listings), special coverage check studies can be conducted to try to find households that are not on the lists. The coverage improvement studies attempt to improve list coverage by using more experienced staff or more rigorous methods to fill in the missing units on the frame. Due to the cost of coverage improvement methods, such studies are usually done on a sample basis in order to estimate the list frame coverage error and, possibly, adjust for it. The coverage error is then the difference between what has been accomplished by means of the regular methodology and the more advanced methods.

Example 3.2.2   The data in Example 3.2.1 provided a fairly precise point estimate of the coverage bias for the telephone frame since information is widely available on the noncoverage rate for the telephone frame as well as for numerous characteristics of telephone and nontelephone households. However, in some cases, very limited information is available on these quantities. It is then useful to consider a "what if" analysis using the foregoing formulas.

Suppose that for some area, or subpopulation of interest, the frame noncoverage rate is not known exactly except that it is not larger than 12%. Further, the difference between the mean income for the covered and noncovered populations is known to be between $10,000 and $12,000. The formula for relative bias can be used to compute whether the coverage bias in these situations would be problematic by computing the maximum possible bias under this scenario. For example, we might assume that the noncoverage rate is 12% and that the difference between the two means in the formula is $12,000. This would produce an estimate of the noncoverage bias of $1440 using formula (3.2). If the population mean is $18,268, the relative bias is $1440/$18,268 = 0.079 or 7.9%.

If this bias is acceptable, the frame is acceptable, since we assumed the worst-case scenario in this example. If the bias is not acceptable, more information is needed in order to get a better assessment of the potential effects of noncoverage bias. Such "what if" analysis allows the survey designer to determine whether coverage error may be problematic for a particular population characteristic and application of the survey data. Figure 3.3 is provided for this purpose. In the figure the relative coverage bias is plotted as a function of the difference between the covered and noncovered parts of the population. That is, on the x-axis we have plotted the quantity

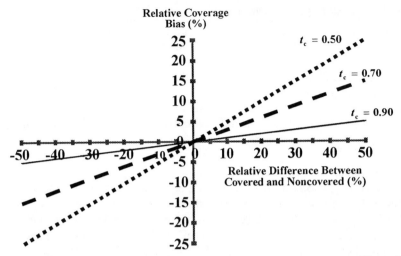

Figure 3.3   Coverage bias as a function of t_c and the relative difference between $\overline{Y}_c$ and $\overline{Y}_{nc}$.

$$\frac{\overline{Y}_c - \overline{Y}_{nc}}{\overline{Y}_p}$$

and on the y-axis the quantity RB_{nc}. The lines in the graph correspond to different levels of the coverage rate, t_c. The first line is t_c equal to 0.9 or 90% coverage; the second, 70% coverage; and the third, 50% coverage. The x-axis is the difference between the means of the covered and noncovered parts of the population expressed as a percentage of the total mean. The following example illustrates the use of this graph for conducting "what if" analysis in situations where precise information on the frame quantities is not known.

Example 3.2.3   Suppose that we believe the difference between $\overline{Y}_c$ and $\overline{Y}_{nc}$ is as high as 30% of the total, $\overline{Y}_p$. Then, finding 30 on the x-axis, we move up from this point to the line that comes closest to the coverage rate. For example, if the frame covers 90% of the population, the relative bias, which is read on the y-axis, is around 2.5% (here we interpolated between 0% and 5%). However, if the coverage rate is closer to 70%, the relative bias is much higher, about 9%. Then at 50% coverage, the relative bias goes up to about 17%. By using this graph, it is possible to examine a number of scenarios of differences between the covered and noncovered parts of the population for various coverage rates as well as various relative differences between the characteristics of covered and noncovered populations. Of course, the less certain we are about the components of coverage bias, the wider our bounds on the possible coverage bias. As we have just seen, if we know only that the frame coverage is somewhere between 70 and 90%, the relative coverage bias may be as small as 2.5% and as large as 9% for the same relative difference. For this reason,

Table 3.1   NASS List Frame Coverage

	NASS List Frame (%)	1977 Agricultural Census (%)
Total farms	56.3	89.2
Land in farms	77.6	98.6

Source: National Agricultural Statistics Service data.

we may need to do further research to determine whether a particular frame we are planning to use is acceptable.

Example 3.2.4   Consider the list frame the U.S. National Agricultural Statistics Service (NASS) used for selecting farms in the agency's farm surveys. In Table 3.1 we compare the NASS list frame with the frame developed for the 1977 U.S. Agricultural Census. For farms, the NASS list frame covers 56.3% of the population. The actual farm acreage represented on the NASS list frame, however, is much higher, 77.6%. The Agricultural Census covers many more farms, 89.2%, and almost all the land, 98.6%. Given the low coverage rate of farms (56%), it is important to consider that there is a problem with coverage bias in using only this frame for agricultural surveys. First, note that while 44% of total farms are missing from the NASS frame, the coverage of land in farms is still about 78%, suggesting that only 22% of the land in farming is not represented on the frame. Thus, it appears that what is missing from the frame is primarily small farms, and most of the major agricultural operations are represented on the NASS frame.

Let us now apply the formulas for coverage bias to determine the potential bias for the NASS sampling frame. Since the coverage bias is a function of both the coverage rate and the difference in characteristics under study between the covered and noncovered parts of the population, we wish to determine whether small farms are different from large farms with respect to the most important survey characteristics. For characteristics such as gross income, sales, total inventory, and production, the answer is certainly "yes." However, for characteristics such as total yield of wheat, corn, and so on, for a particular growing season, it may not be important that these smaller farms are missing from the frame if collectively, they contribute very little to total yields. For other items, however, frame noncoverage could have an important effect on agricultural estimates, and NASS has conducted studies to evaluate these effects. The Agricultural Census has a much smaller risk of incurring bias due to noncoverage because it covers a large fraction of the farms and the land in agriculture. Only about 1% of the land in farms is missing in the Agricultural Census. Thus, since t_c is very small, the coverage bias will be small even if the difference between $\overline{Y}_c$ and $\overline{Y}_{nc}$ is very large.

Example 3.2.5   Now consider an example from the U.S. Health Interview Survey. In the mid-1980s, the U.S. National Center for Health Statistics (NCHS) experimented with RDD for conducting the Health Interview Survey.

At that time, the RDD frame covered only about 93% of the households in the United States. So one question that NCHS had to answer in evaluating whether to move the Health Interview Survey to RDD was the effect of coverage bias on key estimates produced by the survey. One of the items on the survey was whether respondents are currently smoking. For this item, t_c is 92.8%, $\overline{Y}_c$ (the proportion of persons currently smoking among telephone households) is 28.8%, and $\overline{Y}_{nc}$ (the proportion of smokers in nontelephone households) is 49.6%. Then applying the formula for $\overline{Y}_p$ in (3.1), we obtain a value for the total population mean:

$$\overline{Y}_p = 0.928 \times 0.288 + 0.072 \times 0.496 = 0.30$$

The bias is then computed from (3.2):

$$B_{nc} = 0.072(0.288 - 0.496) = -0.015$$

and hence the relative bias, RB_{nc}, is $-0.015/0.30 = -0.05$ or -5.0%. This suggests that the proportion of persons currently smoking is underestimated by 5% using RDD, a difference that cannot be ignored in discussions of data collection mode.

Example 3.2.6 Another example concerns mental stress on the job. The percentages t_c, $\overline{Y}_c$, and $\overline{Y}_{nc}$ are 92.8, 17.6, and 17.0, respectively. Here we can see that there is not much difference between covered and noncovered populations in terms of whether jobs cause mental stress. Persons in telephone households seem to be equally susceptible to stress as persons in nontelephone households, and therefore the relative bias due to noncoverage is very small, in this case just 0.0025.

Example 3.2.7 Finally, consider an example concerning private health insurance. NCHS wants to estimate the proportion of persons who have private health insurance. The agency wants to do this for groups both below and above the poverty level (Table 3.2). Below the poverty level, telephone coverage is only 72.6%. Above the poverty level, telephone coverage increases to 96.3%. The relative bias for the "below" subpopulation is 16.3%, and for the "above" subpopulation 1.2%, which tells us where the coverage problems are.

Table 3.2 Relative Coverage Biases in a Health Insurance Study (Percent)

	t_c	$\overline{Y}_c$	$\overline{Y}_{nc}$	Relative Bias
Below poverty	72.6	34.5	16.8	16.3
Above poverty	96.3	87.1	59.3	1.2

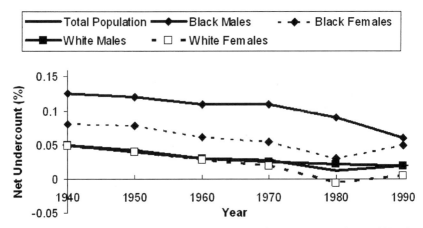

Figure 3.4   Estimated percent net undercount in the U.S. Census of Population and Housing.

The U.S. Decennial Census frame is used quite often in U.S. survey work to help target samples, decide on the allocation of samples to various strata, and so on. In addition, many U.S. survey organizations conduct surveys of households, so the experience of the U.S. Census Bureau in covering the population in the decennial census is very much a key issue for those who conduct household surveys. The data in Figure 3.4 represent the net undercount in the census from 1940 through the 1990 census. The second (bold) line from the bottom on this chart represents the total population. Over the years, the net undercount, or people who are missed in the census, has decreased as a result of the U.S. Census Bureau's very intensive efforts to reduce the undercount. In the 1990 census, however, the net undercount increased a little. White women appear to have been overcounted in 1980 as the net undercount became negative, which in effect means an overcount. As we can see from this figure, the black population seems to be missed more than the white population, black men being missed more than black women, and men in general being missed more than women in general.

These experiences of the U.S. Census Bureau for the decennial census are not unlike what any U.S. survey organization would experience in conducting a survey of the population of households. Survey organizations would miss approximately the same people, perhaps at even higher rates than the Census Bureau misses these people in the census. Blacks will be missed more than whites, and males will be missed more than females in these surveys. The procedures that are being employed in survey organizations are not that different from what the Census Bureau does in collecting the decennial census, so these results are indicative of coverage error in all demographic surveys. Of course, all countries have their own patterns of population groups that tend to be missed more often than others.

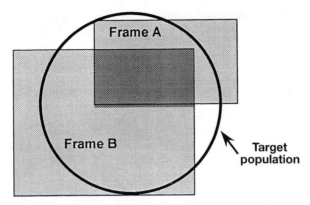

Figure 3.5 Population coverage using two frames. Neither frame A nor frame B covers the entire population; however, the combination of the two frames covers almost the entire population. It is essential to address the overlap of the two frames in either the construction of the combined frame or in the estimation process; otherwise, estimates will be biased by the unequal probabilities of inclusion in the sample.

> The *coverage bias* is a function of the noncoverage rate and the difference in characteristics studied between covered and noncovered populations.

3.3 REDUCING COVERAGE BIAS

In this section we discuss a few methods for reducing or eliminating the coverage error associated with various sampling frames. As mentioned previously, perhaps the most effective bias reduction approach is to try to repair the frame by removing duplicates and erroneous inclusions and improving coverage by fieldwork to identify units missing from the frame. This is a costly procedure, but many survey organizations try to build systems for their demographic and business surveys, and with such a strategy, costs are amortized across many surveys over a long period of time. For a smaller survey organization this might not be a cost-effective or even possible way of decreasing the total mean squared error. *Multiple frames* (see Figure 3.5) is an option that might be more cost-effective for both small and large organizations.

In Figure 3.5, the target population is represented by the circle, and the two rectangles overlaying this circle depict the coverage of each frame. In this example there are two frames, frame A and frame B. Neither of these frames alone does a good job in covering the target population. As illustrated, however, both frames in combination do a much better job than either frame by itself. The small rectangle in the middle illustrates the overlap between the two frames. If the frames are combined without unduplicating the units in this overlap region, the duplicated units would have a higher probability of selection than the units outside this region. Since the units in the overlap region are represented on both frames, they essentially have a double chance of being selected, whereas units that are on only one of the frames have only one chance of selection. This is not necessarily a problem if at some point during the survey we determine whether units that we are investigating belong to frame A or to frame B or to both. That way, we know their probabilities of inclusion and can weight the responses according to the inverse of their probabilities of inclusion appropriately.

For example, suppose that frame A is an RDD telephone number frame. Thus, every household that is selected from this frame is a telephone household. Suppose that frame B is some other kind of frame not based on telephone ownership, such as a mailing list. Further suppose that information on telephone ownership is not available on this frame. If we use both frames in combination to select the sample, we will have to determine telephone ownership for all the units on frame B in order to know the probabilities of selection for all the units in the sample. Otherwise, the sample cannot be weighted properly (see Chapter 9).

A problem arises when the combination frame is formed in such a way that it is difficult to know to how many source frames a particular unit selected for the sample belongs. Usually, when a number of list frames are combined, it is best to try to unduplicate the units among the various frames. This can often be done fairly easily with a good computer program that can match names, addresses, and other identifying information between the frames and identify the duplicates. However, in this process some errors will occur and some duplicates will remain. There is also a possibility that different units on the frames are matched erroneously and some units are mistakenly eliminated from the frame. Still this may be a much more cost-effective and error-proof way of reducing the coverage error for the survey than using exhaustive efforts to complete a single frame.

There are also various techniques for estimating the noncoverage rate for a frame and adjusting for the noncoverage bias. These methods have been developed primarily to check the census coverage errors. In a *record check*, an external record is checked to see if members of the target population that are on the external record have been missed in the census. Such records might exist for subgroups of the population, such as for children below 1 year of age. Suppose that a current birth record is available and we sample n children among those N who are 1 year or less and check whether they were enumerated in the census. Then the proportion of children on the record who cannot be found in the census is an estimate of the proportion missed in the census.

When a better aggregate estimate of the study population is available, a common method of comparing the two estimates is the *coverage ratio*, calculated as an estimate from the survey divided by a "preferred" aggregate estimate (i.e., an independent population control total). An example of the use of coverage ratios is shown in Table 3.3. The comparison of a survey estimate to population controls is a fairly crude check on coverage error since deviations

Table 3.3 U.S. Current Population Survey Coverage Ratios

Age	Nonblack		Black		All Persons		
	Male	Female	Male	Female	Male	Female	Total
0–14	0.929	0.964	0.850	0.838	0.916	0.943	0.929
15	0.933	0.895	0.763	0.824	0.905	0.883	0.895
16–19	0.881	0.891	0.711	0.802	0.855	0.877	0.866
20–29	0.847	0.897	0.660	0.811	0.823	0.884	0.854
30–39	0.904	0.931	0.680	0.845	0.877	0.920	0.899
40–49	0.928	0.966	0.816	0.911	0.917	0.959	0.938
50–59	0.953	0.974	0.896	0.927	0.948	0.969	0.959
60–64	0.961	0.941	0.954	0.953	0.960	0.942	0.950
65–69	0.919	0.972	0.982	0.984	0.924	0.973	0.951
70 or more	0.993	1.004	0.996	0.979	0.993	1.002	0.998
15 or more	0.914	0.945	0.767	0.874	0.898	0.927	0.918
0 or more	0.918	0.949	0.793	0.864	0.902	0.931	0.921

Source: U.S. Bureau of the Census (1992).

between an estimate and a control number may be due to factors other than coverage.

The frame is a very important part of the survey process. If the frame is not adequate, large coverage errors might occur and it may be costly or impossible to reduce or compensate for them. Many government or other large survey organizations try to coordinate their frame developments for entire programs of surveys. In those cases the starting point is a business register, farm register, or person register. These registers form a frame database from which frames for individual surveys of, say, agriculture are generated, thereby simplifying frame maintenance and updates of auxiliary information related to individual register units.

Auxiliary information, which is often obtained from the sampling frame, is an important factor to consider when estimation and other design features are decided. Auxiliary information includes data on the population units that can be used to improve estimates of population parameters, for example, by ratio estimation (see Chapter 9). In fact, to be of practical use, a frame should contain more information than just a listing of population units. There is need for data that can be used for efficient stratification and estimation, as described in Chapter 9. In addition, who to contact within a unit is an important piece of information that may be included on the frame, particularly for business surveys or mail household surveys. Business survey frames may also contain information on the size of a business and its primary industry or product line. This information may be important not only for drawing a sample, but also for making an initial contact for an interview.

There are a number of more advanced methods for the reduction and estimation of coverage errors. Some of these may be found in the following: multiplicity sampling (Sirken, 1970; Johnson, 1995), multiple frame methods

(Hartley, 1974; Kott and Vogel, 1995), mechanisms for linking frame elements to others in the target population not covered by the frame (Groves, 1989), and frames and registers (Colledge, 1995). Over the years the U.S. Census Bureau has conducted very extensive research on coverage errors in connection with the U.S. decennial censuses of population and housing. Coverage error models for census data are discussed in Wolter (1986) and postenumeration surveys, whose purpose is to evaluate and adjust for potential coverage error, are discussed in Hogan (1993).

SEVERAL METHODS FOR PREVENTING OR REDUCING COVERAGE BIAS

- Removing duplicates and erroneous inclusions from a frame
- Using more than one frame
- Checking an external record to see if members of the target population are missing on a frame

---

3.4 UNIT NONRESPONSE ERROR

In this section we consider the second type of nonobservation error: nonresponse. Nonresponse occurs when we are unable to conduct data collection on a unit selected for the sample. When an entire interview or questionnaire is missing or not obtained for the sample unit, we refer to *unit nonresponse*. However, even when a completed questionnaire is obtained for a unit, not all items on the questionnaire may have been completed that should be completed, a type of missingness referred to as *item nonresponse*.

There are no clear guidelines to say when a unit has answered enough of the items on a survey to be considered as a responding unit with extensive item nonresponse, as opposed to a unit nonrespondent. Technically speaking, if the unit has provided an answer to a single question on the questionnaire, it could be called a respondent. However, in practice the requirement for a unit response is usually much more stringent than that. Usually, criteria are developed to determine how many or which key items on the questionnaire need to be completed before a questionnaire can be classified as completed. Then a questionnaire that does not meet these criteria is called incomplete and the unit is classified as a nonrespondent. It is up to the survey researcher to decide when a case having considerable item nonresponse has provided sufficient information to be considered as a responding unit. Item nonresponse is discussed further in Section 3.6.10.

Unit nonresponse occurs because sample units cannot be contacted or because they refuse to participate in a survey when contact has been established. These cases are referred to as cases of *noncontacts* and *refusals*, respectively, and the distinction is important since the two groups are treated differently in nonresponse follow-up operations and may be treated differently in the estimation process as well. The distinction between noncontacts and refusals is not always clear, however. For instance, in a mail survey some of those who do not respond and who have not been contacted are, in fact, refusals. The survey organization has perhaps sent copies of the questionnaire several times or even tried to telephone the sample unit without success. Many organizations classify those as noncontacts, but obviously some of the noncontacts are aware of the contact attempts but choose not to react, which in effect means that they refuse to participate.

In a household telephone survey, the answering household member might tell the interviewer that the sample person is not at home when that is not true. This is also a case of hidden refusal. The problem with these scenarios is that we cannot say when we have a genuine noncontact or a hidden refusal. Item nonresponse occurs because questions are skipped intentionally or unintentionally. In a mail survey, questions can be overlooked because of a bad questionnaire layout, because the respondent is unsure about the meaning of the question, or because he or she simply does not want to answer the question. In an interview the interviewer might miss the skip instructions or the respondent might refuse to answer some questions.

As for coverage errors, there are two main strategies for dealing with unit nonresponse. One strategy aims at reducing the nonresponse rate, and the other is to use schemes that adjust for remaining nonresponse. Item nonresponse can be dealt with in a preventive way through work on wording and placement of questions. Various schemes for *imputation* (replacing missing values) are commonly used to adjust for any remaining item nonresponse (see Section 3.6.11 and Chapter 7). Here we discuss the components of unit nonresponse, the meaning of nonresponse bias, and some methods for reducing nonresponse. Adjustment schemes are more demanding technically, and such schemes are not discussed in detail. They usually require the expertise of a statistician with strong technical skills.

To facilitate our understanding of bias due to nonresponse, consider Figure 3.6. This figure is essentially the one that we considered when we discussed noncoverage error (see Figure 3.1). However, parts of the figure are relabeled to correspond to the nonresponse problem. As we shall see, the same type of reasoning we used when discussing noncoverage error applies to nonresponse error as well. As before, the large rectangle represents the population represented on the sampling frame. Note that this population corresponds to the unshaded region in Figure 3.1. For this population, we assume that two types of sample units exist: those that would respond to our survey (the respondents) and those that would not (the nonrespondents). It may seem confusing at first to discuss respondents and nonrespondents at the population level rather than at the sample level since the only units that we ask to respond to the survey are those in the sample (i.e., sample units). This rea-

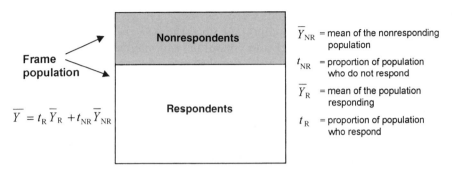

Figure 3.6 Nonresponse problem. The shaded region of the box represents the portion of the population who will not respond to the survey under the current methodology. The survey non-response rate can be considered as an estimate of the area of this region as a proportion of the frame population. The unshaded region represents the portion of the population that will respond. The respondents in the sample are selected from this region. Note the similarity of this figure to Figure 3.1. The concepts of nonresponse bias and coverage bias are quite closely related.

soning is, however, a simplification since it allows us to use the general formulas that we have developed for coverage bias without a lot of new notation or technical development. It will also help us to understand how nonresponse error contributes to survey bias, using the same principles that we used in discussing noncoverage error.

When we sample from the population, the sampling process ignores the division between respondents and nonrespondents since we cannot tell one from the other until we attempt to contact the sample units and ask them to participate. Whenever we draw a nonrespondent in our sample, we will count one nonresponse, since that sample unit will not respond. In the end, the only sample that we have data on are those that we drew from the respondent population. Thus, the nonresponse rate is the proportion of the rectangle in the shaded region.

The conceptual division of the population in these two groups is a very simple deterministic model of a very complex reality. Like other models it is not intended to be an exact representation of reality. In reality, the division between respondents and nonrespondents is not that simple. A more exact representation would have to take into account the fact that different survey procedures would produce different divisions of the population into respondents and nonrespondents. In fact, we rely on this to happen when we try different approaches to minimize the nonresponse rate. A more complex model would include assumptions about sample units' varying response probabilities (i.e., depending on circumstances, a sampled unit might sometimes be a respondent and sometimes a nonrespondent). The simple deterministic model assumes that response probabilities are either 1 or 0. However, the simple model is still very useful for understanding how nonresponse affects survey estimates. The notation should be familiar since we defined a similar notation for coverage error.

In Figure 3.6, let t_R denote the proportion of the population who will respond to the survey (the unshaded region) and let $\overline{Y}_R$ be the mean of the respondent population. Note that t_R can be interpreted as the response rate we would expect (on average) from the survey. Let t_{NR} denote the proportion of the population who will not respond to the survey (the shaded region) and $\overline{Y}_{NR}$ be the mean of the nonrespondent population. Finally, as before, let $\overline{Y}$ denote the mean of the entire population.

Similar to our approach in Section 3.2 we can construct the mean of the entire population by weighting together the respondent and nonrespondent means, where the weight is a function of the size of the respondent population (i.e., the "expected" response rate for the survey):

$$\overline{Y} = t_R \overline{Y}_R + (1 - t_R) \overline{Y}_{NR} \tag{3.4}$$

which in words states the following:

> mean of the frame population = (expected response rate)
> $\times$ (mean of respondents)
> + (expected nonresponse rate)
> $\times$ (mean of the nonrespondents)

Futher, we can derive an expression for the bias and the relative bias due to nonresponse as follows:

$$B_{NR} = (1 - t_R)(\overline{Y}_R - \overline{Y}_{NR}) \tag{3.5}$$

or, in words,

> nonresponse bias = (expected nonresponse rate)
> $\times$ (difference between the mean of the respondents
> and the mean of the nonrespondents)

and

$$RB_{NR} = (1 - t_R) \frac{\overline{Y}_R - \overline{Y}_{NR}}{\overline{Y}} \tag{3.6}$$

which states essentially that the relative bias due to nonresponse is the bias divided by the frame population mean.

Example 3.4.1 Suppose that for a telephone survey, we observe a response rate of 75% and therefore set $t_R = 0.75$ in (3.6). Thus the nonresponse rate is $t_{NR} = 1 - t_R$. Suppose further that the average income for respondents is

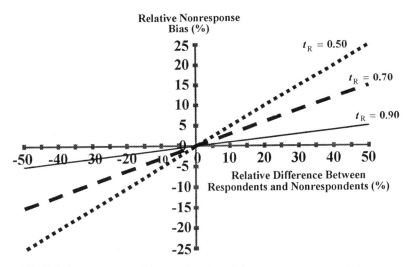

Figure 3.7 Relati̲ve nonre̲sponse bias as a function of the response rate, t_R, and the relative difference between $\overline{Y}_R$ and $\overline{Y}_{NR}$.

$\overline{Y}_R = \$107,000$ and the average income for nonrespondents is $\overline{Y}_{NR} = \$89,000$. Then $\overline{Y} = 0.75 \times \$107,000 + 0.25 \times \$89,000 = \$102,500$. Using the relative bias formula in (3.6), the relative bias due to nonresponse is

$$\text{RB}_{NR} = \frac{0.25(107,000 - 89,000)}{102,500} = 0.044 \quad \text{or} \quad 4.4\%$$

Thus we say that there is a 4.4% relative bias as a result of the 25% nonresponse rate.

As before, if we do not have information on the mean of the nonrespondents for the characteristics of interest or otherwise cannot compute the relative difference between respondents and nonrespondents, we can, if we wish, perform a "what if" analysis, as we did in the case of insufficient information for coverage bias. We might also want to evaluate the bias if the response rate could be increased or if it were lower, or if the difference between the incomes of respondents and nonrespondents were greater.

For these purposes we can use the relationship between the relative bias due to nonresponse and the relative difference between respondents and nonrespondents shown in Figure 3.7. As we did in Figure 3.3, in Figure 3.7 we have plotted the relative difference between respondents and nonrespondents on the x-axis and the relative bias on the y-axis. Each line on the graph corresponds to a different response rate. As an example, let us determine from the graph the relative bias due to nonresponse for the situation where the response rate is 70% and the relative difference between respondents and nonrespondents is believed to be between 20 and 30%. For the lower relative difference, we find the 20% point on the *x*-axis and read up to the 70% response rate line. Reading over at a right angle to the *y*-axis, we estimate the relative bias to be approximately 6%. Similarly, to obtain the upper bound on the relative bias, we repeat this process for 30% for which we estimate a relative bias of about 9%. Therefore, our range on the relative bias is between 6 and 9%.

Figure 3.7 also serves to emphasize that the bias due to nonresponse is not just a function of the response rate. There has been much focus on response rates in surveys, and some surveys with low response rates have been discontinued on the basis of the response rate without evaluating the bias, which also depends on the difference in the characteristics under study between respondents and nonrespondents. In reality, the bias may be acceptable even though the response rate is lower than expected.

As we can see from the nonresponse bias formula, nonresponse bias depends both on the difference between respondents and nonrespondents and the nonresponse rate. If the nonrespondents are not much different from the respondents for the characteristics we are measuring, the response bias might be quite small even though the response rate is low. So in evaluating the nonresponse bias, we need two pieces of information. We need to know the response rate, and we also need to know something about the differences in the characteristics between respondents and nonrespondents. In many surveys not much is known about the latter, and if that is the case, the value of surveys with large nonresponse rates can be questioned.

Nonresponse bias is a function of the nonresponse rate and the difference in characteristics under study between the respondents and the nonrespondents.

3.5 CALCULATING RESPONSE RATES

From the preceding discussion, it might be concluded that the concept of nonresponse rate is universally understood and agreed upon. However, there is no universally accepted definition of response rate, primarily because the eligibility of every unit in the sample cannot always be determined due to nonresponse. This suggests that caution is needed when comparing response rates between surveys.

A number of measures, other than response rates, can be computed using data on the outcomes of the data collection contacts which are useful for monitoring different aspects of the data collection process. Some of these measures relate to different aspects of the data collection process and serve as measures of key data collection process variables. By having measures of these, it becomes possible to control the data collection process, which in turn

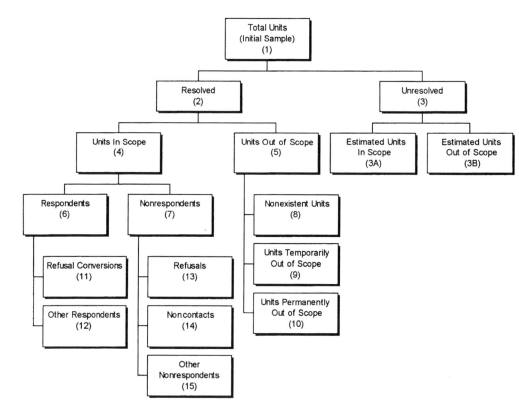

Figure 3.8 Final disposition of the sample with components of response and nonresponse. [Adapted from Hidiroglou et al. (1993).]

can improve the process, resulting in decreasing nonresponse rates, and in many cases, decreasing nonresponse bias. Therefore, it is useful to discuss some of these rates and how they should be computed.

Examples of key process variables of interest include contact rates, cooperation rates, refusal rates, and refusal conversion rates. Figure 3.8 is adapted from Hidiroglou et al. (1993), who discuss a model for computing response rates and other data collection process statistics that are used at Statistics Canada. In this figure, the total sample represented by the first box at the top of the flowchart is divided into two categories of units: units that are resolved and units that are unresolved. *Resolved* means essentially that we have determined whether the unit is either in scope (it is a member of the target population and should therefore be included in the survey) or out of scope (it is not a member of the target population). For example, in a survey of farms, households not in agriculture are out of scope for the survey since they are not members of the target population. In some cases, we do not have enough information to know whether a unit is in or out of scope. For example, a household that may, by the survey criteria, be classified as being in agriculture is classified as unresolved because the household is unwilling to provide information needed to make that determination. This situation is also quite common in RDD telephone surveys in which many of the numbers called never answer. Since we never talk to anyone at the number and have no other information about the number (e.g., whether it is associated with a household or establishment or a pay telephone or some other kind of telephone), we are not able to classify it as a household or as out of scope for the survey. Therefore, the number is designated as unresolved.

Since the denominator of any measure of response rate is the number of units in scope, to compute a response rate we need to estimate what proportion of the unresolved units is really in scope. A number of methods have been developed for this, depending on what information is available for the unresolved units. If no information is available other than that the unit appears to be occupied or the number appears to be active, we can estimate (albeit crudely) the number of unresolved numbers that are in scope from the proportion of *resolved* numbers that are in scope. Let p_{resolved} denote the proportion of resolved units that are in scope, and let $n_{\text{unresolved}}$ denote the number of unresolved numbers. Then the number of unresolved units that are in scope is the number of unresolved units times the proportion of resolved units that are in scope (i.e., $p_{\text{resolved}} \times n_{\text{unresolved}}$).

Now, considering the units in scope, these also can be classified as to two major types: respondent units or nonrespondent units. Further, respondent units may be subdivided into units that never refused and units that initially refused and were ultimately converted to respondents, referred to as *refusal conversions*. The nonrespondent units can be further subdivided into three groups: final refusals, units never contacted, and other types of nonrespondents. This "other" group consists of units that were contacted but could not participate because of illness, unavailability during the interview period, language barriers, and so on. The out-of-scope units can also be subdivided into three groups. One group comprises *nonexistent units* (i.e., units that appear on the frame but are no longer in existence). One example of a nonexisting unit is a housing unit that has been demolished or, in an RDD survey, a telephone number that is no longer in service. If the sampling frame is quite out of date, the number of nonexistent units may be substantial.

Temporarily out-of-scope units are units that at one point were in scope but now are out of scope. These could be farms that satisfied the definition of a farm at the time the frame was constructed, but because their sales have dropped, no longer satisfy the definition. At some time in the future, these units may again satisfy the definition of a farm. Finally, *permanently out-of-scope units* are units that are out of scope and there is no prospect they will ever be in scope. These could be farms that have sold out their operations and never intend to become part of agriculture again.

During the course of a survey project, a data collection manager may be interested in various rates to monitor the progress of data collection—response rates certainly, but also cooperation rates, language barrier rates, noncontact rates, unresolved rates, and so on. In Chapter 1 we discussed the importance of monitoring rates that reflect the key process variables for a survey in order to improve the data collection process continuously. The survey designer can choose variables and rates that fit any monitoring and improvement ambitions.

Regarding the response rate, however, there is a need for standardized treatment, since for better or worse, it is the most widely reported and compared process statistic for judging the quality of a survey. However, progress toward adopting standard methods of computing response rates has been slow, and consequently, it is still very important to understand precisely the methods used in computing various response rates when comparing them across surveys. Many national statistical agencies have standards for calculating response rates based on conceptualizations such as the one from Statistics Canada that we just discussed. But the conceptualizations differ across agencies and countries, due to varying administrative resources to conduct surveys.

Quite recently, the American Association for Public Opinion Research (AAPOR) (2001) provided guidelines to help researchers calculate outcome rates for RDD telephone surveys and in-person household surveys. Six different formulas for calculating response rates are provided in the AAPOR guidelines. This alone suggests the importance of understanding the computation of particular rates that are being compared. Currently, there is growing support to convince editors of survey journals to adhere to AAPOR guidelines when publishing articles on nonresponse issues related to RDD telephone surveys and in-person household surveys.

Let us consider a few useful rates that can be computed from Figure 3.8, beginning with the response rate for the survey. Recall that the response rate is intended to be an estimate of the quantity t_R, the expected response rate. If we are able to estimate t_R accurately, we have the best information available for accessing the *potential* for nonresponse bias. Of course, as noted previously, full knowledge of the bias requires knowing the difference between the estimates for respondents and nonrespondents. Nevertheless, in the absence of this information, estimation of t_R should be the goal.

Simply stated, the survey response rate is just the number of respondents divided by the number of sample members that are in scope for the survey. This is the best estimator of t_R for samples where each sampling unit is selected with equal probability. Other estimators, referred to as *weighted estimators*, are required to estimate t_R for unequal probability samples, as discussed subsequently.

One of the principal reasons that standards for calculating response rates are not accepted universally is the difficulty in computing the denominator of the response rate. The problem arises due to the existence of unresolved units in the sample. The appropriate way to compute the *response rate* is to take the number of responding units, box (6), divided by the total number of in-scope units, including those that are known to be in scope and the estimated number of in-scope units; in the figure, that would be boxes (4) and (3A). Thus the

response rate is [6]/([4] + [3A]). One way in which response rates are calculated misleadingly is to leave out the estimated number of in-scope units, box
(3A). What often happens is that the person computing the response rate
makes the assumption that all the unresolved units are out of scope, which
may not be true, of course. For example, in an RDD telephone survey, some
telephone numbers unresolved at the end of the survey may have been to
persons who were on vacation during the survey period or were left uncon-
tacted for other reasons.

If no other information is available to help estimate the number of unresolved units that are in scope, a simple and widely accepted approach is simply
to estimate the proportion of resolved units that are in scope and then multiply the number of unresolved units by this fraction. In other words, the number
of cases in box (5) divided by the number of cases in box (2) is the in-scope
proportion among the resolved cases. Then this quantity times the number in
box (3) is the number to enter in box (3A). This can be expressed as

$$[3A] = \frac{[5]}{[2] \times [3]}$$

and therefore [3B] = [3] − [3A].

Another rate often computed in survey work is the *cooperation rate*. The
cooperation rate reflects the degree to which units who have been contacted
in the survey agree to participate in the survey. So we would take the number
of responding units, box (6), and divide that by the total number of units that
were contacted, which would be the sum of the numbers in boxes (6) and (13).
Thus, the cooperation rate is [6]/([6] + [13]). Other rates that can be computed
are the *refusal rate*, *nonresponse rate*, and *noncontact rate*. A summary of the
rates is given below.

Response rate:	$\dfrac{[6]}{[4]+[3A]}$
Cooperation rate:	$\dfrac{[6]}{[6]+[13]}$
Refusal rate:	$\dfrac{[13]}{[4]}$
Nonresponse rate:	$\dfrac{[7]+[3A]}{[4]+[3A]}$
Noncontact rate:	$\dfrac{[14]+[3A]}{[4]+[3A]}$

So far, we have been discussing only unweighted response rates; however, as noted previously, when the sample is selected using unequal probabilities of selection (see Chapter 9), weighted response rates may be more appropriate estimates of t_R and thus a better measure of the potential effect of nonresponse bias on the mean squared error. When the probabilities of selection for the sample vary considerably from unit to unit, weighted and unweighted response rates can be quite different. For example, suppose that we draw a sample in which persons who have been arrested in the past year are overrepresented. For example, this subpopulation may account for only 3% of the entire population, but it comprises 50% of the sample. Suppose that the response rate for this group is 40% and the response rate for the remainder of the sample is 80%. Then the unweighted response rate is just the average of these two response rates, or 60%. However, the weighted response rate would weight the arrestee response rate by 3%—its prevalence in the population—and the nonarrestee group by 97%. Thus, the weighted response rate is $0.03 \times 40 + 0.97 \times 80 = 78.8\%$, a difference of 18.8 percentage points. Computations of weighted response rates are essentially equivalent to estimating the weighted proportion for a sample, discussed in Chapter 9.

The unweighted response rate is still an important process variable since it is an indicator of the success of the fieldwork in obtaining responses from sample members. In that regard, the unweighted response rate is more useful to monitoring the progress of work in the field and for identifying problems with nonresponse that can be addressed while the fieldwork is ongoing. However, it is not necessarily an indicator of the degree to which nonresponse could be affecting the accuracy of estimates. For that, the weighted response rate is more relevant. This is because the weighted response rate is an estimate of t_R, the proportion of the population in the shaded region of Figure 3.6. As we noted in that discussion, the nonresponse bias is a function of t_R and the difference between the means of the respondents and the nonrespondents. Thus, t_R can be viewed as an indicator of the effect of nonresponse on the bias components. In unequal probability samples, the weighted response is an estimator of t_R, whereas the unweighted response rate is not.

The rates we have discussed are based on one possible set of case codes, the one used by Statistics Canada and discussed by Hidiroglou et al. (1993). There are, however, a number of such sets of case codes developed and used by different organizations (such as the one developed by AAPOR), and attempts to set uniform standards across organizations have just begun. Also, while admittedly the number of different possible rates is a sign of a complicated survey world, we have here avoided to discuss some even more complicated scenarios. For instance, in a school survey, the sample design may be hierarchical, which in this case means that in a first step, say, schools are chosen, and in a second step, teachers within selected schools are chosen. Here nonresponse can occur in both steps. There is a resulting increase in nonresponse rates associated with such multiple survey stages.

Another example of complexity is the longitudinal survey, where sample units are followed over time and each follow-up or *wave* generates nonresponse or *sample attrition*, typically with a steady increase in attrition for each wave. Also in a longitudinal survey sample, units might change (such as changed household sizes) over time, complicating the picture further. The U.S. Federal Committee on Statistical Methodology (2001) has reviewed nonresponse rate calculations for some of these complex scenarios.

3.6 REDUCING NONRESPONSE BIAS

As we have seen, the nonresponse rate is one of two determinants of nonresponse bias. The nonresponse rate is usually easier to reduce than the second determinant, the difference between respondents' values of the study variables and those of nonrespondents. Consequently, much effort has been devoted to developing survey practices aimed at increasing the response rate as much as possible and thereby minimizing the risk of a response bias even if differences of the type mentioned are present.

Quite often, data users and producers ask: What is an acceptable response rate? Having analyzed the expression for nonresponse bias, the answer should be simple: It depends. This answer is not very instructive, but if we have only the response rate and do not know enough about the second term in the expression for nonresponse bias, there may be no satisfactory answer to that question. A low response rate may be quite acceptable, whereas a higher response rate may contribute substantially to total bias. Nevertheless, since data on the characteristics of nonrespondents are typically not available, many organizations have adopted operational standards for acceptable response rates which are both achievable and effective in limiting the risk of nonresponse bias.

Nonresponse bias arises whenever there is nonresponse and $\overline{Y}_R$ and $\overline{Y}_{NR}$ differ. Therefore, the goal of nonresponse bias reduction should aim to reduce $t_{NR}(\overline{Y}_R - \overline{Y}_{NR})$, not just t_{NR}. Consequently, unless we know something about $\overline{Y}_R - \overline{Y}_{NR}$, it is usually not meaningful to speak about acceptable nonresponse rates. However, the nonresponse rate is a useful indicator of the *potential* for nonresponse bias.

3.6.1 Causes and Effects of Nonresponse

A number of factors can affect nonresponse rates. In surveys of individuals and households, examples of factors include lack of motivation; lack of time; fear of being registered or documented by the authorities; unavailability due to illness, vacation, work, or bad timing; answering machines; and language problems. Many of these factors are also present in establishment surveys, since establishment respondents are people, too. But in establishment surveys, additional factors come into play. Examples of such factors are difficulties determining a designated respondent and obtaining permission to collect data; staff changes; restrictive establishment policies concerning survey participation; lack of knowledge; low priorities on the part of the respondent; and the fact that responding might be associated with considerable costs for the establishment.

There are also factors within the survey organization that can have an effect. Examples of such factors are interviewer workload, data collection periods that are too short, and questionnaires that are burdensome and difficult to use. The list of factors can be made much longer, but obviously, some factors are more easily controlled than others. Internal factors should be easier to handle than, for instance, locating sample units. Factors that cause nonresponse are closely connected with the common classification of nonresponse: inability to contact, refusal, and other nonresponse (illness, living in institutions, language problems, etc.).

Experience suggests that a classification of nonresponse into different categories can be quite useful for data collection since methods to reduce nonresponse rates differ depending on the type of nonresponse. For example, units difficult to contact are addressed by effective tracing procedures and by using efficient patterns for repeated contact attempts. Refusals are, for the most part, handled by the use of persuasive arguments and by making the response situation as unburdensome as possible. Other cases of nonresponse, such as language barriers, illnesses, and so on, may be handled by the use of proxy respondents (i.e., persons who can speak on behalf of the sample person). Such categorizations make it easier to allocate resources properly for reducing nonresponse rates.

Another reason that the nonresponse taxonomies are helpful is that if recorded over time, they can tell us something about changes in the *survey climate*, a notion used to describe a state of various populations' willingness to participate in surveys. As this notion suggests, the climate can change over time, depending on specific circumstances, such as public debates about invasion of privacy and respondent burden. Such factors can temporarily affect the willingness to comply to survey requests. In Table 3.4 we have listed some of the most common reasons for nonresponse and the type of nonresponse these reasons are likely to generate.

Let us expand a little on the two main nonresponse categories: *unable to contact* and *refusal*. As an example, in a field interview survey, dealing with noncontacts means returning to the household or farm or establishment repeatedly at different times of the day and days of the week until contact has been made. If the interviewer in a household survey is not able to find anyone at home who can answer the survey questions, it may be possible to find a neighbor who could at least give some information about who lives in the

Table 3.4 Common Reasons for Nonresponse and the Type of Nonresponse It Is Likely to Generate

Reason for Nonresponse	Typical Nonresponse Category
Not motivated	Refusal
Lack of time	Refusal
Fear of being registered	Refusal
Traveling, vacation	Unable to contact
Not a good time	Refusal
Unlisted telephone number	Unable to contact
Answering machine, telephone number display	Refusal and unable to contact
Wrong address or wrong telephone number	Unable to contact
Illness or impaired	Other
Language problems	Other
Business staff changes	Refusal and unable to contact
Business owner change	Refusal and unable to contact
Business restructured	Refusal, unable to contact, and other
Survey too difficult	Refusal
Business policy not to participate in surveys	Refusal
Low priority	Refusal
Too costly	Refusal
Sensitive questions	Refusal
Boring topic	Refusal
Heavy interviewer workload	Refusal and unable to contact
Data collection period too short	Refusal and unable to contact
Screening	Refusal
Bad questions or questionnaire	Refusal
Moved	Unable to contact

household and perhaps some information on the characteristics of the missing household. For a telephone survey it is much the same thing, calling the telephone number until no more time is available or until you have exhausted just about all the possibilities for times of the day and days of the week to reach the respondent. To reduce the contribution of no-contact cases to the nonresponse rate, and therefore to nonresponse bias, the main strategy is simply persistence, continuing to attempt contact with the sampling unit until you finally have to give up.

Refusals, on the other hand, can be a much more difficult problem to resolve. There have been many studies to try to understand why people refuse to participate in surveys. If we could understand those reasons well enough, we might be able to do better at persuading people to participate in surveys. There is much information on why people do not want to participate in surveys. Many surveys collect such information on a regular basis (i.e., interviewers ask sample units at the time of refusal why they refuse). Even if people refuse to participate, they seem to be willing to provide a reason why. For instance, in a study conducted by the Research Triangle Institute, 42.7% of refusers said that they were not interested. Studies such as these are interesting, but they are not very useful, since there is not much that a survey designer can do to counteract this if, in fact, the topic is boring or tedious to large portions of the population. Furthermore, there is often a more "sophisticated" underlying reason that sample units choose to refuse. Is the sample unit not interested because the interviewer did not make it seem interesting, or because the interviewer sounded boring? Was the sample unit not interested because he or she did not fully understand what the survey was about? Or is "not interested" just an easy way to refuse to participate in a survey?

There is much more that we can do to develop an understanding of why people do not want to participate in a survey and how we can get them to change their minds. During recent years, important research has addressed this issue which aims at developing a theory for survey participation. Important contributions in this regard are those of Groves et al. (1992) and Groves and Couper (1998). Their research is related to surveys of individuals and households, but some of their findings are also relevant to establishment surveys, since in most surveys people are supposed to convey the information sought by the survey researcher. (Exceptions are surveys where readings are taken by field-workers or where survey data are recorded using mechanical devices such as TV viewing meters.) To a large extent, people react similarly to stimuli whether they respond for themselves or on behalf of an establishment. However, establishment surveys differ in many other respects from surveys of individuals and households. An informative discussion of these differences may be found in Riviere (2002).

Groves et al. (1992) suggest a number of factors that influence refusals, factors that have been discussed in the vast literature on survey methodology. The survey design, the mode of data collection, the respondent rule, the length of the interview, the length of the interview period, the survey topic, and the questionnaire design can all influence refusals. Respondent characteristics also have a great influence on refusal rates. The age, gender, income, and health of respondents, whether they live in an urban area or rural area, the crime rate of the area they live in, and whether or not the respondents are literate will influence their propensity to respond. Interviewer characteristics also influence response rates. The age, gender, race, and perceived income of an interviewer can affect response rates, as well as their experience, the way in which they handle reluctant respondents, their self-confidence, and their expectations and preferences. Do they believe that people, if approached appropriately, will respond to the survey, or do they think it is hopeless in many cases? Other factors that come into play are the attitude that they display in the interview and their recent experiences. Perhaps they have had a bad day with a lot of refusals and this has temporarily eroded their confidence. Perhaps their motivation to try to get participation is low, because of low payment or because the importance of high response rates is not stressed sufficiently by their supervisors. Finally, societal factors such as social responsibility and the legitimacy of the survey objective can influence refusal rates.

With each factor or family of similar factors that generate nonresponse there is associated one or more methods or techniques to reduce the nonresponse. We review some of these techniques; however, it should be emphasized from the outset that a single measure is rarely sufficient to get a reasonable response rate. A collection of methods should be in place where each method in the collection addresses one or more of the many factors thought to contribute to nonresponse. Moreover, this collection of methods will not be the same across surveys and thus should be tailored to the needs of the specific survey. Another reason that several methods should be used is that different methods attract population groups differently. In theory, emphasis on using a single measure to obtain a very large response rate among, say, a group of young people or of medium-sized farms, might not reduce nonresponse bias if at the same time the difference in characteristics between respondents and nonrespondents increases.

Despite the ambiguity regarding the effect of a high nonresponse rate on data quality in the absence of information on the difference between respondents and nonrespondents, it is still quite clear that nonresponse should never be taken lightly in the survey process. First, nonresponse means that the achieved sample size is smaller than planned or smaller than it could have been if the response rate were higher. As discussed in Chapter 9, this smaller sample size will result in a reduction in the precision of the estimates, which in turn will lead to wider confidence intervals for the parameters of interest. Second, there is often a difference in characteristics between respondents and nonrespondents, for at least some of the items in the survey. For example, in 1991, a Swedish Labor Force Survey study showed that the number of unemployed was underestimated by 8%, due to nonresponse bias. (A new estimator using auxiliary information has since been developed, thereby reducing the bias considerably.) Third, nonresponse is disruptive to the statistical production process, which could lead to increased costs and delayed survey results. Fourth, nonresponse is a quality feature that many survey users and sponsors have heard of, and they know intuitively that a high nonresponse rate is not in line with good survey practices. Data users may know very little about how the data were collected and all the steps that have been taken to improve data quality. Therefore, the survey response rate is viewed as an indicator of the quality of the entire survey process, including all five sources of nonsampling error discussed in Chapter 2. Indeed, to these users, the nonresponse rate may also be indicative of the competence of a survey organization. High response rates become synonymous with efficient, high-quality data collection operations. So to stay in business, there is a practical reason to keep nonresponse rates low. In the long run, however, the survey community should vigorously educate sponsors and users regarding the totality of survey quality indicators.

> A single measure is rarely sufficient to obtain reasonable response rates. A *collection* or *battery of methods* should be in place, where each method is supposed to address one or several factors contributing to nonresponse.

3.6.2 Theories of Survey Participation

Two main schools of thought dominate the development of theories for survey participation. One concerns interview surveys and uses specific psychological concepts to explain survey participation. Also included is social skills analysis, which has been shown to improve interviewer–respondent doorstep interaction (Morton-Williams, 1993). For mail and other self-administered modes of data collection there is another line of research led by Dillman (1978, 1991), based on social exchange theory, which has some implications for interview surveys as well.

A basic theory of survey participation was developed by Cialdini (1990) and elaborated on by Groves and Couper (1998). This theory integrates the influences observed for sociodemographic and survey design factors with the less observable effects of the psychological components of the interaction between interviewer and respondent. Factors at work include societal levels of interviewers and sampled units, attributes of survey design, characteristics of the sample persons, attributes of the interviewers, and respondent–interviewer interaction. Figure 3.9 describes factors thought to affect survey participation (see Groves and Couper, 1993). This conceptual framework for survey participation in some ways contradicts traditional ideas that interviewer behavior should be highly standardized. The practical implications of their theory are that tailoring techniques should be part of interviewer training and that survey materials should be tailored to specific attributes of sample persons in interview surveys of individuals or households. By *tailoring* we mean that interviewers adapt their approaches to specific sample units by the characteristics of those sample units. Interviewers try to focus on characteristics that may be related to some basic principles thought to facilitate compliance with the survey request.

Cialdini (1984) has identified six psychological factors that may have a large influence on refusal rates. The first of these psychological factors is *reciprocation*. Essentially, what Cialdini suggests is that persons will be more inclined to respond to a survey if they see it as a repayment for a gift, payment, or other concession made to them, or if they think it benefits them. This is the scientific basis for giving incentives to respondents. More detailed research has shown that prepaid incentives are more effective than promised incentives (Berk et al., 1987). As a consequence, many mail surveys will enclose a $2 to $5 incentive along with the questionnaire. This is the meaning of *prepaid incentive*: The respondents receive compensation before they respond.

The literature on incentives seems to indicate that promised incentives are viewed by respondents as a payment for services (see Singer et al., 1999).

Factors out of researcher control   *Factors under researcher control*

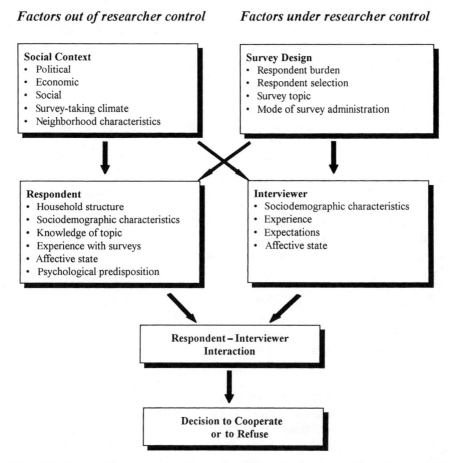

Figure 3.9 Factors affecting the decision to cooperate or to refuse to participate in household surveys. [Adapted from Groves and Couper (1993).]

Promised incentives tend not to work as well as prepaid incentives because some respondents may view the payment for their time as inadequate. The reciprocation theory also suggests that sample members will respond to a survey if they feel that it is beneficial to them. Successful interviewers sometimes employ this device by telling the respondent how the survey is important to some group of which the respondent is a member. For example, in a U.S. survey of the elderly regarding Medicare payments, the interviewer can inform a respondent how this survey will benefit Medicare recipients, perhaps by giving information to the agency in charge of the Medicare system about how to improve services, payments, or other features of the program.

The second psychological factor is *consistency*. This factor suggests that compliance with the survey request is consistent with a sample member's announced position, beliefs, attitudes, or values. For example, if in the course of interviewer–respondent introductions, respondents say that they are very concerned about the environment, and if there is some aspect of the survey that deals with the environment, the interviewer can use the consistency technique to inform the respondents how participation in the survey will benefit the environment. It is obvious that this technique could be effective given that sample members have already expressed their concern about the environment. Many people are just concerned about their general well-being. In this case, it may be beneficial to inform respondents as to how survey results will be used to make their lives and society better.

The third psychological factor is *social validation*, which states that sample members will show more willingness to comply if they believe that others will also comply. This is a well-known factor in, for instance, the sales business and the upbringing of children. For example, a door-to-door salesperson trying to sell a product to a householder might say that many others in the neighborhood are buying the product, in an effort to convince you that the product is acceptable to many others. Thus, even though the householder is not aware of the product's features, he or she may now be interested in knowing just what makes the product so popular. As another example, a small boy may be unwilling to wear a bicycle helmet, but decides to do so when he is informed that an older boy in the neighborhood is wearing one. The same type of psychology can be used effectively to convince respondents to participate in a survey. The interviewer might say that many other respondents in the neighborhood have participated in the survey because they are convinced of the importance of the survey and want to be a part of it.

The fourth psychological factor is *authority*. That is, compliance is more likely if the request comes from a legitimate authority. We have known for quite some time from reports in the survey literature that in some countries government-sponsored surveys have lower refusal rates than do surveys conducted by private organizations. Market research surveys have the lowest cooperation rates among any surveys conducted. Authority would seem to explain why this is so. The government is viewed as a legitimate authority.

For example, in the United States, the Census Bureau, the National Agricultural Statistics Service (NASS), and other federal agencies have as their mandate the collection of survey data to benefit the general good. NASS collects data dealing with agriculture. So when NASS approaches a farm operator to ask for his or her cooperation, NASS is viewed as an agency that has legitimate authority. For that reason, cooperation rates with NASS surveys are fairly large, just as they are for the Census Bureau and other government agencies conducting surveys.

When a private survey organization conducts a survey for some government agency, it might try to use the letterhead for that government agency or a letter signed by a prominent person at the agency as a sign of legitimate authority. Even marketing research firms not conducting government surveys have caught on to the importance of legitimate authority and getting sample members to comply with their survey request. Some marketing research organizations develop letterheads that look and read like those of government agencies. It should be noted, however, that in some parts of the world, government agencies are not viewed as particularly authoritative, but rather, are viewed with suspicion.

The fifth psychological factor is *scarcity*. Rare opportunities are generally considered more valuable than others. Sample members may be more willing to comply with a survey request if they think it is a rare opportunity and they need to take advantage of it. Interviewers sometimes use this factor when they tell sample members that there are only a few days left to make their voices heard, that this is the last day of the survey, that only one person in 2000 is contacted, and that we really need the sample units' responses now or they will lose their opportunity to participate.

Liking is the sixth psychological factor. Sample members are more willing to comply if the interviewer is appealing to them or if the interviewer appears and sounds like he or she is similar to them with regard to values, opinions, language, social background, or personal style. This is one of the reasons why many interviewers have learned to dress down in low-income areas and to dress up when they are interviewing in high-income areas. Some interviewers may try to compliment the sample person or may have a complimentary remark about their home, child, pet, flower garden, and so on. They may even imitate the respondent's way of speaking.

In Table 3.5, these six psychological factors and their implications are summarized. It is important to know that the heuristics mentioned could be more or less powerful, depending on where in the world the survey is conducted. Most of the psychological factors that we have just discussed require that we know something about the sample members before we ask them to participate in a survey. For example, consistency requires that we know something about the sample members' beliefs or attitudes or values. The liking principle requires that we know something about what the respondents would like in an interviewer. To apply reciprocation, we may need to know something about how a respondent views a survey and what we may be able to bring out about the survey that would benefit the respondent. How do we apply these and the other principles if we are interviewing strangers?

Here the idea of *prolonged interaction* can often be used to great advantage. Prolonged interaction means maintaining a conversation with the sample member for the purpose of identifying cues that allow the interviewer to tailor his or her approach to that particular respondent: listening to the reasons the respondent is giving for not wanting to participate and responding appropriately using effective persuaders (based on the psychological principles in Table 3.5) to counterargue those reasons. Sometimes, it is wise to postpone the tailoring. For instance, in telephone interviews, when a sample member seems very reluctant, some cues can be obtained for a later attempt to persuade the member to participate, since a later attempt can often increase the chances of obtaining cooperation.

Table 3.5   Six Psychological Factors and Examples of Their Application

Psychological Factor	Meaning	Example of Survey Implication
Reciprocation	Compliance as repayment for a gift, payment, concession, or benefit to the respondent.	At the time of the survey request, an incentive is given to the sample unit. Information brochures are included in mailings.
Commitment and consistency	Compliance is consistent with an announced or fervently held position (belief, attitude, value).	Respondent: Tax money should not be used for surveys. Interviewer: It is important that decisions be based on data. Surveys generate data.
Social validation	More willingness to comply if one believes that similar others would also comply.	Interviewer: Most people participate in this survey and seem to enjoy it.
Authority	One might comply with requests from well-known government agencies and others having legitimate authority.	If applicable, interviewers should stress agency reputation.
Scarcity	One is willing to comply with requests perceived as rare opportunities.	Interviewer could mention opportunity to: • Represent thousands of others. • Participate now because data collection period is soon over.
Liking	One is more willing to comply with requests from interviewers who are liked or appealing.	Interviewers can use this heuristic by choice of clothes, attitudes, way of speaking, and general style.

Apparently, some of these principles have been understood by survey designers and interviewers for a long time, even before the explicit discussions provided by Cialdini, Groves, Couper, and others. Much of that understanding has been based on common sense, and the principles have been applied intuitively in developing methods to gain cooperation. For example, the reciprocation principle has led to the use of incentives and information brochures and other material that might be valued by sample members. Reciprocation can also be a concession by the interviewer. In some cases, the interviewer might ask a respondent for a complete one-hour interview, and the respondent refuses. If the design allows, the interviewer could ask for a concession—a 15-minute interview for the purposes of obtaining measures of just a few, key survey variables. There is some research showing that following a larger request with a smaller request will make compliance with the smaller request more successful.

The consistency principle can be used when developing interviewer instructions regarding survey introductions. Making a connection between a survey and the respondent's values could mean that interviewer introductions or the advance letter contain sentences such as: "These surveys are conducted so that decisions can be based on facts." Authority and scarcity principles are commonly used to improve survey introductions.

The principles outlined above can be used in a more consistent fashion than in common practice. One reason why interviewers differ when it comes to gaining survey participation is that their social skills and their ability to look for cues and react to those cues quickly and successfully can differ dramatically. Examples exist of interviewers who are extremely successful at gaining cooperation even though they are working in difficult areas. The only explanation for their success is that they are skilled in recognizing, interpreting and addressing cues provided by the sample members and responding to them effectively.

Many experienced, veteran interviewers apply these principles in their everyday work. However, few of them have thought about the techniques they employ as part of a structured theory of survey participation. Thus, experienced as well as inexperienced interviewers can benefit from a discussion of the psychological factors we just covered, particularly the strategy of prolonged interaction and tailoring. Interviewers with high cooperation rates should be monitored and observed more often so that more is known about how they actually carry out their task. Good interview practices could then be incorporated into interviewing training together with training in prolonged interaction and tailoring.

Many of the principles can also be used in mail surveys to improve the wording of the advance or accompanying letter. More advanced procedures based on social exchange theory have been used to develop combinations of steps to maximize response rates in mail surveys. We discuss these combinations below.

3.6.3 Using Combinations of Methods

Since the causes of nonresponse are many and varied, a complete strategy for reducing nonresponse must be multifaceted, with specific techniques targeted to each cause. However, there are some general rules for any survey involving respondents, as follows: (1) the questionnaire, whether administered by the respondent or by an interviewer, should be easy to complete; (2) the survey should be introduced in an interesting and professional way by means of either an advance letter or an interviewer; (3) there should be a plan and resources set aside for follow-up procedures, including reminders, refusal conversion attempts, and tracing activities; and (4) interviewers should be thoroughly trained in techniques for avoiding refusals and for contacting hard-to-reach sample members.

More specific methods depend on the mode of data collection. For mail surveys, Dillman (1978, 2000) has developed a strategy built on social exchange theory that is a standardized step-by-step combination of methods to obtain high response rates. The strategy has been very successful in many applications, and the current version consists of nine steps.

DILLMAN'S NINE-STEP STRATEGY FOR IMPROVING MAIL SURVEY RESPONSE RATES

Step 1:   An advance letter is sent to the sample units.

Step 2:   The questionnaire is sent using stamps as postage, if feasible.

Step 3:   After some time, depending on the survey, a combined thank you/reminder card is sent.

Step 4:   A second copy of the questionnaire is sent to those who have not responded following step 3.

Step 5:   A third and final copy of the questionnaire is sent to those who have not responded after step 4, using special mail services, or follow-up is conducted by telephone.

Step 6:   In all mailing rounds the sample units receive stamped return envelopes, if feasible.

Step 7:   In all correspondence with sample units, personalization is used (i.e., the sample unit's name is typed on the advance letter letterhead).

Step 8:   The questionnaire is designed to be easy to complete by all respondents.

Step 9:   A small monetary incentive is used.

Source: Adapted from Dillman (1978, 1991, 2000).

Studies of an earlier, seven-step version have shown that if any of the steps are omitted, nonresponse rates will increase (DeLeeuw and Hox, 1988). The final result of the strategy will vary, depending on country, topic, and population. Also, individual steps and the time between follow-up rounds might have to be modified, for the same reason. For instance, in a business survey, one important step is to identify the "right" informant in the company or organization. Another modification might be to omit the use of stamps, since in most countries that is not feasible in government surveys.

In interview surveys it is important that interviewers be well trained and motivated. They should be familiar with methods available to gain cooperation, which means that the implications of theories for survey participation should be clear to them, that efficient sources for locating sampled units be used, and that an efficient call-back strategy be used. At Statistics Sweden, experiments have shown that interviewers not using all means available usually achieve lower response rates than do interviewers who use all means available.

In extensive diary surveys, where respondents are supposed to keep records of activities, purchases, and similar things, it has been shown that although the diary is self-administered, intense follow-ups by interviewers or other staff are very effective. For instance, in a two-week household budget survey, Dillman (2000) found that five to seven contacts with diary keepers is not unreasonable. Also, it might be worthwhile to send a note telling the diary keeper that the period is soon ending, to counter any fading interest.

Sometimes, general efforts and strategic methods related to the data collection might not be sufficient. Some surveys might require the use of special methods because the causes of nonresponse are special for those surveys. The only way to decide what methods and combination of methods are appropriate is to collect data on the causes of nonresponse. Such data can result in design adjustments, such as modifications of the information material, special assistance to large companies, or the use of more than one data collection mode.

3.6.4 Privacy and Confidentiality

Privacy and confidentiality are concepts related to the protection of survey data. *Privacy* is the right of individuals and businesses to decide what information about them can be collected. *Confidentiality* is the extent to which data already collected are protected from unauthorized use. Most countries have data acts and other legal frameworks that regulate these issues. Nevertheless, concerns about privacy and confidentiality are problematic for statistical organizations and other data collection organizations, since privacy and confidentiality are linked to issues related to measurement errors and nonresponse.

Privacy is linked to item nonresponse as a result of question sensitivity but also as a result of rights to privacy for individuals and other data providers. What is sensitive can vary over time, between cultures and other subgroups, and between individuals. But as long as a significant portion of the population under study believes that a question is sensitive, the question should be treated as sensitive to avoid both item and unit nonresponse. The best way to avoid nonresponse problems as a result of asking sensitive questions is to avoid asking them in the first place. However, most policy needs of governments require asking sensitive questions. Avoiding such questions is therefore seldom an option but should be considered when possible. It is more realistic to counter some of the effects in the questionnaire design phase. For some topics, open-ended questions are better than questions with fixed response alternatives.

For example, the question "How often do you drink alcohol?" may be less threatening to some people than "Do you drink alcohol every day, every other day, ... ?" since the first question is more general. For a variable such as *income*, the more general question may be more sensitive. Fixed response alternatives as wide as possible have been shown to achieve higher item response rates than a direct question on income. By extending the question introduction, it is sometimes possible to eliminate some of the sensitivity by using wordings such as "Years ago, the issue of '_____' was not discussed openly, but today such topics are openly discussed," or: "It is not unusual that . . ." A question such as "How many cigarettes do you smoke per day?" is less likely to imply that smoking is a somewhat stigmatized behavior than the question "Are you a smoker?" followed by "How many cigarettes do you smoke per day?" (see Dalenius, 1988). Questioning strategies like these may have a positive effect on item response rates, but the trade-off may be increased measurement errors. When everything has been done regarding question wording, it is generally recommended that sensitive questions be placed near the end of the questionnaire, to avoid unit nonresponse for the items that otherwise would follow the sensitive questions.

If the sensitive questions are very important or perhaps the entire survey can be considered sensitive, a mode of administration should be chosen that provides the greatest level of privacy for the respondent. As discussed in Chapter 4, mail surveys tend to be less threatening and prone to social desirability bias than face-to-face or telephone interview modes. Social desirability bias is the survey error resulting from a reluctance of sample units to reveal that they possess socially undesirable traits. Instead, they report in a more socially desirable fashion or not at all. There are various ways to combine regular interviews with self-administered modes for those parts that are sensitive (see Chapters 4, 5, and 6).

A few special modes have been developed for the collection of sensitive information. One method used in the past, now replaced by computer-assisted methods, is to use self-administered questionnaires for that part of a face-to-face interview that deals with sensitive questions. To avoid discovery by the attending interviewer, the respondent is asked to place the questionnaire inside a sealed envelope provided by the interviewer. Another method that has been used is *randomized response* (see Chapter 4).

Weaknesses in pledges of confidentiality seldom cause large effects on response rates. It is the fear itself that is the problem—whether or not there are pledges affects only a small portion of individuals or establishments. Even in the absence of a strong measurable link between confidentiality assurance and decreased nonresponse rates, statistical and other data collection organizations have an ethical and legal obligation to protect data from being disclosed improperly. Therefore, a lot of research has been conducted toward the development of methods that limit the possibilities of such disclosure. Among methods designed to protect data on individual units, some of the more common are described below. All these methods aim at reducing the risks of disclosing individual data for specific sample units. A disclosure of individual data could have substantial negative consequences for a survey organization, with reduced respondent, public, and user confidence that could take years to rebuild. No serious survey organization can afford to take these issues lightly, and therefore most organizations have a program to control invasion of privacy that is properly balanced against sociatal needs for data. For a review of confidentiality methods, see Fienberg and Willenborg (1998).

SOME COMMONLY USED METHODS FOR ENSURING CONFIDENTIALITY OF SURVEY DATA

De-identification or anonymization. Anything on a record that can serve as a direct identifier of a person or establishment (e.g., social security or other identification number) is removed.

Suppression. A variable value is deleted from a date file for some or all sample members. One reason for this might be that the number of units in a table cell is so small that an association can be made between individual units in the cell and their variable values.

Interval values. The original value is replaced by an interval encompassing that value. A variant of this approach is *top coding*, replacing all very large values with a maximum value. For example, incomes of $100,000 or more are coded simply as >$100,000.

Adding noise. An original value X is replaced by a value $X + a$, where a is a small random number.

Encryption. Data are scrambled using an encoding algorithm that is very difficult to decode. This is commonly used to transmit data from the field to data processing offices, either by mail or electronically.

3.6.5 Choice of Data Collection Mode and Response Rates

Errors associated with the data collection mode are treated in Chapter 6. However, the choice of data collection mode is of importance when trying to reduce nonresponse rates. Therefore, we discuss this relationship briefly here (i.e., data collection mode and nonresponse). The literature is rife with examples, which show that different data collection modes generate different levels of nonresponse. A generally accepted ordering of the modes by their expected response rates is face-to-face, telephone, mail, and other self-administered modes, such as Web surveys and diary surveys, with face-to-face mode producing the highest response rates. Seldom, however, can the choice of mode be driven solely by expected response rates. Other factors, such as cost, coverage, timeliness, and measurement aspects, also come into play (see Chapter 6 for a more complete discussion of these considerations). Knowledge about mode and nonresponse can be used by choosing a main mode based on a combination of all design assumptions and then, if necessary, combine it with a second or even third mode to increase response rates. It is not quite clear why different modes produce different levels of nonresponse rates. Respondents, if they have a choice, will express their preferences regarding the mode of data collection. Some studies have shown that most respondents prefer filling out a mail questionnaire rather than being interviewed. On another level it is generally more difficult for a respondent to turn down a participation request from an interviewer than not to fill out a questionnaire, especially if the interviewer is equipped with knowledge on how to tailor his or her approach to the specific respondent.

In practice, the basic strategy is to start with a main mode that is suitable for all design aspects involved. As for nonresponse, the main mode is used to its full potential with certain constraints, such as a prespecified number of callback attempts. Then another mode is used to increase response rates. This strategy is very efficient because it takes advantage of the fact that respondents' preferences vary and different modes have varying possibilities to locate respondents, but also that it makes respondents aware of the fact that the survey is important. This strategy, called *mixed-mode*, is clearly a compromise.

If there is a main mode considered as most appropriate for the survey, any use of other modes might introduce an additional error. For instance, a questionnaire developed for a mail survey can afford to have quite a few response alternatives for its questions. If that questionnaire is used in a telephone follow-up, there is a risk that the respondent does not hear or cannot distinguish between all response alternatives as read by the interviewer. So, from some perspectives, changing the mode might result in decreased quality. On the other hand, there are other aspects of quality that might be improved by using a mode administered by an interviewer. As usual, we are facing a trade-off situation when it comes to quality. Our interest should be to keep total quality as high as possible given various constraints. If using more than one mode to increase response rates generates a net increase in total quality, a decision should be made to use it if other considerations, such as cost and timeliness, allow it. Most of the time, however, information on the components of this and other trade-offs is not available. As a consequence, most mixed-mode designs are ad hoc with minimal control, although there are examples of very carefully planned and executed mixed-mode surveys.

Common practice in a mixed-mode survey is to start with a less expensive mode, such as mail, and then continue with a more expensive one, such as telephone interview, that will increase response rates. In this situation, cost is usually the major design factor, although sometimes a less expensive mode might be in concordance with an important design factor such as low measurement error.

For example, we know that responses to some sensitive questions may be more accurate if we use a mail survey rather than an interview. In addition, mail surveys are less expensive than interviews, so costs should benefit from its use. But even if the preferred mode is the expensive one, continuation with a less expensive one can sometimes be justified. Carrol et al. (1986) show that three waves of face-to-face interviews in a residential energy consumption study resulted in accumulated response rates of 76.6%, 84.0%, and 89.9%, respectively. Two waves of mail collection added another 4.7%, and about 75% of these were refusers in the face-to-face phase.

3.6.6 Respondent Burden

One important correlate of nonresponse is the burden the respondent perceives in completing the questionnaire and other survey tasks. It is widely accepted that if a survey request is perceived as interesting and easy, the likelihood of obtaining cooperation will increase. However, respondent burden is a very general concept. In surveys of individuals and households it reflects the degree to which the respondent perceives the survey as demanding and time consuming. In business surveys, burden often reflects the cost involved for the responding organization to comply with the survey request. If steps are taken to reduce the burden, response rates will generally improve.

Typically, burden includes questionnaire or interview length, the workload on part of the respondent in terms of time and effort, the pressure the respondent might feel when being confronted with questions, and the number of survey requests the respondent receives within a certain period. Two general methods have been used to address the problem of respondent burden. One is to revise the survey request to one that is perceived as less burden by the respondent, and the other is to try to compensate the respondent by using an incentive.

Reduction of the respondent burden suggests that the time required to complete the questionnaire should be minimized. Questions should not be included unless they are explicitly part of an analysis plan established ahead of time. (Chapter 2 provided a scheme for ensuring that this is the case.) When the questionnaire is developed, the following sequence of considerations should be made for each question, with the sole purpose of trying to reduce the respondent burden.

- Is the question necessary?
- For whom is the respondent answering (him- or herself, the household, the business)?
- Can the respondent answer the question with a reasonable effort?
- Is it likely that most respondents want to provide an answer?
- Is the question answerable? For example, would the designers of the questionnaire have difficulty in providing an answer if they were asked the same question?

If one or more of these issues is problematic, the inclusion of the question should be reconsidered. Common burden-related problems with questions include difficult terms, wordings, and concepts, situations where the respondent is asked to remember events or place them in time, and when the respondent is forced to go to another source or make calculations to come up with an answer. Cognitive research, such as think-aloud protocols and focus groups (see Chapter 8), can disclose many of these problems, and questions can be changed or removed.

It is not only the number of questions and their nature that can pose problems. The organization of the questionnaire is also critical, particularly for self-administered modes. Here, it is important that self-administered surveys are simple to understand, and navigating from question to question is obvious. The phrase *respondent friendly* has been used by Dillman and others to describe such questionnaires (Jenkins and Dillman, 1997; Redline and Dillman, 2002). A new theory is emerging in this field based on psychological and sociological theories, which treats issues such as verbal and nonverbal language aiming at establishing design principles for self-administered questionnaires. (Of course, many of these principles may apply in interview surveys as well, but their effect on nonresponse is probably not that large in interview surveys since the interviewer plays such a major role in the perceptual process.) Examples of questionnaire design principles aimed at making the respondent's job easier and reducing errors in self-administered surveys are provided below. See Dillman (2000) for a more complete set of principles.

DESIGN PRINCIPLES FOR RESPONDENT-FRIENDLY MAIL QUESTIONNAIRES

- Use the visual elements of brightness, color, shape, and location to define a desired navigational path through the questionnaire.
- When established format conventions are changed in the midst of a questionnaire, prominent visual guides should be used to redirect respondents.
- Place directions where they are to be used and where they can be seen.
- Present information in a manner that does not require respondents to connect information from separate locations in order to comprehend it.
- Ask only one question at a time (i.e., avoid *double-barreled* questions).

Source: Adapted from Dillman (2000).

Having done all we can to reduce respondent burden, to increase response rates further may require the use of incentives. The use of incentives has been the focus of much research during the last several decades. It is an important topic and has a bearing on many other design aspects, such as cost, measurement error, and ethics. It might be fair to say that incentives almost always increase response rates. Many marketing institutes use small incentives such as lottery tickets, publications, and calendars on a regular basis to increase sample units' willingness to participate. The typical scenario is that the respondent is promised a gift once he or she has participated. Sometimes the incentive is used only as a means to sway reluctant respondents, and then it becomes an ethical issue. Certainly, a government survey should not use an incentive as a refusal conversion method. Every sample unit should have the same probability of benefiting from an incentive offer. This standpoint may, however, be debatable.

As mentioned the literature on survey incentive research suggests strongly that prepaid incentives are more efficient than promised incentives in terms of resulting response rates. Berk et al. (1987) showed that even a "no incentive" treatment generated larger response rates than a promised incentive. In Section 3.6.2 we discussed the reciprocation principle and other theories that help explain why incentives work, especially why prepaid or predelivered incentives increase response. Based on social exchange theory, the respondent has a tendency to compensate for a gift received before the request to participate. The compensation is in many cases much larger than the value of the gift in terms of value of time spent in the participation. If the incentive is promised, another theory, economic exchange theory, probably comes to play. Here the sample unit sees the survey request as a businesslike proposal that could be accepted or turned down, depending on a judgment of the request/remuneration situation.

There is evidence that incentives can sometimes increase response rates by as much as 10 percentage points, but their usefulness should be weighed against other means of reducing nonresponse. Also, as can be expected, incentives work differently for different populations and population groups, and very little is known about their effect on the measurements per se. In establishment surveys the incentives are usually different from those used in surveys of individuals and households. Stimuli used in establishment surveys include provision of results for the individual establishment, compared to results for other establishments. These comparative results should, of course, concern groups of units that are of particular interest to an individual establishment: for instance, establishments of approximately the same size or that come from the same region as the individual establishment.

3.6.7 Use of Advance Letters

Many survey researchers realize the difficulties obtaining survey participation if sample units are contacted without advance notification. In an RDD survey it is not possible to contact all sample units in advance. There are ways to obtain a reasonable percentage of addresses by matching telephone numbers against address lists even though not all RDD surveys do that. In other surveys we have a better opportunity to obtain addresses. A letter sent out in advance usually has a positive effect on the response rate. Examples of studies confirming the effect are found in Sykes and Collins (1988). There are exceptions to the rule, however. Advance notification can alert some sample units and allow them to prepare a firm refusal when being contacted with the actual request (Lyberg and Lyberg, 1991). In many countries, the legal framework states that sample units selected for official statistics surveys should be notified in advance. Luppes (1994, 1995) has assembled a checklist for what should be contained in an advance letter. His study is based on a review of advance letters used in a number of expenditure surveys across countries. Recent Swedish experiments have shown that advance letters in videotape format have proven effective using a well-known public figure as the narrator (Ahtiainen, 1999).

SOME RULES FOR WRITING EFFECTIVE ADVANCE LETTERS

- The letter should be easily understood, and terminology unfamiliar to the respondents should be avoided.
- The contents of the letter should be presented in a number of short paragraphs.
- The tone of the letter should be friendly.
- There should be a natural order between different pieces of information.
- Potentially controversial wording should be avoided.

3.6.8 Interviewers and Response Rates

In Chapter 5 we discuss errors due to interviewers and interviewing. In this section we highlight findings that relate interviewers and nonresponse.

In interview surveys the interviewer is the link between the questionnaire and the sample unit. The interviewer has two main roles: to gain cooperation and to carry out the interview once cooperation has been established. These roles require two very different skills, and usually, interviewers are not equally skilled at both.

Characteristics of the interviewer and his or her work procedures affect the nonresponse level considerably. Especially crucial is the first contact, even the very first seconds of that contact. The "doorstep" strategies are crucial, and the new ideas based on the psychological theories we have described are important. The interviewer should assess each situation and tailor his or her approach based on prior information, perhaps obtained at the initial contact. Some factors that have proven important are provided below.

Interviewers who use all the means available regarding tracing techniques, tailoring, and refusal conversion methods obtain better results than do interviewers who do not use these means to their full potential. Recent experience suggests that interviewer training can be much improved by taking advantage of the new developments in tailoring. Also, very little has been done to study and incorporate in the training materials the work methods that the best interviewers use in approaching sample units.

SOME INTERVIEWER CHARACTERISTICS RELATED TO HIGH RESPONSE RATES

Experience. Interviewers with considerable experience tend to get lower nonresponse rates than interviewers with less experience.

Interviewer's demographic characteristics. Gender and age play a role, in combination with similar respondent characteristics and the survey topic.

Interviewer's voice, accent, and interviewing style. Voice and accent can be important in telephone surveys, less so in face-to-face surveys. A personal interviewing style may elicit more information from respondents but may also introduce larger interviewer error.

Interviewer expectations and preferences. Interviewers with a positive attitude have a better chance gaining cooperation than do interviewers who think it will be difficult to gain cooperation.

Attitude and confidence. For instance, several refusals in a row might affect an interviewer's ability to approach the next sample unit.

3.6.9 Follow-up

We have seen various efficient ways to recontact sample members who are difficult to contact or who are reluctant or refuse to participate. This follow-up can take various forms. In mail surveys, nonresponding sample units are sent reminders. We have discussed such schemes previously: for instance, Dillman's total design method. These reminders follow a well-established pattern, so that each new reminder does not look exactly the same as the preceding one. Data on the outcome of the reminder rounds ought to be collected so that decisions on when to terminate the efforts are based on these data. Typically, such collection includes data on time between reminder rounds and the outcome of each reminder round and how response is distributed among different sample groups.

In telephone interview surveys and in face-to-face interview surveys with telephone follow-up, interviewers should be persistent. Theories for the placement of calls, *call-scheduling algorithms*, have been developed. If no system controlled by the survey designer is in place to distribute contact attempts over time (which is possible in CATI applications, see Chapter 6), scheduling has a tendency to be based on interviewer preferences. Swires-Hennesey and Drake (1992) have shown that if call scheduling is left to the interviewers, they start with convenient calls on weekdays, and when needed, they continue calling on weekends or evenings. When developing call-scheduling algorithms, the basic assumption is that calls should be made during times when sample units are most likely to be found. Kulka and Weeks (1988) used a conditional probability approach and presented a ranking of three-call algorithms: that is, if three calls are allowed, how should they be distributed? The three-call combinations that ranked highest were:

1. Weekday evening, Sunday, Sunday
2. Sunday, weekday evening, weekday evening
3. Sunday, weekday evening, weekday morning

The worst combinations were:

1. Weekday afternoon, weekday afternoon, weekday afternoon
2. Weekday afternoon, weekday morning, weekday morning

It is easy to see that the worst combinations might very well be among the first an individual interviewer might choose if no restrictions are imposed. If there is not enough time to develop a formal call-scheduling algorithm, one might simply adopt the following general strategy: A majority of the call attempts should be made in the evenings and during the weekends.

Face-to-face follow-ups in the field are less common, due to the costs involved. In principle, the telephone contact strategies also can be applied in the field. In practice, telephone contact strategies are used to set up appointments with respondents. In other cases, interviewers are instructed to follow up on nonrespondents when they conduct interviews in the nonrespondents' neighborhood.

3.6.10 Item Nonresponse

The causes of item nonresponse can often be traced to problems with questions or questionnaires. The reasons that not all items have been answered include:

- The respondent deliberately chooses not to respond to a question because it is difficult to answer or the question is sensitive.
- The questionnaire is complicated, and if the mode is self-administered, the respondent might miss skip instructions or certain questions, or the respondent might simply exit the response process because it is boring, frustrating, or time consuming.
- The questionnaire contains open-ended questions, which increases the risk for item nonresponse.
- The respondent or the interviewer makes a technical error so that the answer to a specific question should be deleted.
- The questionnaire is too long.

In some situations, item nonresponse is closely connected with measurement error. A respondent can answer sensitive questions by responding in an uninformative way. For example, if the question offers response alternatives such as "don't know" or "no opinion," those alternatives can serve as escape routes for respondents who otherwise might have left the question unanswered.

Item nonresponse is best handled by preventive measures associated with the design of questions and questionnaires. Good design strategies include the use of fewer sensitive questions, better skip instructions, and avoiding questions that are too much of a burden to the respondents. The analysis becomes complicated when there is item nonresponse. Therefore, a number of imputation procedures have been worked out over the years where missing data are replaced by data generated in some other way (see Section 3.6.11).

Questions on income and socially stigmatizing behaviors or circumstances are notorious candidates for extensive item nonresponse. As noted by Christianson and Tortora (1995), item nonresponse is not reported completely or systematically by most statistical agencies. When it is reported, the item nonresponse rate is usually calculated as the ratio of the number of eligible units responding to an item, to the number of responding units eligible to have responded to the item. It is useful to calculate item nonresponse rates by key questions, by sample groups, or by interviewer. That way the rates can serve as very important process variables (see Madow et al., 1983; U.S. Federal Committee on Statistical Methodology, 2001).

3.6.11 Adjusting for Nonresponse

Despite all efforts to minimize nonresponse, less than 100% of the sample will typically respond to a survey, and usually, it is considerably less. After data collection is completed and all the data are in, there are still methods to be employed for reducing the effect of nonresponse on the estimates. These methods are referred to as postsurvey adjustments for nonresponse. A number of statistical methods are available for performing these adjustments to the data, but they can be classified into just two general categories: weighting and imputation. The mathematics involved in describing and demonstrating these postsurvey adjustment methods are complex and go beyond the scope of the book. For an up-to-date review, the reader is referred to Groves et al. (2002) and Brick and Kalton (1996).

In many applications, no effort is made to compensate for remaining nonresponse, which is unfortunate. In these cases, the survey estimates are based solely on the respondents with no adjustments to compensate for the differences between the estimates for respondents and nonrespondents. Usually, some information is available on the nonrespondents that can be used as the basis of a nonresponse adjustment. If so, this should be done since not adjusting an estimate for nonresponse assumes that the estimate based just on the respondents represents the entire target population, including nonrespondents. In effect, this means that the analyst assumes that $\overline{Y}_R = \overline{Y}_{NR}$, which is not a sound practice unless there is clear evidence that this is actually the case. To compensate for unit nonresponse, it is standard practice to multiply the survey weights (see Chapter 9) by factors called *weight adjustments*. These adjustment factors attempt to adjust the estimate of $\overline{Y}_R$ so that it is closer to the population mean, $\overline{Y}$.

Most weighting schemes increase the standard errors of the estimates, but the bias reduction resulting from the adjustment for nonresponse is expected to generate a smaller MSE. All weighting procedures are based on a set of assumptions. If these assumptions do not hold, the nonresponse bias will not be reduced to the extent anticipated.

Imputation is used to handle item nonresponse but in some instances may also be used for unit nonresponse. It is a procedure where missing data are replaced by artificial or "modeled" data. Usually, imputation is done only for the key characteristics in a survey, since it requires some fairly extensive statistical work to determine the appropriate model to use for imputing the missing values. A different model is needed for each characteristic.

When imputation was introduced, the technique was a means to construct complete data sets, thereby avoiding calculation problems in an era when computational facilities were less advanced (see Ogus et al., 1965). There were also strict recommendations regarding the allowed level of item nonresponse (at most 5%) for imputation to be applied. To obtain complete data sets is still the goal of imputation, but today there are no computational problems. Also the "5% rule" is no longer applied. Today, imputation is performed on data sets containing much larger nonresponse rates.

There are numerous ways of imputing missing values (see Dillman et al., 2002). Three examples are:

- In *hot deck imputation*, the missing variable values are replaced by the values of one of the respondents. Often, the rule is that missing values are taken from a randomly selected respondent among those that are similar to the nonrespondent on characteristics that are observed for both respondents and nonrespondents. Hot deck imputation is used for imputing income in the U.S. Current Population Survey.

- In *nearest-neighbor imputation*, the imputed value is taken from the respondent that is "closest" to the nonrespondent according to a measure of distance. Obviously, the technique is similar to hot deck, but whereas hot deck is a random procedure, nearest neighbor is a deterministic one.

- In *direct modeling*, a statistical model is developed for predicting the missing characteristic using as the predictor variables other survey items that are known to be highly correlated with the missing characteristic. For example, an imputation model for income predicts a missing income value for a respondent using the respondent's education, age, gender, or other characteristics inquired after in the questionnaire.

The work with adjustment procedures can be quite complicated and should be performed by specialists. Survey managers are strongly advised to incorporate adjustment methods as part of the survey estimation procedures.

There are two principal *methods for nonresponse adjustment*: weighting and imputation.

CHAPTER 4

The Measurement Process and Its Implications for Questionnaire Design

In this chapter, as well as in Chapters 5 and 6, we consider the various components of nonsampling errors that together comprise perhaps the most complex major source of nonsampling error in surveys: the measurement process. In this chapter we consider the interaction of the respondent and the interviewer (if present) with the questionnaire. We begin by introducing a framework for studying the various components of the measurement process, of which the questionnaire is a major part. The focus in this chapter is the process that respondents may use to understand the survey question, retrieve or deduce whatever relevant information is needed to respond to the question, consider how to convey that response given the choices presented by the question response categories, and finally, communicate the response. As we will see, there can be many obstacles confronting the respondent on route to a response.

4.1 COMPONENTS OF MEASUREMENT ERROR

As shown in Figure 4.1, the measurement process is composed of six primary components that contribute to overall measurement error for a survey: the interviewer, the respondent, the data collection mode, the questionnaire, the information system, and the interview setting. The data collection mode refers to the means of communication used for the interview, be it a telephone, face-to-face meeting, or if no interviewer is involved, a self-administered questionnaire. The information system refers to the body of information that may be available to the respondent in formulating a response. For example, it could be physical records, other persons in the household, or a person's own memory. Finally, the setting is the environment within which the interview takes place: a home, a classroom, an outdoor setting, a hospital, and so on.

116

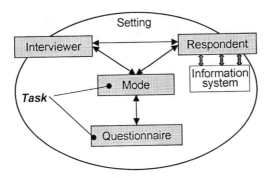

Figure 4.1 Potential sources of measurement error.

A key component of any measurement process is the instrument that will be used to collect the survey information. This may be a mechanical device used for collecting physical measurements of some type (e.g., a blood pressure meter) or a paper or computerized form or questionnaire that is the instrument used for collecting data from people or transcribing information from records. In survey work, the measurement process usually involves a subject, respondent, or other person supplying the information to be gathered in the process and a data gatherer, interviewer, or person who applies the instrument and records the measurements or responses.

The nature of the interaction between the interviewer and the respondent is influenced to a large extent by the *mode of data collection*—the method by which the instrument is applied. For example, an interview may be conducted by telephone or face-to-face and the interviewer may use a paper-and-pencil questionnaire [referred to as *paper-and-pencil interviewing* (PAPI)] or a computerized questionnaire [referred to as *computer-assisted interviewing* or (CAI)]. For self-administration methods, respondents complete the questionnaire without the aid of an interviewer. Mixed-mode surveys are also common. For example, some surveys that ask about sensitive topics may collect responses to those sensitive questions by self-administration, with other parts of the interview collected by an interviewer. Mail surveys may use telephone and/or face-to-face interviewing to follow-up nonrespondents. In other surveys, to encourage their participation, respondents may be given a choice of the mode of data collection.

In *direct-observation* surveys, data are not supplied by a respondent but rather, are collected directly by an interviewer or an observer. For example, in a crop yield survey, an agricultural agent (the observer) will estimate the expected yield of a field of some crop by sight, with some direct physical measurements on the field. No respondent is involved. Similarly, some surveys may require that the interviewer make visual inspections of the respondent, the respondent's dwelling or neighborhood, the behavior of family members during the interview, and so on. In other data collections, the data collector may obtain the required survey information directly from the records of the company or institution by transcribing the information into a paper or computerized form.

The information system refers to the various sources of information that may be used as an aid in completing the questionnaire. For example, a respondent may respond unaided by external information sources or may draw on information from others in the household or business, a company database or records system, or household records such as calendar notes, pay stubs, receipts, and so on. Finally, the interview *setting* refers to the environment within which the survey takes place, be it a noisy office or home, outdoors in a field with the respondent sitting on a tractor, the doorstep of a home, a school cafeteria, or any other place where an interview can take place.

The arrows between the error components in Figure 4.1 suggest that the components are interrelated and interact during the measurement process. Thus, the error contributed by one component is influenced and may be altered by the presence of other components in the process. As an example, interaction between the interviewer and respondent is influenced by the mode of interview, the setting, and the questionnaire design. In telephone interviewing, interactions between the respondent and interviewer may be briefer and less social than in face-to-face interviewing, which tends to be a more social setting with longer interactions between interviewer and respondent. The ability and willingness of the respondent to respond to each question accurately may also be affected by the setting. Respondents may be more contemplative and precise when they are interviewed in a quiet room than when the interview is being conducted in an uncomfortable and noisy environment such as in a busy hallway outside the respondent's apartment.

The record systems that respondents access can have a tremendous influence on response accuracy. Company or household records may be inaccurate, outdated, incomplete, or difficult to access as a result of the setting or interview mode, timing of the interview, and so on. As an example, an establishment survey interview may be conducted at an employee's small workstation where the records needed for the interview are not accessible. Further, the space at the workstation may be too limited and there may not be enough room for the employee to spread out the materials and files that he or she needs to respond to the questions accurately. In a household survey, the interviewer may have to conduct a rather long interview at the doorstep because the respondent is unwilling to invite the interviewer inside the home.

As a result of these complex interrelationships among the components of measurement error, the effect on the response of a particular error source can be quite unpredictable since interactions of that error source with other sources of error can alter its influence on the response process. As an example, in the survey methods literature, there are a number of examples where a particular interview mode can have very different effects on survey response, depending on the population being surveyed, the characteristics of the interviewers, and so on. This suggests that our discussion of the effects on the response of a specific source of measurement error should consider its inter-actions with the other components of error shown in Figure 4.1. Thus, when appropriate, we consider how the effect on data quality of one component in the process may change as a result of changes in the other components in the process.

In the remainder of this chapter, the focus is on three critical components of the measurement process: the respondent, the questionnaire or instrument, and the source of information. Of particular interest in our study is the process that respondents typically use to arrive at a response. Chapter 5 is devoted to the study of interviewer errors. In Chapter 6 we consider errors arising from the data collection mode and setting. Understanding these components of the process will provide valuable insights regarding the design of surveys that facilitate and enhance the measurement process.

4.2 ERRORS ARISING FROM THE QUESTIONNAIRE DESIGN

Roots of Questionnaire Research

Survey researchers have long recognized the potential for alternative wordings of a question to affect the answers that people give, particularly for opinion questions. A famous example that is often cited is given by Rugg (1941). When asked, "Do you think that the United States should forbid public speeches against democracy?," 54% of respondents said "yes," they should be forbidden; when asked "Do you think the United States should allow speeches against democracy?," 75% said "yes," suggesting that only 25% would not allow such public speeches. These results demonstrate that by a simple alter-ation of the question wording—replacing *forbid* with *allow*—the results of an opinion poll can be shifted. As surveys were repeated by different survey orga-nizations using different question wordings, it became apparent that question-wording effects also apply not only to opinion questions but to behavioral questions (see, e.g., Sudman et al., 1974).

An often-used experimental method for testing the difference in responses between two alternative wordings of the same question is the *split-ballot experiment*. In its simplest form, the method involves splitting a sample into two random halves, with each half receiving one of the question versions; however, the same idea can be applied to more than two splits of a sample. Since partitioning of the sample is done randomly, each subsample should yield the same estimates, on average. Therefore, as long as the only difference in the methods applied to each sample split is the questionnaire, any dif-ferences in the estimates can be attributed primarily to the differences in questionnaires.

Two pioneers in our understanding of question response effects in surveys are Seymour Sudman and Norman Bradburn. In the early 1970s, these methodologists conducted a meta-analysis of some 900 split-ballot and other measurement error studies in the survey literature (Sudman et al., 1974). *Meta-*

analysis is a statistical approach to combining the quantitative outcomes from a number of individual studies on a particular, well-defined research question. Its purpose is to provide a general answer to the research issue that is common to and being addressed by all the studies (both published and unpublished) that can be found on the issue. Sudman and Bradburn's analysis considered most of the components of measurement error in Figure 4.1 but focused primarily on questionnaire design variables such as questionnaire length, use of difficult words, open (i.e., a question that obtains a verbatim response from the respondent) versus closed (i.e., a question that provides response categories from which the respondent must choose) question form, the position of the question in the questionnaire, the salience of the question to the respondent, and the use of devices to aid the respondent's recall of information. Their research provided a comprehensive and integrated picture of how various questionnaire design features affect survey response.

An exciting development in the past decade has been the recognition that both cognitive psychologists and survey researchers have made contributions to the other's research. Cognitive theories have already led to a much fuller understanding of the task of a survey respondent (see Chapter 8 for a description of cognitive methods) and how aspects of the questionnaire, especially the context within which a question is asked, affect survey response.

Goal of Questionnaire Design

The goal of questionnaire design is threefold. First, as discussed in Chapter 2, each research question to be addressed by the survey implies a number of concepts to be measured in the population. The questionnaire translates these concepts into survey questions that allow the interviewer or respondent to provide information on the concepts. Thus, one goal of questionnaire design is to write questions that will convey the meaning of the inquiry exactly as the research intended. Second, the questionnaire should provide the preferred manner for eliciting information from respondents and should state the questions in a manner designed to generate the most accurate responses possible. In other words, the questionnaire should be designed to minimize systematic and variable errors within the available budget and other constraints of the survey. In addition, the time required to complete the questionnaire and other aspects of respondent burden should be minimized subject to the analytical goals of the survey. This means that the questionnaire should be designed to allow the most efficient means of providing the information required. Reducing the burden on the respondent is essential since it will usually increase the likelihood that he or she will agree to participate and reduce the chance of error in response due to fatigue and inattention to the response task. Finally, the questionnaire should be designed so that the cost of data collection stays within the data collection budget. For a complex survey with many difficult questions, achieving these goals often will require considerable skill and experience. The following example illustrates a few complications that can arise in the design of a questionnaire.

Example 4.2.1 Consider a study of schoolteachers aimed at, among other things, estimating the number of hours that students receive instruction on various health topics and in physical exercise. These data are needed at three levels of childhood education: elementary school (corresponding to the first six years of education), middle school (corresponding to grades 7 and 8), and high school (grades 9 through 12). For each level, the survey should collect information on the number of hours of "exposure" to certain health topics that students receive who complete a typical 12-year course of education.

For example, one research question in the study asks about the extent to which the current educational system emphasizes in-school physical exercise for students. The survey researcher may specify the concept simply as: "With what frequency do students at each grade level engage in school-sponsored group exercise, and what is the average duration of this in-school exercise?" The task of the questionnaire designer is to translate this research question into a series of survey questions that are appropriate for each education level so that teachers can understand the concepts and can therefore respond to them accurately.

To design questionnaires that work well at all grade levels, questionnaire designers must have some knowledge of the way in which schools function so that the questions reflect or accommodate the various scenarios that interviewers will encounter as they interview teachers at each level. This may require that several questionnaires be developed. In addition, a number of other issues arise that require considerable knowledge of the population and the researcher's intentions. Some examples are:

- Assuming that the survey takes place in the spring, to what period of time should the question refer (i.e., what is the *reference period*)? Should the question refer to the preceding fall semester, the current spring semester (including part of the semester that has not yet occurred at the time of the interview), or both? Or should the question refer to the preceding school year?

- How detailed should the estimates of exposure time for a particular topic be? For example, when attempting to determine the number of hours a teacher taught a topic, is it sufficient to offer the choices "0 or none," "1 to 10 hours," or "more than 10 hours," or is more detail needed?

- What memory aids would be helpful? For example, do teachers follow a lesson plan or syllabus that would help in determining the response to this question? If so, is this followed accurately, or can they deviate considerably from it? Does this vary across elementary, middle, and high school grades?

- For assessing questions about physical exercise, what does the researcher consider to be "exercise," and is this definition likely to be misunderstood by some teachers? As an example, at the high school level, students may take courses such as physical education, whereas at the elementary school level, there may be no formal courses on physical education. There is the potential for teachers at these two levels to have a very different perception of physical exercise, and neither of these may be consistent with the meaning intended by the researcher.

As can be seen from this list of issues, the development of clearly understood and easily answered questions requires close collaboration between survey methodologists and survey researchers or subject matter experts. The subject matter expert may be best at providing the required knowledge of school systems and teaching practices, while the survey methodologist's strength may be in developing survey questions that seek to reduce measurement errors to the extent possible within the constraints of data collection.

Close collaboration between the survey methodologist and the subject matter researcher will usually yield a questionnaire that is less subject to specification error and measurement errors than one developed without this collaboration. However, often a questionnaire that is designed even under ideal conditions will still contain important flaws in the design and would benefit from further revision and refinement. For example, some questions may still be confusing to respondents due to situations that were not anticipated in the design, the question ordering may be awkward or unnatural, or the response categories may be too limiting. The questionnaire length may substantially exceed the length allowable by the budget, requiring that some questions be eliminated. It is not uncommon for questionnaires designed by subject matter and methods experts to still pose serious problems for data collection.

These problems can often be identified by conducting pretests and other evaluations of the questionnaires prior to the survey. One method often used involves conducting the interview with a small number of respondents who represent the target population to be surveyed (e.g., schoolteachers in Example 4.2.1). These interviews can be conducted in the same manner as they would be in an actual survey, or specially designed *cognitive interviewing techniques* can be employed to better identify problems that arise during the interview process. Using these methods, the cognitive interviewer may probe the respondent's understanding of the concepts using such questions as: "How did you come up with that answer?", "What does the term "_____/_____" mean to you?", or "Tell me the process you used to recall that number." Cognitive interviewing techniques and other pretesting methods as well as several non-interviewing methods for identifying questionnaire problems are discussed in some detail in Chapters 8 and 10.

In the last 15 years or so, survey designers have applied cognitive theories to the task of writing questions and designing questionnaires. One theory that has been used extensively in questionnaire evaluation is the so-called cognitive theory of the survey response process. In the next section, we will examine this theory and show how it can be used to identify and correct problems in the design of questionnaires. As we shall see in Chapter 8, this cognitive theory

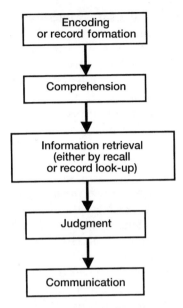

Figure 4.2   Five stages of the response process.

of the response process is the basis for a number of questionnaire evaluation methods.

4.3   UNDERSTANDING THE RESPONSE PROCESS

In this section we consider a model of the survey response process originally proposed by Kahn and Cannell (1957) that has been used widely in the design and pretesting of questionnaires. [See Tourangeau et al. (2000) for a more complete description of the process.] In this model, the respondent proceeds sequentially through five distinct cognitive stages as he or she responds to a single survey question: (1) encoding or record formation, (2) comprehending or understanding the question, (3) recalling or computing a judgment, (4) formatting a response to the question, and (5) editing and communicating the response. These stages are shown as a flow diagram in Figure 4.2. Although this response model reflects a somewhat idealized paradigm for the response process, it is nonetheless quite useful for thinking about the design of questions and for identifying potential problems with the questionnaire. The model is developed primarily for surveys of individuals; however, extensions of the model for use exclusively in business surveys can be found in Edwards and Cantor (1991), and Sudman et al. (2000). The following is a brief description of a model described in Biemer and Fecso (1995) which is applicable to both household and establishment surveys.

In formulating a response to a survey question, the respondent must first have the knowledge, belief, or attitude required to provide a valid response. If the information requested will come from a physical record such as a company database, the record or data entry must exist at the time of the interview in order that the information can be retrieved (*encoding/record formation*). Second, there must be a shared meaning among the researcher, the interviewer, and the respondent with respect to each of the words in the question as well as the question as a whole (*comprehension*). Then, to respond to questions concerning events or behaviors that occurred in the past, the respondent will attempt to *retrieve the required information* from memory. If information requested is some characteristic of a company's past performance, the respondent may attempt to *retrieve the information* from the company's files. Of course, some questions, such as attitude, belief, and opinion questions, require only that the respondent *compute a judgment* during the interview without the need to recall information. However, even in these situations, a respondent may try to recall a previously formed judgment.

Once the information has been retrieved or a judgment has been computed, the respondent must decide how to communicate it to the interviewer. To do so, he or she may need to *format the response* so that it conforms to the answer categories from which the respondent must choose. Finally, the respondent *communicates* the response to the interviewer (or records the response on the form if the mode is a self-administered questionnaire), taking into account risk and benefit of responding accurately and honestly. At this stage, respondents may decide to edit or revise their response and respond in a way they know is not completely accurate, due to social desirability influences, fear of disclosure, or acquiescence (i.e., the tendency to agree with statements using an agree or disagree format). These concepts are discussed in more detail below.

An important benefit of this model is that it decomposes the response process into smaller tasks that may be treated separately in the design and evaluation of surveys. One criticism of the model is that it is too simple to capture the complexities of completing a survey questionnaire. For example, in practice, respondents may respond to a question without trying to understand the terms or to recall the correct information. This may happen because the respondent is fatigued, uninterested, or just being playful. Such short-circuiting of the response process is sometimes referred to as *satisficing* behavior (Krosnick, 1991). When a questionnaire is long and monotonous or when the respondent is not motivated to provide good responses, the respondent may begin to satisfice or answer without really trying to respond accurately (i.e., without *optimizing*). However, as we will see, this simple model, like the other simple models we consider in this book, is still quite useful in that it provides some important insights regarding questionnaire design in many diverse situations. Next, we consider how the response model can be used as an aid in the control and evaluation of measurement errors in surveys.

Encoding and Record Formation

Information encoding is the process by which information is learned and stored in memory. For an event or experience to be recalled or retrieved in the measurement process, a record of it must be created. For example, for a respondent to answer accurately a question regarding the behavior of a household member in a survey, the behavior must first be observed and committed to memory so that it can be recalled during the interview. Similarly, if a business transaction or numerical data item is to be retrieved from an establishment's database during an interview, the information must first be recorded in the establishment's database.

In *encoding or record formation*, knowledge is obtained, processed, and is either stored in memory or a physical record is made.

The encoding or record formation stage of the response process is the only stage of the process that takes place prior to initiation of the measurement process, perhaps by many months or even years prior. Still, it is considered a critical part of the process because respondents cannot be expected to retrieve facts, events, and other data if these data have never been encoded in memory or saved as a physical record.

The failure to encode information is an important cause of error in surveys. Consider, for example, surveys that allow proxy respondents to provide information requested by the questionnaire. A *proxy respondent* is someone who provides a response to a question that is in reference to another person. For example, a survey of very old persons may allow the sample members' caregivers to respond to questions when the sample members are unable to respond for themselves. Proxy responses are allowed in survey work primarily to increase the survey response rates, since persons who are the subjects of the questioning are not always accessible, available, willing, or able to participate for themselves. If the only persons who were allowed to respond to these questions were the subjects themselves—referred to as *self-response*—the amount of missing data in the survey due to nonresponse could be too extensive to provide valid estimates. In addition, proxy responses can be more accurate than self-responses in some situations. As an example, questions about medications taken by an elderly patient in a hospital might be provided more accurately by the patient's nurse rather than by the patient. However, there are also many situations where proxy responses may be less accurate than self-responses. Still, proxy responses may be preferred when the only alternative is nonresponse.

As an example, in a survey of health, the question "How many times in the last 30 days have you visited a physician?" is asked for each family member. If this question is answered by *self-response* (i.e., it is answered by the object of the question), the encoding stage of the response process is unlikely to be a problem since this information should have been encoded into memory during doctor visits. Of course, recall error could still be a problem if the person has had many visits to the doctor and unable to remember them all clearly. However, encoding error could be a problem if the question is answered by a proxy respondent who may not know about the reference person's visits to the doctor, since that information was never encoded. In such cases, perhaps the best outcome is that the proxy respondent admits not knowing this information. Unfortunately, quite often proxy respondents may try to guess or otherwise provide responses regardless of whether they have encoded the information required. Errors of this type are referred to as *encoding errors*.

The proxy reporting in the U.S. Current Population Survey provides an illustration of encoding error. The Current Population Survey (CPS) is a household sample survey conducted monthly by the U.S. Census Bureau to provide estimates of employment, unemployment, and other characteristics of the general U.S. labor force population. For this survey, in a study of proxy reporting, Roman (1981) found that the unemployment rate was much higher for self-reporters than for proxy reporters. A plausible explanation for this is that many proxy respondents may not know whether the reference persons who are without jobs were looking for work in the preceding week. Persons who were looking for work are classified as unemployed by the CPS rules, whereas persons who were not looking for work are classified as not in the labor force. Thus, encoding error could partly explain why a higher proportion of self-respondents are unemployed than proxy respondents.

Moore (1988) provides an excellent review of the literature on proxy reporting. As Moore points out, almost all proxy studies suffer from an important limitation referred to as *selection bias*. That is, when comparing the responses of proxy reporters with self-reporters, many studies simply cross-classify the data by type of reporter and compare the means or proportions for the two groups. However, persons who give self-reports may have very different characteristics than persons for whom proxy reports are obtained. Self-reporters may be more accessible, available, and cooperative than persons whose reports are provided by proxies, so that the true characteristics of the two groups differ. Thus, the measurement error associated with the type of report is entangled or confounded with the true group differences. For this reason, much of the literature comparing the accuracy of proxy reporting with self-reporting is inconclusive.

Encoding error can be problematic for self-reports as well. A study conducted by the U.S. National Centers for Disease Control (CDC) provides a good illustration of encoding error in a survey of childhood immunizations. The U.S. National Immunization Study (NIS) collects information from parents regarding the immunizations received by their children by the age of 2 years. Since children this age should have received at least 14 doses of five different vaccines, even the most conscientious parents have difficulty reporting their child's vaccinations accurately. To investigate why, a series of studies was conducted using the cognitive response model in Figure 4.2 as a guide. Initially, it was hypothesized that recall error was the primary cause of mis-reporting since it was observed in the NIS that parental reports unaided by shot records or other memory aids tend to understate the number of immu-nizations that children receive. However, another hypothesis is that the problem is encoding (i.e., parents' reports may be in error because they know very little about the vaccinations at the time they were administered).

Both hypotheses were tested in separate studies. To test the encoding error hypothesis, a study of children age 7 and younger was conducted at a pediatric clinic. Parents who visited the clinic with their children were asked to com-plete a short interview about their children's medical visit as they were leaving the clinic. Surprisingly, even immediately after the vaccinations had been administered, most parents had little knowledge regarding which vaccinations their children had received that day. The most prevalent error was failing to know that a shot had been given (i.e., a false negative report) rather than reporting shots that had not been given (i.e., a false positive report). The false negative rate was almost 50%, while the false positive rate was only 18%. The study concluded that parental reporting error was encoding error and that the use of recall cues and memory aids to increase reporting accuracy would there-fore not be effective (see Lee et al., 1999).

Another type of encoding error occurs when the respondent has only incomplete, distorted, or inaccurate information regarding a question topic. For example, a survey of farm operators asked farmers to estimate the value of a particular parcel of the land that they own that is used for farming. Some farmers who have no intention of selling their land would not even fathom a guess as to what their land is worth. However, some respondents could decide to supply an estimate even though they have no information on which to base an estimate. They may have heard that land nearby is selling for some amount, say $10,000 an acre, and will suppose that their land is worth that much, too. However, the information they have may be inaccurate or otherwise not indicative of the land's real value. This is an example of how a respondent's response can be distorted by inaccurate or incomplete information on the topic of the survey question.

For establishment surveys, errors resulting from the record formation stage of the response process can occur when an establishment's records are either missing, incomplete, or incompatible with the survey requirements. It is not uncommon that the information requested on the survey form to be similar but still very different from the data that are stored in the establishment's data-base. This incompatibility between the questionnaire and the source of the information causes errors when the respondent simply provides the informa-tion directly from the establishment's database rather than reformatting it to be more compatible with the survey request.

An example of this type of error occurred for the U.S. Current Employ-ment Survey (CES) conducted by the U.S. Bureau of Labor Statistics (BLS).

In an evaluation of the quality of the CES data, Ponikowski and Meily (1989) discovered that 59% of the businesses did not adhere to the definition of unemployment. The main problem was that many companies included employees on leave without pay, although the survey questionnaire requested that these employees not be included in the payroll. When asked why this error was made, approximately 40% of the respondents who committed the error said the cause was incomparability of the survey requirements with their accounting systems. Those establishments did not reconstruct their payroll data to satisfy the survey requirements, but rather, simply gave the number that was more readily available on the company database.

Thus, in designing a questionnaire that asks about individual characteristics and behaviors, a key decision is whether to allow proxy responses. This strategy should be weighed against the risk of obtaining inaccurate information for some items, on the one hand, and missing data, on the other. For some surveys it may be better to obtain data that are inaccurate rather than no data at all. However, there are situations where a proxy response rule should not be used or used only as a last resort to avoid a unit nonresponse. For example, proxy response would not be acceptable for opinion or attitudinal questions. If proxy responses are allowed, a decision rule should be specified for identifying the appropriate informant in the various situations that interviewers will encounter.

For example, for household surveys, the ideal informant (i.e., the person providing the information for the survey) is usually the person in the household who is most knowledgeable about the person who is the object of the question (referred to as the *reference person*). This is usually the spouse or a parent or other caregiver for children in the survey. However, often the ideal informant will vary depending on the topic of the question and the relationship between the household members. It may be impractical to try to interview the ideal proxy for each question and person in the survey. If only one person is to be interviewed in each household, the best strategy may be to identify the informant who is best overall for the key items in the survey.

Similarly, for establishment surveys, the accuracy of the information provided by a company may depend to a large extent on the person providing it. If information is requested on a company's operating costs and other expenditures, the ideal respondent may be the company's chief financial officer. However, the refusal rate for the survey could be quite high if this were the only acceptable respondent. To increase cooperation rates, the respondent rule should be flexible enough to allow other employees in the company to provide this information, within specified limits. This may be difficult to control, particularly for self-administered questionnaires and surveys that request information on a variety of topics, including accounting, personnel, management, and production. Obtaining data on such a wide range of topics may require interviewing or collaborating with not one, but several persons in the company.

Comprehension (Understanding the Question)

The second stage of the response process is comprehension, or understanding the question. At this stage, the respondent reads or hears the statement of the question and attempts to understand what information is being requested. Thus, an important goal for developing good questions is to describe to the respondent precisely what information is needed in words the respondent can easily understand. This stage is essential in order for the respondent to answer the question accurately. Several types of errors can be introduced in the response process at this stage.

> To *comprehend* or *understand a question*, the respondent considers the question and attempts to understand what information is being requested.

First, the wording of the question may be complicated or may involve unfamiliar terms. For example, "In what year did you matriculate at this university?" may not be understood by some students; "In what year did you first enroll at this university?" is more readily understood. Beyond the literal meaning of the question, the interpretation of the question as the researcher intended it must also be conveyed accurately to the respondent. For example, the question "Do you own a car?" contains no unfamiliar or complicated words, yet respondents may still not understand what information is being requested. What constitutes car ownership? Suppose that a person is purchasing a car but is still making payments on it. Perhaps a car is being leased for a three-year period rather than purchased. What about cars that are owned jointly by husband and wife? Or a car that is driven exclusively by a son or daughter, but ownership has never legally been transferred to him or her. Would any of these situations qualify as "ownership?" Without some clarification, respondents may use whatever interpretation comes to mind thus creating variable error or unreliable responses.

Another problem that may arise at this stage is the introduction of context effects. A *context effect* occurs when the interpretation of a question is influenced by other information that appears on the questionnaire, such as the previous questions in the questionnaire, section headings preceding the questions, instructions presented for answering the questions, and so on. Because of the potential for context effects, even the position of a question in the questionnaire can affect the meanings respondents attribute to the question. For example, the question "How satisfied are you with your health insurance?" may elicit very different responses when preceded by the question "How satisfied are you with your doctor?" than when not.

Quite often in questionnaire design, the context implied by prior questions or information in the questionnaire can be quite effective for facilitating question comprehension. When questions that deal with a single topic are grouped together in one section of the questionnaire and the context of the section is made clear to the respondent, question comprehension is enhanced. For example, in an agricultural survey, an entire section of the questionnaire asks about the characteristics of the entire farm operation, while other sections of the questionnaire pertain to only specific segments of land identified within the farm. Thus, it is not necessary to precede each question in the questionnaire with an instruction to indicate which questions deal with the entire farm operation and which questions deal with specific land segments. Due to the grouping of the questions and the context of the sections, respondents understand that some sections are devoted to part of the farm while others are devoted to the entire operation. Questions appearing in each section are then clearly specified by their context.

However, often, context of a question can lead to misinterpretations of questions that result in response errors. Such response errors are also referred to as context effects. A context effect may occur if the respondent mistakenly believes that all questions in a section of the questionnaire pertain to the same entity when they do not. As an example, if in the entire farm section of the aforementioned questionnaire, one of the questions pertains only to a portion of the farm, such as a field or land segment within the farm, the respondent may not notice the change in context and may provide a response for the entire farm.

Context effects such as these can be avoided if the change in context can be made more obvious by the use of transition statements, section headings, boldface fonts, and so on. However, context effects cannot be controlled in all cases because respondents may be influenced by the preceding questions and efforts to prevent such influences are ineffective.

For general population surveys, the use of technical terms or words whose meanings are understood by only a small segment of a population can also lead to comprehension errors in surveys. For example, a question from the U.S. National Health Interview Survey (NHIS) asks: "During the past 12 months, did anyone have gastritis? Colitis? Enteritis? Diverticulitis?" Here the strategy must be that if the respondent does not know the term, he or she must not have had the condition. However, often, the technical term can be replaced with a common term. For example, rather than "otitis media" use "ear infection."

Comprehension errors may also arise in the translation of questions from one language into another: for example, English-to-French translations. If the translations are literal and ignore the cultural and semantic nuances of the French speakers in the population, the translations, although technically accurate, can lead to comprehension errors. Thus, it may be necessary to reword the question in the new language to convey the proper meaning rather than attempting to maintain a strict verbatim translation of the question.

Finally, the response alternatives themselves can lead to comprehension problems. Quite often, respondents use the response options as an aid in interpreting the question. For example, for the question "Do you own a car?"

response options such as "Own outright," "Purchasing," "Leasing," and so on, help to clarify what is meant by car ownership. Another problem occurs when a question is stated clearly but the response alternatives use complicated or ambiguous terminology or may not correspond well to the question. Answer categories that overlap or are not mutually exclusive are also problematic, yet respondents are supposed to select a single category. The following examples illustrate some difficulties that can arise in the comprehension stage of the interview process.

Example 4.3.1   Following the 1977 U.S. Economic Censuses, the U.S. Census Bureau conducted an evaluation of the data quality for the Census of Manufacturers and found some evidence of comprehension error in the survey. In one finding, the bureau discovered that erroneous inclusion and exclusion in the amounts provided for annual payroll totaled $3.7 billion, about 2% of the census total for annual payroll. About one-third of this error was attributed to the exclusion of pay for employees on annual leave or vacation. This might have been an error in the creation of the data record within the establishment since some establishment databases may not have captured annual payroll with the inclusion of vacation pay. However, after further investigation and reinterviews of respondents, the bureau discovered that the real problem was that survey respondents did not understand that vacation pay should be included. They could easily have included it if the question had indicated clearly that vacation pay was to be included.

This example illustrates a fairly common problem in establishment surveys. In reporting of accounting and financial data, respondents often do not understand what information (personnel, expenditures, salaries, etc.) should be included for an item.

Example 4.3.2   This example is provided from a study conducted by Groves et al. (1991). The respondents were asked the following two questions:

1. Would you say that your own health in general is excellent, good, fair, or poor?
2. When you answered the preceding question about your health, what did you think of?
   a. Your health compared to others your age?
   b. Your health now compared to your health at a much earlier age?
   c. Your health in the last few years as compared to more recently?

Table 4.1 shows the percent of respondents who indicated each interpretation of the question. The results suggest considerable disparity in the way that respondents interpret questions about their overall health, and therefore a disparity in their responses to the question (i.e., variable error). However, the

Table 4.1   How Respondents Interpret Questions About Their Overall Health

Compared to others	22%
Compared to an earlier age	43%
Last few years versus more recent times	42%

Source: Groves et al. (1991).

study also found some evidence of systematic error or bias in the response to this question due to the way that males and females may interpret this question differently. For example, in response to question 1, 43% of men said that their health was "excellent" compared with only 28% of women. However, when the gender comparison was restricted to respondents who used the same interpretation (i.e., a, b, or c), the differences were much smaller and even disappeared for interpretation b. This analysis suggests that overall, males and females have different mixes of interpretations to this question and, consequently, comparisons between males and females may be biased.

Information Retrieval

Given the respondent's understanding of the question, the respondent is now ready to retrieve whatever information is necessary to respond to the question. At the information retrieval stage, information that is needed to formulate a response to the question is retrieved by the respondent. This process may involve the recall of information stored in long-term memory at the encoding stage; the retrieval of data from external sources such as computer databases or from household or personal files; or consultation with others in the household or establishment who have the required information. Some questions, such as opinion or attitudinal questions and basic personal demographic characteristics, do not require the retrieval of factual data (i.e., events, dates, autobiographical information). However, information may still be retrieved from memory in the form of feelings, viewpoints, positions on issues, and so on. In addition, this stage includes the process of reflecting on the issues raised by the questions in order to arrive at an attitude, belief, or opinion.

> *Information retrieval* refers to information that is retrieved either from memory or from external sources, such as other family members or co-workers, company databases, or household files.

As indicated in Figure 4.2, this stage of the response process could involve a choice of data sources when the information requested is available from two or more sources. For example, the question may ask about a person's total income received in the preceding tax year. This information may be available from both the preceding year's income tax return and in the individual's memory. If the respondent is motivated to provide the best response, he or she may access the previous year's income tax return rather than rely on memory. Similarly, in an establishment survey, the question may ask about the number of employees who work in the respondent's organization. The respondent may know the number approximately and provide that number or may decide to consult the company's personnel records and provide a more accurate number. In each case, the source ultimately consulted will depend on the burden involved in providing more accurate information, the degree of accuracy requested or implied by the question, the respondent's assessment of how much accuracy is required based on other questions in the questionnaire, and so on.

Errors of omission or recall errors may occur during the process of retrieving information from memory. Two fairly common causes of a failure to recall information are forgetting and telescoping. *Forgetting* may occur for questions that require recall from long-term memory. In general, events that took place in the distant past are more likely to be forgotten than events that took place in the more recent past. Exceptions to this are events that are particularly salient, such as the death of a loved one or the birth of a child.

In *telescoping error*, the event is remembered but the date of the event is inaccurate. *Forward telescoping error* occurs when events are remembered as occurring closer to the interview date, and in *backward telescoping*, events are remembered as occurring further from the interview date. *External telescoping error* refers to reporting erroneously events that occurred outside the reference period as occurring within the reference period. For example, a respondent may report that he or she went to the doctor within a two-week reference period when in fact the visit was prior to the reference period. In this case, the respondent telescoped the event forward in time so that it is counted erroneously as occurring during the reference period. *Internal telescoping error* occurs when the timing of the events that occurred within the reference period is in error due to telescoping them either forward or backward in time. For example, a trip out of town that occurred during a one-month reference period is reported as occurring longer ago than it really occurred (i.e., it is telescoped backward in time) (see Figure 4.3).

Forgetting will usually lead to the underreporting of events and may therefore be classified as a systematic error. Forgetting is quite common when events occur frequently during the reference period and the respondent is asked to count up the number of events. A consequence of forgetting is underreporting of the number of events. For example, suppose that respondents are asked to count up the number of trips of any length they took in their cars within the last month. If a respondent who took many trips during the reference period attempts to count them up one by one, he or she is likely to underreport the actual number as a result of forgetting. This effect of forgetting is perhaps smaller for a respondent who took only a few trips than for a respondent who took many trips. Thus, the effect of forgetting is that estimates of the average number of trips per month respondents take may be negatively biased.

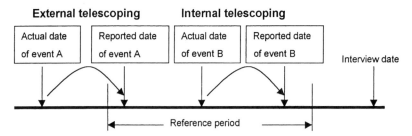

Figure 4.3 External and internal telescoping. Event A occurs outside the reference period but is reported within the reference period. This is an example of forward external telescoping. Similarly, event B is an illustration of forward internal telescoping. Backward internal telescoping occurs when events that occur closer to the interview date but within the reference period are reported as occurring further in the past but still within the reference period.

External telescoping error may also result in a bias; however, the direction of external telescoping bias is usually in the opposite direction from forgetting bias since now the number of events reported will tend to be larger than what actually occurred. As a result, estimates of the total number of events that occurred may tend to be overestimated as a result of external telescoping. External telescoping is a particular problem for highly salient, emotional, or rare events that may leave a lasting impression on the respondent, such as being a victim of a violent crime or witnessing a bad automobile accident. Respondents may remember these events as occurring more recently than they actually occurred. Especially for rare events, the respondent might also report an event occurring outside the reference period as having occurred within the reference period because they feel the event should be documented, to try to help the interviewer or to "tell their story."

An alternative to counting the number of events in a reference period is to estimate the number. For example, rather than asking a respondent to count the number of trips in a car in the last 30 days, the survey could simply ask about the number of trips during the last week, or during a typical week, and multiply this number by approximately 4 to arrive at the monthly estimate. Indeed, respondents may resort to this type of estimation rather than counting when there are many events to count because it is cognitively easier than trying to remember every event. Indeed, if the number of trips is large, respondents may decide that trying to count the number of events is futile and may resort to estimation as a way to arrive at a more accurate answer.

Whereas counting events often results in systematic error from forgetting and telescoping, estimating the number of events often results in variable error since the estimates may be higher than actual for some respondents and lower than actual for others. Thus, with estimation, there is a tendency for respondents who underestimate the number of events in the reference period to offset the overestimates of other respondents. Across the sample of respondents the average or total of the estimates may have little or no bias. However, the estimated counts obtained from respondents will tend to be more variable than the actual counts as a result of the error in the estimation process. Since variable error is usually less damaging than bias to estimates of means, totals, and proportions (see Chapter 2), estimation may be preferred over counting when the bias from counting is expected to be quite large.

The estimation error in extrapolating an estimate for a short time interval to a longer one may be quite severe if the frequency of events is not somewhat uniform across the smaller time intervals. For example, estimating the number of cigarettes smoked monthly in a population by extrapolating a daily or weekly rate may be reasonably accurate. However, estimating the annual frequency that certain types of foods are eaten in a population by the foods eaten in the preceding month could be quite variable if the consumption of some food items tends to be seasonal. Also, a longer recall period than monthly may be required for events that occur less frequently than monthly. For example, extrapolating the number of trips on a commercial airline during a given month to estimate the number of annual trips will yield an estimate that is subject to considerable variable error.

Respondents may also simply decide to guess or provide a rough estimate of the number of events rather than counting or estimating. This type of behavior is another form of satisficing. Respondents' answers may be close to accurate, but with more cognitive effort, they could provide even more accurate responses. As mentioned previously, satisficing occurs when respondents are not motivated to provide accurate responses or are overburdened by the survey request. Like estimation, guessing can lead to increased amounts of variable error in the estimates. However, the variable is likely to be much larger with guessing than with estimating.

Finally, another problem in the information retrieval stage that is quite common, particularly in establishment surveys, is the use of outdated or otherwise inaccurate records. For example, the survey may ask the respondent to supply the number of current employees on the company payroll, and the respondent may provide information that is several months old. Consequently, an erroneous figure is reported by the respondent. In the following, we provide some real examples of errors in the retrieval process.

Example 4.3.3 The first example is from the Census of Retail Trade, which is a census conducted by the U.S. Census Bureau of all retail establishments in the United States. For the 1977 Census, the Census Bureau conducted a reinterview study to evaluate the quality of the census results. In this study, professional staff from the Census Bureau revisited a sample of establishments to obtain information from them that would help evaluate the census error. For example, where possible, the reinterviewers asked respondents to check the establishment's files to get a "book value" for question items that required the retrieval of information from records. One finding from this study was a considerable amount of measurement error in the number of employees reported. Further analysis determined that approximately 75% of the error in the reports was due to respondents estimating or guessing the number of

Table 4.2   Rotating Panel Design of the NCVS[a] for a Typical Month

Sample Component	Month of Interview		
	6 Months Prior	Typical Month (M)	6 Months Later
Subsample 1	Not yet activated	Interview 1 ⟶ (bounding)	Interview 2 ⟶
Subsample 2	Interview 1 ⟶ (bounding)	Interview 2 ⟶	Interview 3 ⟶
Subsample 3	Interview 2 ⟶	Interview 3 ⟶	Interview 4 ⟶
Subsample 4	Interview 3 ⟶	Interview 4 ⟶	Interview 5 ⟶
Subsample 5	Interview 4 ⟶	Interview 5 ⟶	Interview 6 ⟶
Subsample 6	Interview 5 ⟶	Interview 6 ⟶	Interview 7
Subsample 7	Interview 6 ⟶	Interview 7 ⟶	No longer interviewed

[a] The design of the NCVS is such that each month, seven independently selected samples (shown in the first column) are interviewed. Each sample has been previously interviewed a different number of times. For example, in a typical month denoted by M in the table, subsample 1 is interviewed for its bounding interview (first interview), subsample 2 is interviewed for its second interview, subsample 3 is interviewed for its third interview, and so on. Each sample is interviewed a total of seven times, including the bounding interview at six-month intervals. For example, subsample 2 was introduced for its bounding interview six months prior to month M. In month M it is interviewed for the second time, six months after month M it is interviewed for the third time, and so on. Note that subsample 7 is interviewed for the last time in month M. This pattern is repeated every month of the year. At each interview, questions regarding the crimes and victimizations occurring during the prior six months are asked.

employees rather than consulting their records to obtain an exact count. Perhaps the burden of checking the company's records to obtain an accurate figure was more than respondents were willing to assume. Consequently, they resorted to satisficing: providing a figure that is "close enough."

Example 4.3.4   A second example of retrieval error is provided by an evaluation study conducted for the U.S. National Crime Victimization Survey (NCVS). The NCVS is a periodic survey conducted by the U.S. Census Bureau for the U.S. Bureau of Justice Statistics. The survey design is a monthly rotating panel survey where respondents are interviewed at six-month intervals. That is, each month a new sample of households is added to the survey and interviewed for the first time. In addition, households that were interviewed six months earlier are also interviewed. Once a household has been interviewed seven times at six-month intervals, the household is "retired" from the survey, meaning that it is no longer interviewed. A tabular representation of this design appears in Table 4.2.

At each interview, respondents are asked to recall events related to criminal activity they witnessed or experienced (as victims), such as assaults, personal theft, burglary, auto theft, and so on, that have occurred during the preceding six-month period. The data are the basis for the crime victimization reports published by the U.S. Bureau of Justice Statistics. Because of the length of the recall period, forgetting, telescoping, and other memory errors can be a problem, distorting the reports of crime victimizations and imparting systematic and variable errors to the estimated victimization rates. In high-crime areas, where the frequency of thefts, burglaries, and other types of crime is high, remembering exactly when a crime took place may be quite difficult. Then, too, some crimes, such as a petty theft and minor assaults, may be difficult to recall even if they are relatively infrequent.

To eliminate much of the external telescoping in the survey, the NCVS uses the first interview in the sequence of seven interviews as a *bounding interview*. That is, the first interview is used to establish the beginning of the recall period for the second interview. In the first interview, the respondents are asked about victimizations that occurred within the last six months. However, due to external telescoping, victimizations that occurred seven or more months earlier could also be reported.

Since victimization estimates based on the first NCVS interview are known to be considerably biased upward, the Census Bureau decided many years ago that the victimization data based on the first interview are unusable for purposes of estimating the victimization rate. Instead, the victimizations reported in the first interview can be used to eliminate telescoping in the second interview, for example, by matching the victimizations between the two interviews, eliminating the crimes in the second that were reported previously. Similarly, the second interview can serve as a bounding interview for the third, the third for the fourth, and so on, for all remaining six interviews. The second through the seventh interviews are referred to as *data interviews*, emphasizing that unlike the bounding interview, the data from these interviews are used in the estimation of national crime victimization rates. Thus the second interview is actually the first data interview, the third interview is the second data interview, and so on.

Although it can reduce telescoping, the bounding interview does not address the potential for forgetting in the victimization reports. One way in which forgetting can be reduced is to shorten the reference period by conducting the NCVS interviews at more frequent intervals. In the early 1980s, the Census Bureau conducted a study to evaluate the effects on data quality of using a three-month rather than a six-month recall period in the NCVS. For this study, they used a split sample design in which a fraction of the NCVS sample was interviewed at three-month intervals and the remaining fraction of the sample was interviewed at the usual six-month intervals. Thus, by totaling the crimes reported in two consecutive three-month recall interviews, an estimate can be constructed that is directly comparable to the number of crimes reported in a single six-month recall period. For example, the total number of crimes recorded for January–March and April–June for the three-month recall design should equal the total number recorded for January through June for the six-month recall design.

Table 4.3   Comparison of Three- and Six-Month Recall for the NCVS (per 100 Persons 12+ Years Old)

Type of Crime	6-Month Recall	3-Month Recall	Difference
Total personal crimes	12.8	15.5	−2.7*
Crimes of violence	3.5	4.3	−0.8*
Crimes of theft	9.4	11.2	−1.8*
Total household crimes	23.0	26.4	−3.4*
Burglary	8.5	9.7	−1.2*
Larceny	12.7	15.1	−2.4*
Auto theft	1.8	2.1	−0.3

Source: Bushery (1981).

* Statistically significant difference at the 5% level of significance.

The victimization rates for the two designs are compared in Table 4.3. Entries with an asterisk indicate that the differences are *statistically significant*: that is, the magnitude of the difference is larger than could reasonably be expected to occur by chance. The table shows clearly that three-month recall always provides higher reporting, usually significantly higher reporting, than a six-month recall period. If the only error in the estimates was forgetting, the fact that the three-month recall period yields higher estimates indicates that the three-month recall error is less subject to forgetting error and is therefore less biased. However, there may still be a small amount of external telescoping error present in the data despite the bounding interview design. Thus, both the six- and three-month recall estimates may be somewhat positively biased, due to telescoping error. However, there does not seem to be any plausible reason why the three-month recall design should have larger telescoping bias than the six-month design. Therefore, we conclude from these results that the three-month recall estimates are generally larger and so are less biased than estimates based on six-month recall.

Although three-month recall is less subject to recall bias for crime victimization estimates, for a fixed-cost design, estimates based on six-month recall may still be more accurate when the entire mean squared error is considered. Since the three-month design requires more frequent interviews with respondents, the sample size for the three-month design must necessarily be smaller to maintain the same survey costs as the six-month design. Thus, although measurement bias is reduced, the sampling variance for the three-month design may be as much as twice that of the six-month design. Hence, the total mean squared error, which is the sum of the squared bias plus the variance could actually be larger using the three-month recall.

Besides cost, there are other considerations in the decision to switch to a three-month recall period. For example, how would the user community react if the standard errors of the victimization rates were dramatically increased, even if the overall mean squared error were reduced for the estimates? Since biases in the victimization rates are not reported (the recall bias cannot be

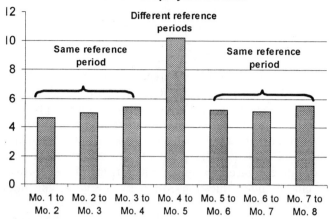

Figure 4.4   Seam effect. The percentage who changed employment status between adjacent months is between 4 and 5% of the population except for months 4 and 5, where this percentage jumps to 10.2%. The reason is the seam effect. Months 1–4 employment statuses are obtained in one interview, while months 5–8 statuses are obtained in the next interview four months later. Thus, months 4 and 5 are collected at two different points in time approximately four months apart.

estimated without special evaluation studies as described earlier), any reduction in the bias may be unnoticed and unappreciated by data users. Indeed, to data users, data quality may actually appear to worsen with the increase in standard errors of the estimates. These considerations have led the Census Bureau to maintain the six-month recall design in the NCVS, despite the advantages of the three-month recall for reducing recall bias.

Another problem often encountered with panel surveys is the *seam effect*, a phenomenon whereby many more month-to-month changes are observed between adjacent months within the same reference period than between adjacent months that straddle two different reference periods. Figure 4.4 illustrates this effect as measured in the Survey of Income and Program Participation (SIPP) conducted in the United States. The most obvious explanation for this effect is recall error. Respondents are more likely to remember their employment status in the month preceding the interview than in the month four months prior to the interview. In that case, the changes between the seam (i.e., months 4 and 5 in the figure) are erroneous and the real change is likely to be between 4 and 5% as in the other nonseam months.

However, another explanation is that changes between the months within the reference period are understated. That is, respondents may say that there was no change in their employment status during the entire four-month reference period as a form of satisficing. For example, they may want to avoid additional questioning about the change. The most likely explanation for the seam effect is that it is due to a combination of factors that act both to reduce the within-reference-period changes and to increase the between-reference-period changes.

Formatting the Response

Following the information retrieval stage, the next stage of the response process is referred to as a *judgment and response formatting*. It is at this stage of the response process that information retrieved in the preceding stage is evaluated and a response is formulated according to the format requested in the question. Quite often, survey questions are closed-ended, meaning that the question requires that the respondent choose a response from a list of response alternatives. Since appropriate answers are already provided for the respondent, closed-ended questions often save time in the interview and reduce respondent burden.

Open questions ask respondents to phrase answers in their own words and interviewers to record the answers. A closed-ended question might also be used when it is feared that respondents with low verbal skills will not provide useful responses to an open-ended question. However, when the required form of the question is obvious, such as in the case of the question "How many times have you been to the doctor in the last year?," an open-ended question is usually preferred.

As an example, Sudman et al. (1996) recommend the use of open-ended question formats for obtaining behavioral frequencies. This is because, as will be seen in Example 4.3.5, respondents sometimes use the range of numeric response alternatives as a frame of reference in estimating their own behavioral frequency, which can result in systematic bias. Since the responses are numeric, there is no particular difficulty in coding such responses by computer, if desired. However, as noted in the preceding discussion of different recall and estimation strategies, open-format questions may still be biased by other sources of error in the response process.

A compromise between open and closed questions is to use an "other" answer category on a closed question to allow respondents to volunteer an answer when the response alternatives are not sufficient. This approach is recommended for questions where it is not clear what response categories to provide to cover all possible responses. Also, it can be used as a catch-all category to cover an unknown number of possible responses that could be provided by a relatively small minority of respondents who would not choose from among the response alternatives provided.

Thus, for closed-ended questions, the response formatting stage is where the respondent formats the information accessed in the prior stage according to the response choices provided. For open-ended questions, the respondent would try to determine how to construct a response that addresses the question.

For example, the question may ask: "Are your company's computer support services centralized or decentralized?" The response formatting stage involves the process of deciding which of these two response choices best fits the company's computer support services. If the question is open-ended, such as "What is your age?" or "What is your income?," the respondent must decide how accurately to report that information; for example, whether to provide a fairly precise number or a rough or rounded figure.

> When *formatting the response*, information is evaluated and a response is formulated according to the format requested in the question.

A number of errors can occur at this stage of the response process. For closed-ended questions, one common error results when the response categories do not describe adequately what the respondent wants to convey by his or her response. For example, suppose in the preceding example that only two response categories are available: centralized services and decentralized services. However, for the responding company, some computer support services, such as Internet support and e-mail, are centralized, whereas other services, such as PC support and technical software support, are decentralized. The correct response in this case would be some sort of combination of centralized and decentralized computer support. Thus, since neither response category is appropriate, the respondent's response cannot describe reality accurately no matter how he or she answers.

In addition to being too constrained, the response categories can sometimes suggest a typical response or a different interpretation of the question. For example, the question "How often does your company engage in strategic planning?" may take on a different interpretation if the response categories are "Never," "Annually," "Once every two years to three years," and "Every four years or less often," than if the response categories are "Never," "Every month," "Several times a year," "Once a year or annually," and "Less often than annually." In the former case, the respondent may interpret strategic planning to mean companywide planning involving many departments meeting together in large planning sessions. In the latter case, the response categories may convey that the strategic planning of interest is a smaller-scale planning, involving fewer people meeting together and, therefore, more often. Another example of this type of problem is described subsequently.

A third problem arises in the response formatting stage when respondents are hurried and pressured into giving quick responses which are not well considered—called *top-of-the-head responses*. Although this error can occur in any mode, it tends to occur more frequently in telephone surveys than in mail or face-to-face surveys. In telephone surveys, respondents may feel uncomfortable when there are long pauses in the conversation, and perhaps for that reason, they feel some pressure to answer rapidly. In face-to-face surveys, visual communication provides information about what is happening during the long pauses, so there is less pressure on respondents to fill the silence. At any rate, research evidence suggests that in general, telephone surveys are more prone to result in top-of-the-head responses than are other modes of interview. In addition, responses to open-ended questions tend to be shorter over the telephone than in person. This may be an indication that respondents also tend to be less conversational in a telephone interview than in a face-to-face interview.

Respondents may satisfice in choosing a response category from a list of categories, particularly if the categories are unordered or nominal. For example, in a self-administered survey of elementary school teachers, the teachers were presented with a list of 10 instructional aids they might use in the classroom. They were asked to select the aid that they would find most useful in teaching elementary school children. Although the aids were listed in no particular order, those in the top half of the list were selected almost twice as often as those in the bottom half of the list. This could suggest respondents satisficed in that they stopped reading through the list once they found an acceptable response rather than reading through the entire list and selecting the best response.

Satisficing can be a serious problem for open-ended questions as well. For example, the question "What types of activities do you usually engage in on your job?" can present a challenge to a respondent. A person whose job involves many activities would find it difficult to recall these, sort through them to decide which ones are typical, and communicate the list in the response. There is a risk that the information provided is inadequate for purposes of the research.

Example 4.3.5 Response alternatives can sometimes inform the respondent about the researcher's perceptions of the population or the typical responses expected for a question. This information can then be used by the respondent to formulate a response. In some cases the respondent may choose to edit his or her response to conform to the researcher's assumptions about the real world as revealed through the response alternatives.

For example, research on the use of response alternatives to assess the frequency of certain behaviors has determined that respondents may assume that the response alternatives reflect the distribution of the behavior in the population. Specifically, values in the middle range reflect typical behavior, while alternatives at the extremes of the scale reflect rare or "abnormal" behaviors. These assumptions influence responses in various ways. In some cases, respondents may use the range of response alternatives as a frame of reference in estimating the frequency of their own behavior. If they view their behavior as typical, they may select a point near the middle alternative without attempting to assess more accurately the true frequency of their behavior.

To illustrate, Table 4.4 provides the results of a study on TV viewing by Schwarz et al. (1985). In this study, a random half-sample of the respondents were presented the low-frequency alternatives on the left in the table and the

Table 4.4 Responses to TV Viewing Question for Two Response Set Options

Low-Frequency Alternatives		High-Frequency Alternatives	
Response Alternatives	Percent Reporting	Response Alternatives	Percent Reporting
Up to ½ hour	7.4	Up to 2½ hours	62.5
½ to 1 hour	17.7	2½ to 3 hours	23.4
1 to 1½ hours	26.5	3 to 3½ hours	7.8
1½ to 2 hours	14.7	3½ to 4 hours	4.7
2 to 2½ hours	16.2	4 to 4½ hours	0.0

Source: Data from Schwarz et al. (1985). Reprinted with permission from the University of Chicago Press.

other half were presented the high-frequency alternatives on the right. As shown in the table, 16.2% of the respondents who were presented the low-frequency alternatives reported daily viewing of 2½ hours or more, while 37.5% did so when presented the high-frequency responses. That is, the question with the high-frequency alternatives resulted in an estimate that is more than two times the estimate obtained with the same question using the low-frequency alternatives. There are several possible explanations for this.

One explanation is that rather than trying to recall how often they view TV, many respondents simply estimated this frequency using the information about "typical" TV viewing frequency presented in the response alternatives. For the low-frequency alternatives, they assumed that the typical person watches between 1 and 2 hours (i.e., the middle of the low-frequency scale). For the high-frequency alternatives, they assumed that the typical person watches between 3 and 4 hours per day. In either case, the responses reflect their perceptions of how the TV viewing compares to that of the average person.

Another explanation for the effect is the presence of social desirability bias. In this study, the sample was composed of college students who may associate excessive TV viewing as a characteristic of unpopular persons who lead dull social lives. Thus, it is socially undesirable to watch TV excessively. To avoid the appearance of a socially undesirable lifestyle, respondents may select a response category in the middle range of the scale, assuming that this frequency is consistent with typical and thus more socially acceptable behavior in the population.

A third explanation is that respondents are confused by the question. If the question asks, "On average, about how many hours per day do you watch television?," respondents may interpret the term "watch television" differently depending on the response alternatives that are provided. When the high-frequency alternatives are presented, respondents assume that the researcher means being in the same room with the TV while it is on, regardless of how attentively they are viewing the TV. When the low-frequency alternatives are presented, respondents may interpret the term "watch television" to mean active, attentive TV viewing.

Regardless of which explanation is correct, it is clear from this example that the response alternatives provided for a question can have a profound effect on the responses to the question by the information they convey about typical behavior.

Editing and Communication

Finally, the last stage of the response process is editing and communication. In the preceding stages, the respondent understood the meaning of the question, retrieved the information needed to respond to the question, and determined the response category, value, or reply that best describes his or her response to the question. Now at this final stage, the respondent decides whether to edit his or her response, that is, whether to provide the most accurate response or one that has been altered out of social desirability or fear of disclosure concerns, and then communicates this response to the interviewer or selects the appropriate response category. Several types of errors can occur at this stage—social desirability error, fear of disclosure error, and acquiescence—which are discussed below.

In the final step, *editing and communication*, the response is communicated to the researcher, either as it was formulated or after undergoing editing by the respondent.

As described above, social desirability error occurs when a respondent determines that his or her response may not be socially acceptable and changes it to one that is more socially acceptable. For example, respondents who drink excessive amounts of alcohol may deliberately underreport their consumption to an interviewer to avoid possible interviewer disapproval of the true amounts. As a result, a systematic error occurs in the data, and alcohol consumption in the population is underestimated. This bias, referred to as *social desirability bias*, occurs quite frequently in the collection of sensitive data such as socially unacceptable sexual behaviors, drug use, underpaying taxes, and other illegal activities where respondents may be too embarrassed to reveal their true behavior to interviewers.

Since interviewers are the primary catalysts for social desirability error, this error is usually much larger in interview surveys than in self-administered surveys. Therefore, self-administered data collection is usually preferred over interviewer-administered modes as a more accurate method of collecting sensitive data in surveys. Further, there is some evidence in the literature that telephone surveys are slightly better at collecting data that are subject to social desirability bias than face-to-face interviews, although this is not always the case, as we shall see in Chapter 6.

Fear of disclosure error occurs when respondents fear the consequences for providing accurate survey reports and thus edit their responses. For example, a business establishment respondent may be concerned that the company's competitors may somehow gain access to the proprietary information that is being requested in the survey. A high-income earner who cheated on his or her income tax forms may fear that telling the truth in the survey will cause trouble with the tax authorities. Thus, fear of disclosure error is not necessarily affected by the presence or absence of the interviewer. Rather, it is caused by a concern that the information provided may not be kept anonymous and confidential.

Like social desirability error, fear of disclosure error usually leads to systematic error in the data and thus bias in the estimates. One means of avoiding fear of disclosure bias is to assure the respondents that their responses will be kept anonymous and confidential, if indeed this is the case, and, if possible, to take extra precautions in the survey to ensure that the survey responses cannot be linked to the respondent's identity. However, in some cases, these measures are not adequate. As an example, farmers may be hesitant to report seasonal field-workers who do not have appropriate immigration documentation on an agricultural labor survey. Although they may have trust in the assurances of confidentiality of their individual reports, they may still fear that collectively, the survey results will show an increase in the use of undocumented workers by farmers which could lead to increased measures by the authorities to prevent this practice. Thus, they may fear that honest disclosure of these workers will ultimately lead to increases in the price they pay for farm labor.

Acquiescent behavior is a potential problem that can occur during the response editing and communication stage. This error occurs when respondents report as they believe the survey designer or interviewer wants them to rather than as they should report to be accurate. As an example, customer satisfaction surveys tend to provide a more positive assessment of respondents' opinions toward products and services than is true in reality. Respondents are well aware that responses that indicate satisfaction are wanted and therefore tend to acquiesce in that direction. To avoid this type of bias, survey designers should strive to design customer satisfaction survey questionnaires that are neutral in wording and tone and balanced with regard to positive and negative statements. Further, since respondents can be influenced by the partiality of the interviewers or the survey sponsors, satisfaction surveys often use self-administration and survey sponsorship that is viewed as neutral and impartial.

Example 4.3.6   An example of a survey that was designed to minimize the risk of social desirability and fear of disclosure bias is the U.S. National Household Survey on Drug Abuse (NHSDA). The NHSDA is a household survey designed to measure the population's current and previous illicit and abusive drug use activities. The target population includes all persons living in households who are 12 years old or older. Drug and demographic data are collected from each respondent during the interview phase using a combination of inter-viewer- and self-administered instruments. On average, the interview takes about an hour to complete. The interview begins with a set of interviewer-administered questions designed to collect data on the respondent's current and previous use of cigarettes and other forms of tobacco. These initial ques-tions allow the respondent to become familiar with the format of the NHSDA questions.

The remainder of the questionnaire is divided into sections corresponding to each drug of interest: alcohol, the nonmedical use of sedatives, tranquiliz-ers, stimulants and analgesics, marijuana, inhalants, cocaine, crack, hallucino-gens, and heroin. For each section, the interviewer gives the respondents an answer sheet and asks them to record their responses on it. Depending on the complexity of an answer sheet, the interviewer will either read the questions to the respondent or, if preferred, the respondent can read the questions. Upon completion of an answer sheet, the respondent is requested to place the answer sheet in an envelope without allowing the interviewer to see the responses. The motivation for conducting the interview in this manner is to ensure that the respondent understands the questions and does not erroneously skip over major parts of the questionnaire and, more important, to guarantee response confidentiality.

Most of the answer sheets are designed so that even respondents who have never used a particular drug still need to answer each question about the drug. Since both users and nonusers of a drug are asked to respond to essentially the same number of questions, the interviewer is less likely to guess that the respondent is a user or nonuser based on the time the respondent takes to complete an answer sheet. This is another feature of the survey that is designed to protect the privacy of the respondent. In addition, some respondents who indicate under direct questioning that they never used a drug will later answer an indirect question about the drug in a way that implies use of the drug. This redundancy in the questionnaire provides additional information regarding drug use that can be used to compensate for underreporting for the direct question.

Example 4.3.7   Table 4.5 illustrates the risk of fear of disclosure or social desirability biases on various topics that might be included in a survey. Bradburn et al. (1979) conducted a study to identify topics that respondents feel are sensitive and may be too personal as a survey topic, including drugs, alcohol consumption, income, sexual activity, gambling, drinking, and sports. They had respondents rate topics on a four-point scale according to how uneasy they would make "most people": very uneasy, moderately uneasy, slightly uneasy, or not at all uneasy. Table 4.5 provides the list of the items that were presented to respondents, along with the percentage of respondents who said they would feel "very uneasy" discussing the topics in a survey. As we see from the table, sexual behavior and drugs are ranked at the bottom of list, which is understandable since the former is often embarrassing or, in some

Table 4.5   Percentage Who Would Feel Uneasy Discussing Various Topics in a Survey

Topic	Make Most People Very Uneasy (R's Rating)
Sports activities	1
Leisure-time and general activities	2
Social activities	2
Occupation	3
Education	3
Happiness and well-being	4
Drinking beer, wine, or liquor	10
Gambling with friends	10
Income	12
Petting and kissing	20
Getting drunk	29
Using stimulants or depressants	31
Using marijuana or hashish	42
Sexual intercourse	42
Masturbation	56

Source: Data from Bradburn et al. (1979).

cases, socially unacceptable, and the latter is illegal. Sports and leisure activities appear to be topics respondents actually enjoy talking about.

Example 4.3.8   We conclude this chapter with an example of a technique for counteracting both social desirability bias and fear of disclosure bias, referred to in the literature as the *randomized response technique*. One variant of this method asks the respondent two questions, such as: "Were you born in the month of January?" and "Did you report all your income in last year's taxation process?" Note that one question is not sensitive and the other is, potentially. The respondent is asked to respond with "they are the same" if the answers to the two questions are the same (i.e., both correct responses are "yes" or both are "no"). Otherwise, the respondent is asked to respond with "they are different." If the probability of being born in January can be determined for the population (it can in most cases from census data or other population records), the extent of tax cheating can be estimated using an innovative statistical estimation approach. In this way, respondents can avoid revealing their true response to a direct question regarding tax cheating.

The randomized response method, first published by Warner (1965), was initially considered a breakthrough in survey collection of sensitive data. Danermark and Swensson (1987) provide an example of a successful application of a variation of this method on estimating drug use in schools, and there are dozens of other applications discussed in the literature, but the method has not reached the level of use in practical survey work that was anticipated in the late 1960s. For example, respondents do not always understand that their answers are indeed protected. Furthermore, the randomization devices that sometimes have been used (miniature roulette wheels, decks of cards, etc.) have not been perceived as part of serious survey research by some sample members in various applications. Also, to administer a randomized response method for just a few survey questions can be disruptive and impractical.

For comprehensive treatments of questionnaire design issues and methods that implement many of the principles described in this chapter, we recommend Converse and Presser (1986), Sudman and Bradburn (1982), Bradburn et al. (1979), Dillman (2000), Schwarz and Sudman (1996), and Tanur (1992). In addition, for books that deal comprehensively with measurement errors in surveys, we recommend Biemer et al. (1991), Groves (1989), Groves et al. (1988), Lyberg et al. (1997), Rossi et al. (1983), and Turner and Martin (1984).

CHAPTER 5

Errors Due to Interviewers and Interviewing

Survey questionnaires can be designed either for interviewer administration, where an interviewer asks the respondent the survey questions and enters the respondent's responses on the questionnaire, or for self-administration, where the respondent reads the questions and enters his or her responses directly onto the questionnaire without assistance from an interviewer. Combinations of these two types of survey administration are not uncommon. For example, in the U.S. National Household Survey on Drug Abuse (NHSDA), interviewers administer part of the interview, and the remainder of the interview, because it involves collecting highly sensitive information on drug use, is self-administered. In Chapter 6 we discuss a number of methods or modes of data collection which use various ways of communicating survey questions to the respondent, and consider the advantages and disadvantages of each. In this chapter the focus is on surveys that are administered by a survey interviewer who is either communicating face-to-face with the respondent or who communicates by telephone. We discuss the interviewer's role in a survey, the types of errors that he or she may commit in performing that role, factors that may affect the magnitude of the errors, and methods for evaluating and controlling interviewer errors.

The first question that one might ask about interviewers is why they are needed to collect survey data. Indeed, in many situations an interviewer is not needed. For example, for modes of data collection such as mail, Internet, and e-mail, the questionnaires are sent to respondents, who complete the questionnaires without the assistance of an interviewer. Even in surveys that require interviewers to visit respondents, the interviewer's role in the survey could be minimized, due to the sensitivity of the questionnaire content. For example, in the NHSDA of the 1990s, the interview role is essentially to deliver the survey questionnaire, wait until the respondent completes it by self-administration, and then deliver the completed questionnaire to the research organization for processing and analysis. As computer technology advances, "virtual" interviewers which are created by computer software might conduct interviews with no need for human contact with the respondents. The question of whether or not to use an interviewer is one that should be addressed in the early stages of the design process, as described in Chapters 2 and 10.

However, in many data collections, interviewers are an essential part of the survey process and serve a very valuable role. In fact, interviewers do much more than simply interview respondents. For example, interviewers may assist in the sampling process by creating a list of the housing units in a neighborhood so that a sample of housing units can be drawn from it. In many face-to-face surveys, an important role of the interviewer is to find the sample member since the frame from which the sample was drawn may contain many old addresses. With highly mobile populations, determining the current address of the sample member can be quite difficult.

One of the most critical duties of a survey interviewer is to contact the sample members and persuade them to participate in the survey. After gaining cooperation with a household, the interviewer may need to conduct a short screening interview to determine whether anyone in the household is eligible for the survey. When the interview begins, the interviewer may read the questions to the respondent, interpret or clarify the meanings of the questions as necessary, ask probing questions when the responses are ambiguous or unclear, and record the responses in the instrument. The interviewers may also make and record observations about the respondents, the households, or the neighborhoods they visit. If these functions of the interviewer are important to achieve survey objectives, self-administered modes should be ruled out in favor of an interviewer-assisted survey protocol.

The style or manner that interviewers use for interviewing respondents has been the topic of some debate in the literature. Traditionally, interviewers have been trained to be neutral agents of the researcher and to present questions to respondents in a very standardized manner (Fowler and Mangione, 1990). However, Suchman and Jordan (1990) and a number of subsequent researchers have questioned the standardized approach in favor of a more interactive or "conversational" interviewing approach. The latter approach gives the interviewer much more freedom to communicate with the respondent as necessary in order to obtain the most accurate responses. In the next section we discuss these two approaches and provide some guidance regarding the use of each.

5.1 ROLE OF THE INTERVIEWER

Since Suchman and Jordan's article was published in 1990, survey methodologists have debated the role of the interviewer in the survey interview. At one extreme of the controversy is the *standardized interviewing* approach, which attempts to standardize the interviewer–respondent interaction as an experimenter might standardize the treatments in a response–stimulus experiment.

At the other extreme is *conversational interviewing*, which requires a much higher level of interaction between the interviewer and respondent in an attempt to standardize the meaning of the questions to the respondent. To emphasize the differences between these two views, we describe each technique in its most extreme form. However, in practice most survey methodologists advocate an interviewing technique that is to some extent a compromise between these two extreme views.

Standardized Interviewing
The standardized interviewing perspective has been used widely in survey research since the 1950s. This perspective holds that the interviewer's role in the interview is to read the questions exactly as worded to the respondent, making every attempt to be completely neutral in the delivery of the question and regarding the information sought. If the respondent asks for an interpretation of the question, the interviewer is not allowed to provide one, even though he or she may be very capable of doing so. The interviewer may repeat the question, read the response categories again, or encourage the respondent to make his or her own interpretation. Definitions of terms may be provided if they have been "scripted" as part of the interview for all interviewers. The interviewer may probe for clarification neutrally but may not provide feedback to the respondent unless instructed to do so by the survey procedures.

Thus, the goal of standardized interviewing is to present the same question in exactly the same manner to every respondent in order that responses are in no way influenced by the interviewer. Theoretically, if implemented as designed, the interviewer errors occurring in a survey using standardization should be similar to errors that would result if one well-trained interviewer conducted all the interviews for the survey using the same approach for all respondents. In other words, standardized interviewing is intended to eliminate the variation in errors that may be introduced when many interviewers conduct a survey.

The standardized interviewing technique assumes that the questionnaire is well designed and works well in almost all types of survey situations, that the questions are well worded and will not need to be rephrased for some respondents, that the situations that interviewers encounter during the interview typically map well to the definitions of concepts used in the questionnaire, and that few respondents will need any clarification of the concepts covered by the questionnaire. It further assumes that exceptions to these conditions, for the most part, can be anticipated in the design of the questionnaire so that special instructions, clarifications, probing, and so on, can be scripted and delivered consistently by the interviewers. Thus, with standardized interviewing, the interviewers should have few reasons, if any, to deviate from the script, and interviewers can be trained to handle any unanticipated problems with the questions in a completely consistent manner. An important benefit of this degree of standardization is an absence of variation in the error due to interviewers, or *interviewer variance*. As we shall see, interviewer variance can be quite damaging to the survey estimates, so any attempts to eliminate it are certainly worthy of consideration.

Unfortunately, this idealistic goal is frequently not attained in standardized interviewing, for a number of reasons. First, for long, complex instruments, questions are often poorly worded and confusing even after considerable pretesting and revision. Seemingly simple question concepts are often quite complex and are open to interpretation, and the respondent's circumstances do not map easily onto the official definitions of terms used in the survey. In these cases, respondents are confused, and with no assistance from the interviewer, are left to guess at an interpretation of the question. Since the interpretation may be incorrect, so may be the response.

In addition, for some types of question sequences, such as collecting rostering information or event histories, the progression of the interview can be quite unstructured. The respondent may offer information about household membership or recall events such as work histories in a very haphazard manner and not necessarily in the order assumed by the structure of the standardized instrument. In these circumstances, it is best to abandon the structured approach of standardized interviewing in favor of a more flexible approach for recording the information being provided.

Standardized interviewing is an interviewing protocol that requires interviewers to ask questions as worded, to probe, provide feedback, and interact with the respondent in a manner which is consistent across all interviews. It attempts to standardize the behavior of interviewers to reduce the measurement error in responses due to interviewers.

Conversational Interviewing

As an alternative to standardized interviewing, a number of survey methodologists have recently advocated using a more flexible or conversational interviewing approach. In its most extreme form, conversational interviewing essentially abandons the standardized interviewing approach in favor of an approach that is more conversationally natural and similar to a normal social interaction between two strangers. With this approach, the interviewer may alter the wording of the question: for example, to tailor it to the respondent's situation. Morever, the interviewer is free to assist the respondent in any way necessary to clarify the meaning of the question and how the question applies in the respondent's particular situation. Thus, rather than attempting to standardize the behavior of the interviewer, conversational interviewing attempts to standardize the meaning of the questions by providing definitions, clarifications, and other information the respondent might need to understand what the researcher behind the questions is asking. Proponents of this technique claim that conversational interviewing reduces the error variation due to respondents rather than that due to interviewers.

As an example, the written question may ask "How many persons live in the household?" The standardized interviewer would read the question exactly as worded, with no deviations. An interviewer using the conversational approach may reword this question as: "Besides yourself and your husband, are there any other persons living here?" or any other way that seems appropriate. Conversational interviewers are trained to look for cues that the respondent may have misunderstood the question and attempt to clarify the question as necessary. For example, the respondent may reply, "Well, there is my sister who has been living with us for three months, so I guess that would be one person besides me and my husband." The standardized interviewer might accept this answer unless it were part of the standardized interview procedures to probe further into the nature of the sister's living arrangement. The conversational interviewer has complete autonomy to decide whether probing on this issue is necessary to provide the most accurate response. For example, even if not instructed to do so in the questionnaire, the interviewer might ask: "Did your sister live anywhere else for any part the year?" and "Does she have another place she calls home?"

Conversational interviewing assumes that interviewers can be trained to understand the researcher's intent for each question in the survey well enough to convey that intent to the respondent. In this regard, the technique places a much higher expectation on the interviewer than does standardized interviewing. The conversational interviewer must know the intentions of the researcher that underlie the question. For example, consider the question "Does your organization plan to change its method for evaluating employee performance within the next year?" The respondent may have many doubts about what is being asked here: What is meant by organization? My department? My division? My entire company? What is meant by change? Do minor changes count, or do the changes have to be big, as in a complete overhaul? What aspects of the performance evaluation process are of interest to the researcher? Is the question referring to the annual review process, or do interim reviews matter? The conversational interviewer should be able to answer these questions from the perspective of the researcher, whereas there may be no such expectation for the standardized interviewer.

Part of the debate regarding the role of the interviewer centers on whether it is realistic to expect interviewers to be knowledgeable enough about every question on the questionnaire so that they can interpret to the respondent the meaning of the question as the researcher intended it. Some survey methodologists believe that with interviewer salaries being near the bottom of the pay scale in most survey organizations, it is not realistic to expect that hundreds of interviewers can be hired who have the necessary research background and commitment to master the complex nuances of the many questions in a typical social survey. There seems to be agreement that interviewer salaries would have to be increased substantially to attract sufficient numbers of workers capable of mastering the conversational interviewing approach. However, training interviewers in the proper technique for conversational interviewing will also be challenging and the time required to impart the required knowledge regarding the purpose behind each question could be considerably longer than that required for standardized interviewing.

Another cost issue with the approach is administration time. Studies have shown that interviews using the conversational approach can take as much as one and a half to three times longer than interviews using the standardized approach, depending on the complexity of the concepts surveyed and the difficulty of determining how the respondent should respond best given the specifics of his or her circumstances.

Some proponents of conversational interviewing argue that the increase in costs is the price of collecting accurate data. However, opponents of the method argue that the risk of interviewer variance is even greater for the method since the interviewer is given so much control over the way questions are presented and thus in the responses to those questions.

Conversational interviewing is an interviewing protocol that allows interviewers to interact freely with the respondent, to modify and adapt questions to the respondent's situation, and to assist respondents in formulating a response. It attempts to obtain the most accurate response by minimizing all measurement error sources, not just the interviewer's error.

Other Interviewing Styles

The recent literature on the role of the interviewer suggests that a combination of the two approaches may be best in terms of cost and data quality. As an example, the approach proposed by Schober and Conrad (1997) standardizes delivery of the question but allows the interviewer to follow up in an unscripted fashion as needed after reading the question as worded. The interviewer is encouraged to probe to clarify the response, ask follow-up questions to determine the respondent's understanding of the question, correct apparent misinterpretations of the questions, and to make commonsense inferences based on the respondent's response.

To illustrate, consider the following example from the U.S. Current Population Survey (CPS), as discussed in Schober and Conrad (1997):

I'er: *Last week, did you have more than one job, including part-time, evening, or weekend work?*

Res: *Well, it depends . . . I babysit for different people—is that one job or more than one?*

If the interviewer is using the standardized approach, the interviewer would not answer the respondent but rather require that the respondent interpret the question by himself or herself. However, under the *conversationally flexible interviewing approach*, as they refer to it, the interviewer may aid the respondent by explaining that for this question, the U.S. Bureau of Labor Statistics counts babysitting for more than one employer as only one job. Thus, with this approach, the interviewer reads the question as worded, but then the interviewer and respondent work together to ensure that the respondent interprets the question as the survey designer intended.

Conrad and Schober provide some evidence that conversation flexibility gives essentially the same results as standardized interviewing when the questions are easy to answer. However, for more difficult questions, for example, questions that require the respondent to map very complex situations to equally complex definitions in the survey questions, the conversationally flexible approach gives more accurate data. However, this method is subject to many of the same limitations regarding interviewer recruitment, training, and interviewing time as the conversational interviewing approach proposed by Suchman and Jordan.

Another possibility for combining the best elements of standardized and conversational interviewing is to use standardized interviewing for most survey questions and the conversational approach for a few key questions where the standardized approach would be awkward. Examples of the latter are the collection of household roster information, the reporting of trip dates, durations, destinations, and so on, for the past week, the reporting of job history for the last 10 years, and other information that is best recorded in a matrix or grid format in a questionnaire. Since the collection of such information is usually quite unstructured, standardizing the sequence and wording of questions would be very difficult and might even hinder the respondent's ability to recall the information. This type of data collection may be an ideal application of the conversational or conversationally flexible interviewing approach.

This hybrid approach to interviewing standardizes the delivery of the question but allows considerable interviewer flexibility to obtain the best response from the respondent. In that regard, the demand on the interviewer's knowledge of the research objectives is about the same as in conversational interviewing. However, the risk of interviewer variance is supposedly smaller since in most cases, respondents may be able to answer the question as originally worded, without assistance from the interviewer. It is still debatable, however, which of these approaches to interviewing works best in various types of situations.

Dijkstra and van der Zouwen in Holland (Dijkstra, 1987; Dijkstra and van der Zouwen, 1987, 1988) have experimented with two styles of interviewing, which they refer to as formal and personal. In the *formal* or *task-oriented interviewing style*, the interviewer behaves essentially detached from the interview and is trained to refrain from emotional reactions and personal interchanges during the interview. In the *personal* or *person-oriented style*, interviewers are allowed to engage in natural, person-oriented behaviors. For example, the interviewers can make comments such as "Oh, I am so sorry to hear that," "That is nice for you," or "I have similar feelings."

In both styles, interviewers are trained to ask the questions as worded, as in the standardized approach. However, one style is designed to be clinical, unemotional, and in some respects, unnaturally business oriented, and the other style is intended to be much more human and friendly. In split-sample experiments, Dijkstra and van der Zouwen coded the interactions between the interviewer and the respondent for both styles and found that the personal style tends to stimulate respondents to give more adequate responses relative to the formal style. However, it also gives interviewers more freedom to engage in behaviors that may bias responses, such as directive probing and interpreting inadequate answers. Thus, it is important to train interviewers in the personal style, not to add their personal views to the question–answer sequence. On the other hand, a problem with the formal style is the need to unlearn natural person-oriented behavior.

5.2   INTERVIEWER VARIABILITY

5.2.1   What Is Interviewer Variability?

Various terms have been used to describe the errors that are attributable to interviewers. Some of the terms that one encounters in the literature include interviewer variability, interviewer variance, correlated response variance, correlated interviewer error, intra-interviewer correlation, the interviewer effect, and the interviewer design effect. In this section we define these terms, discuss the underlying causes of interviewer variability, and provide some examples and illustrations of the interviewer effect on survey estimates.

In Chapter 2 we defined systematic and variable errors and how they arise in survey data. Interviewer error is related to both of these concepts. An interviewer can make variable errors (i.e., errors that vary from respondent to respondent and, when summed together to form an estimate, cancel one another) (see Chapter 2). An interviewer can also make errors that are somewhat systematic and tend to influence in the same way the responses across all respondents in the interviewer's assignment. Thus, the errors are not offsetting, so, when summed, a bias results.

Variable interviewer errors may arise due to direct observations that interviewers sometimes make in the course of conducting a survey. For example, an interviewer may be asked to estimate the value of each house in a neighborhood. For some houses, the interviewer sometimes overestimates and sometimes underestimates the market value. However, the average value of houses in the neighborhood may still be very close to the actual average value. In this case, the errors tend to cancel out one another. However, another interviewer may not be aware of the current value of housing. It may be some years (if ever) since the interviewer bought or sold a home, so this interviewer's assessments may all tend to be too low. Consequently, the errors in the interviewer's assessments of home values in the entire neighborhood will generally be too low, so the average value of the housing units in the neighborhood will be biased downward. This is an example of *systematic interviewer error*.

Systematic interviewer biases can also vary from interviewer to interviewer. For example, interviewer A may underestimate the value of the housing units in his or her assignment by −5%, interviewer B may overestimate these values by +7%, interviewer C may overestimate by +20%, interviewer D may underestimate by −12%, and so on. This variability between the systematic biases of interviewers is sometimes referred to as *interviewer variability* or *interviewer variance*.

5.2.2 Effect of Interviewer Variability on Population Estimates

In Figure 5.1 we return to the analogy of the marksman and the bull's-eye that we used in Chapter 2 as an aid in understanding the concepts of bias, variance, and mean squared error. It is also useful to understand the concept of interviewer variance and why interviewer variance increases the variability of the sample mean and other statistics. To illustrate, suppose that five interviewers are available to conduct a survey in an area and they are referred to as interviewers A, B, C, D, and E. Each interviewer is assigned the same number of households or establishments, and therefore each interviewer collects data for one-fifth of the sample for the area.

In a typical survey operation involving visits to the sample units, the assignment of the sample units to interviewers would be done to try to minimize travel costs and other costs of collecting data. That is, typically, the interviewer assignments would be constructed as geographic clusters where each unit is geographically proximate to others in the assignment. Then a cluster would be assigned to the interviewer living closest to it, so that travel costs and time to and from the sample units would be saved. However, this method of assignment does not allow us to estimate the interviewer variance, since clusters could be quite different with regard to the characteristic of interest. For example, some of the clusters may comprise mainly high-income households; others, medium- to low-income households. Thus, the differences in average income among the clusters is both a function of the geographic area or neighborhood for the cluster as well as any possible interviewer bias.

For this reason, we need to change the way that assignments are constructed in field studies, to observe interviewer variance. Rather than forming the assignments as clusters, we form the assignments at random. That is, we construct the interviewer assignments by randomly choosing one-fifth of the units for interviewer A, one-fifth for interviewer B, and so on, until all five interviewers have an assignment that is a random sample from the area to be surveyed. This method of assignment is often referred to as *interpenetration*.

Interpenetrated interviewer assignments are constructed by randomly assigning the sample units in an area to each interviewer working in the area so that each assignment constitutes a random subsample of the area's sample. This type of experimental design was used originally by Mahalanobis (1946) as an approach to estimating the variance for field-workers.

With interpenetrated samples, we can observe the effect that the interviewers may have on one particular variable of interest in the survey—say, personal income. The targets in Figure 5.1 are visual conceptualizations of the possible results for income for the five interviewers. The target on the left (Figure 5.1*a*) depicts the results when there is a considerable amount of interviewer variance in the data and the target on the right (Figure 5.1*b*) depicts the situation where there is none or very little interviewer variance. When there is no interviewer variance, the sample values form one cluster on the target. If there are no other biasing factors, the cluster is centered at the bull's-eye.

However, when interviewers influence responses to the income question for the units in their assignments, the values in the sample tend to form clusters

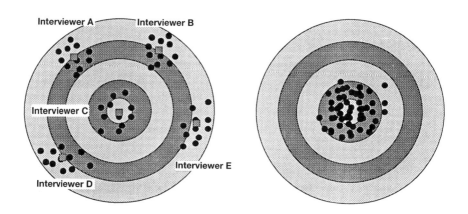

(*a*) **With interviewer variance**   (*b*) **Without interviewer variance**

Figure 5.1   Distribution of sample values with and without interviewer variance. The hits on these bull's-eyes represent the values in the sample for some characteristic. Part (*a*) shows the high level of variation in the sample caused by interviewer variance. Each cluster represents the values in an interviewer assignment. The small squares within the clusters represent the average value of the cluster. With interviewer variance the cluster means differ considerably due to systematic interviewer error. This causes the clusters to be scattered across a bull's-eye, resulting in high interviewer variance. Without interviewer variance, all clusters have the same mean as in part (*b*). The result is a much reduced variance of the sample values.

around the target. As shown in the figure, the values for an assignment are not centered at the bull's-eye, indicating that the values are biased. This is bias due to the interviewer for the assignment. The difference between the centers of two clusters is equal to the difference between the biases of the two interviewers. Thus, if two interviewers have the same bias, their clusters would be essentially overlaid one on top of the other. So, as shown in Figure 5.1a, the values in interviewer A's assignment is on one side of the target, while interviewer E's is on the opposite side, indicating that these two interviewers have very different influences on the responses to the income question. Note that the values plotted on the target in Figure 5.1a are quite dispersed. Comparing the dispersion of values or "hits" on the target for the case of no interviewer variance and that of with interviewer viariance in the figure, it is obvious that the variation in interviewer systematic biases (i.e., interviewer variability) increases the variance of the responses to the income question.

Thus, we see that one effect of interviewer variability is to increase the variance of the observations in the sample. This increase in the variance of the sample observations means that the variance of estimates of population parameters is also increased. As we shall see in the next section, the amount of increase in the estimates of means, totals, and proportions is directly proportional to the amount of variation in the observations that is induced by the interviewers.

5.2.3 Quantifying the Effect of Interviewer Variability on Precision of Survey Estimates

Simple Model for Interviewer Variance

Figure 5.1 suggests that interviewers can influence the responses in ways that displace the observations away from the bull's-eye, or equivalently, add biases to the observations. Interviewer variance arises when these biases differ between interviewers. This concept suggests a simple mathematical model that can be used to quantify the effects of interviewer variability on the observations and the estimates derived from the observations. In this section we present such a model and consider what it suggests about the effect of interviewer error on the estimates.

Figure 5.1 suggests a very simple model for survey error as follows. Let y denote an observation from a sample member and let the index i denote the sample member. Therefore, y_i denotes the observation from the ith sample member where i can take on any value from 1 to n, the total sample size. Thus, we can think of y_i as denoting one hit on the target. Now, denote the mean of the population or bull's-eye by the Greek letter μ (mu) and the deviation of y_i from μ as e_i. Thus, we can write

$$\text{observed value} = (\text{true pop'n mean})$$
$$+ (\text{deviation from the true pop'n mean})$$

or, in terms of the notation we just defined,

$$y_i = \mu + e_i \tag{5.1}$$

for $i = 1, \ldots, n$. For both targets in Figure 5.1, the sum of the deviations which are denoted by e_i is approximately zero. That is, the values y_i are centered around the bull's-eye μ so that the mean of the y_i over all n sample units is approximately μ.

However, for the target in Figure 5.1a, it appears each interviewer has displaced the observations in the assignment by some amount that is approximately the same for all the units in the interviewer's assignment. Let b denote this displacement quantity (or systematic interviewer bias) and let j denote the interviewer (i.e., $j = A, B, C, D,$ or E), so that b_j is the systematic bias for interviewer j. Further, let k denote the unit within an interviewer's assignment. For example, if there are 100 units and each interviewer is assigned 20 units, then k takes on the values $1, 2, \ldots, 20$. Finally, let the Greek letter ε (epsilon) denote the difference between the observed value and the sum of the true mean and the interviewer bias. Thus, y_{jk} denotes the observed value for the kth unit in interviewer j's assignment and ε_{jk} denotes the deviation of y_{jk} from $\mu + b_j$. Therefore, we can write

$$\text{observed value} = (\text{true pop'n mean}) + (\text{interviewer bias}) + (\text{deviation for this respondent})$$

or in terms of the symbols we defined,

$$y_{jk} = \mu + b_j + \varepsilon_{jk} \tag{5.2}$$

for $j = A, B, C, D, E$ and $k = 1, \ldots, 20$.

Let us discuss the interpretation of (5.2) in terms of Figure 5.1a. Consider the cluster of hits labeled interviewer A in the figure. Then each hit could be labeled y_{Ak} for $k = 1$ to the number of hits. For each hit in the cluster, the deviation from the hit to the bull's-eye is $b_A + \varepsilon_{Aj}$ since from (5.2), $y_{Ak} - \mu = b_A + \varepsilon_{Ak}$. In the figure, the center of each cluster of hits is designated by a square (■). This may be interpreted as the bias for the interviewer. Thus, the deviation of the bull's-eye to the square for interviewer A is b_A. Finally, the deviations between the individual hits within cluster A and the square at the center of the hits in cluster A are the variable errors denoted by ε_{Ak}. These interpretations are summarized in Table 5.1.

As a further illustration of the model, consider the situation where interviewers ask farm operators to estimate the market value of various segments of land on the farm. Interviewers are supposed to obtain the operator's estimate of the market value without capital improvements and assuming the farm would be sold as agricultural land and not for development. However, some interviewers misunderstand the instructions and ask farmers to provide a

Table 5.1   Model Components in Terms of Figure 5.1

Model Component	Interpretation	Symbol
Observation within interviewer A's assignment	Hit within interviewer A cluster	y_{Aj}
True population mean	Center of the bull's-eye	μ
Interviewer bias	Center of interviewer A cluster	b_A
Variable	Deviation between hit within interviewer A cluster and the center of the cluster	ε_{Aj}
Total error	Deviation between bull's-eye and hit within interviewer A cluster	$b_A + \varepsilon_{Aj}$

Table 5.2   Value of the Components of the Interviewer Model

j	y_{jk}	μ	b_j	ε_{jk}
1	4800	5100	500	−800
2	6200	5100	500	600
3	4400	5100	500	−1200
4	6700	5100	500	1100
5	5900	5100	500	300

value that includes capital improvements or its value if the land were sold to a land developer. Further, some respondents may never have thought about selling their land and have no idea what the land is worth. In the interest of getting a response rather than an item nonresponse, some interviewers may try to lead the respondent to an estimate. Consequently, the land value estimates may be considerably influenced by what the interviewers think the land is worth and the risk of interviewer variability for the land values question is very high.

Suppose that the values of the land segments in an area are about $5100 per acre. Further, suppose we know that one interviewer obtains values that are biased upward by approximately $500 per acre on average. In reality, the value of the interviewer's bias would not be known; however, we assume it is for illustration purposes. For the five farm operators in his or her assignment, the interviewer obtains the values $4800, $6200, $4400, $6700, and $5900. Thus, the value of the components of the model in (5.2) are as given in Table 5.2. One can verify that the model holds for these values (i.e., $y_{jk} = \mu + b_j + \varepsilon_{jk}$). Further, the variable error, ε_{jk}, sums to zero.

This model of interviewer error somewhat oversimplifies the complex ways in which interviewers can influence survey data. For example, it assumes that an interviewer adds the same constant bias to all the responses that he or she obtains to a particular question. Obviously, this model applies to continuous data items such as age and income rather than categorical data items (e.g., "yes" or "no" responses). However, even for continuous items, it may be too simple since an interviewer may influence some responses more than others. For example, later we will see that the degree to which a respondent is influenced by an interviewer can depend on the question topic, the respondent's characteristics, and the interviewer's characteristics. There are also other ways in the which the model oversimplifies reality. Still, the model can be very useful for providing insights regarding the effects of interviewer variability. It can also be used to guide the design of studies that attempt to estimate the magnitude of the interviewer effect. Thus, even though the model is simple, it is still a very important tool for the study of interviewer error.

Interviewer Design Effect

A number of measures have been proposed in the literature to summarize the degree to which interviewer variance affects the mean of a sample. The most widely used of these measures, due to Kish (1962), is referred to as *the intra-interviewer correlation coefficient*, denoted by the Greek letter ρ (rho) with a subscript "int" denoting "interviewers." Kish defines ρ_{int} mathematically; however, since we are trying to minimize the mathematical content in this book, we will define ρ_{int} in terms of the targets in Figure 5.1.

Kish defined ρ_{int} as the ratio of two variances or measures of variability. The numerator is referred to as the between-interviewer variance and the denominator is the total variance, which is the sum of the between-interviewer variance and the within-interviewer variance. Now in terms of Figure 5.1a, the interviewer means are indicated by the squares (■). The *between-interviewer variance* is simply a measure of the scatter of these squares around the bull's-eye. Recall that the distance of these squares from the bull's-eye for the *j*th interviewer is b_j in the interviewer model. Therefore, the numerator of ρ_{int} is also the variance of the b_j's.

The denominator of ρ_{int} is the sum of the quantity in the numerator and the within-interviewer variance. *The within-interviewer variance* is just the variation among the hits within each of the interviewer clusters in Figure 5.1a. It is also the variation in the scatter of hits in Figure 5.1b, that is, the variance of the sample values after removing interviewer biases from the data. Recall that Figure 5.1b is essentially Figure 5.1a with the interviewer effects (i.e., the b_j) removed. Finally, a third way to compute the denominator of ρ_{int} is to compute the variation of all the hits in Figure 5.1a across all interviewers. Thus, ρ_{int} is the variation in the means of the interviewer assignments shown in Figure 5.1a divided by the sum of that quantity plus a measure of the variation in the hits in Figure 5.1b.

Example 5.2.1 Suppose that the b_j values for the five interviewers in Figure 5.1a are 430, 445, 10, −435, and −450. The variance of these numbers can be calculated using the following formula for the variance of a sample of errors:

$$\text{variance} = \frac{b_1^2 + b_2^2 + b_3^2 + b_4^2 + b_5^2}{5}$$

$$= \frac{430^2 + 445^2 + 10^2 + (-435)^2 + (-450)^2}{5} = 154,150$$

This formula is just the well-known variance formula encountered in elementary statistics courses, simplified to the sum of the squares of the biases divided by 5 since the mean of the five biases is zero. Further suppose that the average within-interviewer variance is 935,000, which is the variance of the numbers in the y_{jk} column in Table 5.2. Assume for the purposes of this illustration that the within-interviewer variance for interviewer A is the same as the average within-interviewer variance across all interviewers. Then

$$\rho_{int} = \frac{\text{between-interviewer variance}}{\text{between-interviewer variance} + \text{within-interviewer variance}}$$

$$= \frac{154,150}{154,150 + 935,000} = 0.142 \tag{5.3}$$

Although it is beyond the scope of this book, it can be shown that under the interviewer error model above, ρ_{int} is the correlation between any two observations within an interviewer's assignment. Another interpretation of the interviewer effect is the correlation between responses that is induced by interviewer error. Thus, we say that the intra-interviewer correlation coefficient is 0.142. Alternatively, we can say that the ratio of interviewer variance to the total variance is 0.142. Although the theoretical values of ρ_{int} are between 0 and 1, the estimates of ρ_{int} can be negative, as we will see. The usual way of dealing with these negative estimates is to replace them by zero and interpret them as the absence of interviewer variance.

The symbol ρ_{int} for a particular survey item denotes the *intra-interviewer correlation coefficient* for the item. It is often used as a measure of the degree to which the interviewers influence survey responses for a survey item. The larger the value of ρ_{int}, the larger the interviewer variance. A value 0 for ρ_{int} indicates no interviewer variance.

Now that we have calculated ρ_{int}, the question arises: What do we do with it? As we saw in comparing Figures 5.1a and b, interviewer variability increases the total variance of the sample responses. It is shown in Kish (1962) that the amount of this increase is related to the *design effect for interviewers*, denoted by deff_{int}:

$$\text{deff}_{int} = 1 + (m-1)\rho_{int} \tag{5.4}$$

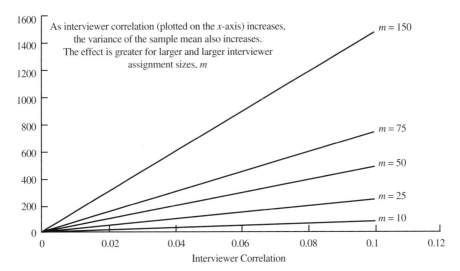

As interviewer correlation (plotted on the *x*-axis) increases, the variance of the sample mean also increases. The effect is greater for larger and larger interviewer assignment sizes, *m*

Figure 5.2   Increase in variance of the sample mean due to correlated error as a function of intra-interviewer correlation, ρ_{int}, and average assignment size, *m*. [Reprinted by permission from Biemer and Trewin (1997), Figure 27.1.]

where *m* is the average interviewer workload for the survey in terms of number of interviews. The quantity $\text{deff}_{int} - 1$ or $(m - 1)\rho_{int}$ is the increase in the variance of a mean, total, or proportion as a result of interviewer variability.

For example, suppose that $\rho_{int} = 0.142$, as we computed above. Further suppose that interviewers, on average, interview 50 persons in the survey. The value of deff_{int} is then

$$\text{deff}_{int} = 1 + (50 - 1) \times 0.142 = 7.96 \qquad (5.5)$$

and $\text{deff}_{int} - 1$ is 6.96. That is, as a result of influences on the survey responses, the variance of the mean of a characteristic having a $\rho_{int} = 0.142$ is increased by almost seven times! This is a tremendous increase in variance, especially considering that the intra-interviewer correlation coefficient is only 0.142, which seems like a small correlation. As we will see through some examples, a ρ_{int} of 0.1 or larger is a very large ρ_{int}. In fact, for most questionnaire items, ρ_{int} is typically in the range of 0.0 and 0.05.

Note that deff_{int} is an increasing function of *m*, so that the larger the average interviewer assignment size, the larger the increase in the variance of an estimator due to interviewers. Typically, *m* ranges from 20 to 80 for face-to-face surveys and from 80 to 150 or more for telephone surveys. In Figure 5.2, we have plotted $(\text{deff}_{int} - 1) \times 100\%$, which is the percent increase in total variance as a function of *m* for values of *m* ranging from 20 to 150 and ρ_{int} ranging from 0.0 to 0.1. This plot clearly shows considerable increases in variance as a result of correlated interviewer error.

> The *interviewer design effect*, deff_{int}, is a measure of the increase in the variance of the mean of a simple random sample due to interviewer variance. For example, a deff_{int} of 1 indicates no increase in variance, while a deff_{int} of 2 indicates that the variance is doubled as a result of interviewer variance (i.e., an increase in variance of 100%).

Example 5.2.2 Consider a telephone survey of 6000 persons and suppose that the survey data were collected by 40 telephone interviewers who split the workload approximately equally. Therefore, m for this survey is approximately 6000/40, or 150 interviews. Suppose that the value of ρ_{int} for a question on the survey is estimated to be approximately 0.013. What is the increase in variance for the estimate of the mean of this characteristic as a result of correlated interviewer error?

Applying the formula for deff_{int} in (5.4), we see that

$$\text{deff}_{\text{int}} = 1 + (150 - 1) \times 0.013 = 2.94$$

Thus, the variance is increased by the factor $(2.94 - 1)$, or 194%.

Effective Sample Size

While the actual sample size for a survey is n, the amount of information we obtain from the sample for the survey characteristic of interest may be much smaller than the sample size suggests, due to interviewer variance. For example, if there were no interviewer variance, the variance of the sample mean for a simple random sample is given by the well-known term σ^2/n. However, as we have seen, interviewer variance increases the variance of the mean by the factor deff_{int} (i.e., the variance of the mean is σ^2/n_{eff}, where n_{eff} is $n/\text{deff}_{\text{int}}$). In other words, when we compute the mean for a characteristic having a deff_{int} value larger than 1, the mean has the same variance as the mean from a sample of size n_{eff}. We say that the sample size, then, is effectively n_{eff} and that n_{eff} is the effective sample size for the characteristic.

> The *effective sample size* for a survey item, denoted n_{eff}, is the sample size, n, divided by the design effect for interviewers, deff_{int}. It is a measure of the loss in precision (or equivalently, the loss of sample information) as a result of interviewer error.

In the example above, the telephone survey of 6000 persons provides the same information for the survey item as a survey of 6000/1.94, or 3092, persons that is completely devoid of interviewer variance. So if we could find a way of eliminating the interviewer variance for this characteristic (i.e., making $\rho_{int} = 0$), the gain in precision for the estimate of income would be approximately equivalent to adding almost 3000 persons to the sample!

Another important finding regarding interviewer variance is that as the number of interviewers for the survey increases, the size of $deff_{int}$ decreases and the effective sample size increases. For example, if the number of interviewers for our telephone survey were increased from 40 to 100, the average assignment size is then reduced from 150 persons to 60 persons. Recomputing $deff_{int}$ for this situation using (5.5), we see that $deff_{int}$ is reduced from 2.94 to 1.77.

Thus, if interviewer variance is expected to be a problem for some important topics in a survey, a survey that employs 100 interviewers will have better precision than a survey that employs 40 interviewers. However, this result ignores some practical problems in increasing the number of interviewers in a survey. For example, the cost of hiring, training, and supervising more interviewers may be such that the reduction in variance obtained by increasing the number of interviewers is not worth it. This is particularly true if each interviewer is given only a few cases to interview. In addition, if the average size of an interviewer assignment is too small, interviewers may not gain much experience in interviewing before they have completed their work. Thus, ρ_{int} may be much higher for the survey with 40 interviewers than with 150 interviewers. Note that in recomputing $deff_{int}$ we assumed that the same ρ_{int} applied in both situations. This may not be the case in practice.

Note that the maximum number of interviewers for a sample of 6000 units is 6000, since this would mean one sample unit per interviewer (i.e., $m = 1$). This is essentially the situation that arises for a self-administered survey since for self-administered surveys, each respondent plays the role of an interviewer in completing the survey. So, for a self-administered survey, we have $m = 1$ and $deff_{int} = 1$ and the effective sample size is $n = 6000$ (i.e., no increase in variance as a result of interviewer variance).

Similarly, the minimum number of interviewers for a survey is one interviewer. For example, suppose that in the telephone survey example, only one interviewer conducted the interviews for the entire sample of 6000 persons. Of course, this is unrealistic, but it is interesting to see what happens to the interviewer variance in that case. Here we see that by (5.4), $deff_{int}$ is at its maximum value 77.7. Again we assume that ρ_{int} does not change as we reduce the number of interviewers to 1, which is probably not likely. Nevertheless, even if ρ_{int} were reduced to one-tenth of its original value (i.e., from 0.013 to 0.0013), $deff_{int}$ is still larger: $deff_{int} = 1 + 5999 \times 0.0013 = 8.8$.

In general, it is a very poor survey design to have only one interviewer collect the data for the entire survey since the effect of interviewer error is maximized in that design. However, deciding on the optimal number of interviewers for a survey should be based not only on interviewer variance concerns but also on survey costs and other logistical factors related to field operations, such as interviewer recruitment, training, and supervision.

Interviewer Variance and the U.S. Census of Population and Housing

Interviewer variance was the focus of a number of studies at the U.S. Census Bureau in the 1950s through the 1970s. In the 1950 Census of Population and Housing, Morris Hansen and his colleagues conducted an experiment to evaluate the effect of interviewer variance on the census results. They found that interviewer variance for many census items was quite high, with ρ_{int} values exceeding 0.1 for some items. This result was an important factor in the decision to conduct a mail-out/mail-back census in 1960 (see Hansen et al., 1961).

Estimating ρ_{int} in Surveys

As discussed previously, for field interview surveys, the usual practice for constructing interviewer assignments is to concentrate an interviewer's workload geographically and near the interviewer's home, if possible, to avoid excessive costs due to traveling to and from sample units. For example, for a survey in a large city employing two interviewers, an interviewer living on the south side of town may be assigned units concentrated to the south and an interviewer living on the north side would be assigned cases near that side of town. This assignment of cases makes it impossible to estimate any interviewer biases since any differences between the average characteristics of their samples could be attributed to the differences of north- and south-side city dwellers.

Suppose instead of the normal assignment that we interpenetrated the assignments of the two interviewers. Recall from our previous discussion that this means that each interviewer's assignment is a random sample from the same population. For example, the south-side interviewer will have approximately half of the north-side sample, and vice versa for the north-side interviewer. In this way, any differences in the population characteristics for the two assignments are eliminated. Then, if the two interviewers perform in a similar manner, we can expect the two assignments will have similar response patterns (within the limits of random sampling error). If the mean characteristics, response rates, or other summary measures for two assignments are significantly different, we can attribute the differences to something related to interviewer performance (i.e., population differences can be ruled out as a cause of the difference).

In experiments for estimating interviewer variance in face-to-face surveys, interpenetrated assignments are used for at least part of the sample, while the usual assignment allocation approach is used for the remainder of the sample. This is due to the extra costs entailed by interpenetrated samples, due to increased travel costs and greater complexity in the coordination of the field staff. In telephone surveys, interpenetrated designs are quite feasible since travel costs are not an issue and random assignment of telephone numbers to interviewers can be managed quite naturally and easily using even the most basic automated call management system.

Given the importance of controlling interviewer error for surveys and the relatively low cost associated with estimating it, should not the assessment of interviewer effects be a routine part of every centralized telephone survey operation? In fact, correlated interviewer error is seldom estimated in telephone survey operations. Part of the reason for this lack of attention to this important component of the mean squared error is lack of information in the research community regarding the damaging effects of interviewer error. In addition, the assessment of interviewer error is a postsurvey quality measure (i.e., it is a measure of interviewing quality that is computed after the data collection is completed). Hence, the value of estimating ρ_{int} for one-time surveys may be less than its value for continuing surveys, since in the latter case, it can be built into an ongoing quality improvement process.

For both modes of interview, obtaining precise estimates of interviewer variance may be impossible for some surveys since the estimates of ρ_{int} are notoriously unstable. In fact, it is usually infeasible to obtain good estimates of ρ_{int} in surveys employing fewer than 20 interviewers with an average workload of 50 cases or less—the standard errors of the estimates would be too large. Since estimates of ρ_{int} are computed by taking the difference between two variance estimates, the estimates can be negative. However, as mentioned previously, ρ_{int} should theoretically take on only positive values since it is the ratio of two positive quantities. Thus, a negative ρ_{int} cannot be interpreted under this model and is usually a sign of instability in the estimates.

Despite these problems, correlated interviewer error has been the focus of a number of studies in the literature, and the results of many of these have been compiled in Groves (1989, Chap. 8). Table 5.3 reproduces some of the results of two tables presented by Groves. It reports the mean values of ρ_{int} for 10 face-to-face surveys and nine telephone surveys. The values of ρ_{int} in this table range from a low of 0.0018 for the Health in America Study to a high of 0.102 for the World Fertility Survey in Lesotho. Most of the values of ρ_{int} are in the range 0.005 to 0.06, with a median value for the table of approximately 0.01. Most of the values of ρ_{int} that are below the median are from telephone surveys, while the majority of ρ_{int} values above the median are from face-to-face surveys. The average value of ρ_{int} for face-to-face surveys is about 0.03, and the average for telephone surveys is about 0.01.

Thus, it appears from available studies in the literature that telephone surveys are less prone to correlated interviewer error than are face-to-face surveys. This result is not surprising given that continuous supervision of interviewers is one of the primary attractions of centralized telephone interviewing. Since the early days of centralized telephone interviewing, this feature was recognized for its potential to reduce interviewer variance through the monitoring and enforcement of standardized interviewer behavior. In addition, centralized telephone interviewing offers many more opportunities for interaction among the interviewing staff than does face-to-face interviewing, which is believed to contribute to greater homogeneity of interviewing behavior.

For face-to-face interviewing, interviews are generally not monitored or observed except in special situations where the supervisor may occasionally accompany an interviewer on his or her rounds to evaluate the interviewer's performance. The face-to-face mode encourages respondent–interviewer

Table 5.3 Values of ρ_{int} from Interviewer Variance Studies in the Literature

Studies Reporting ρ_{int}	Interview Mode	Average Value of ρ_{int}
Study of Blue-Collar Workers (Kish, 1962)	Face-to-face	
Study 1		0.020
Study 2		0.014
Canadian Census, 1961 (Fellegi, 1964)	Face-to-face	0.008
Canadian Health Survey (Feather, 1973)	Face-to-face	0.006
Study of Mental Retardation (Freeman and Butler, 1976)	Face-to-face	0.036
World Fertility Survey (O'Muircheartaigh and Marckwardt, 1980)	Face-to-face	
Peru, main survey		0.050
Peru, reinterview		0.058
Lesotho, main survey		0.102
Consumer Attitude Survey (Collins and Butcher, 1982)		0.013
Interviewer Training Project (Fowler and Mangione, 1985)	Face-to-face	0.005
Average ρ_{int} for face-to-face surveys		**0.0312**
Study of Telephone Methodology	Telephone	0.0089
Health and Television Viewing	Telephone	0.0074
Health in America	Telephone	0.0018
1980 Post Election Study	Telephone	0.0086
Monthly Consumer Attitude Survey	Telephone	
November 1981		0.0184
December 1981		0.0057
January 1982		0.0163
February 1982		0.0090
March 1982		0.0067
Average ρ_{int} for telephone surveys		**0.0092**

Source: Groves (1989), Chap. 8.

interaction, which can be biasing. In addition, when communication can be visual as well as aural, there is greater potential for interviewer influence on responses than in the telephone mode. Body language, facial expressions, and other gestures can convey much to the respondent.

However, larger interviewer workloads are more typical in telephone surveys than in face-to-face surveys. For example, the U.S. Current Population Survey (CPS), which conducts all first-time interviews using face-to-face interviewing, has an average interviewer workload of approximately 50 households. Assuming an average face-to-face value $\rho_{int} = 0.03$ for this survey, the CPS would have an average $deff_{int}$ of about 2.5. For centralized telephone interview, assume that $\rho_{int} = 0.01$ and compute the average telephone interviewer workload corresponding to a $deff_{int}$ of 2.5. It is 150 interviews (i.e., the $deff_{int}$ for a face-to-face survey with $m = 50$ and $\rho_{int} = 0.03$ is equivalent to the $deff_{int}$

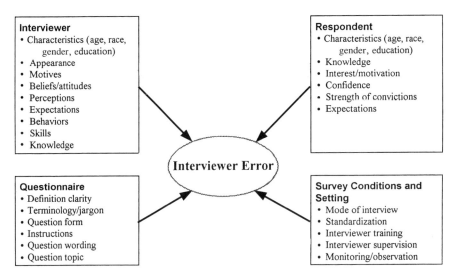

Figure 5.3 Design factors that may explain interviewer effects.

for a centralized telephone survey with $m = 150$ and $\rho_{int} = 0.01$). Thus, telephone surveys having interviewer workloads of 150 or fewer interviews per interviewer should generally have smaller interviewer contributions to variance than face-to-face surveys with average workloads of 50 cases or more.

Given the potential for interviewer error to reduce dramatically the precision of the survey estimates, survey methodologists have devoted much effort to identifying the causes of interviewer errors, the types of respondents and survey situations that are most prone to interviewer influences, and methods for reducing and controlling interviewer effects through better questionnaire design as well as interviewer training and supervision, monitoring, and evaluation. In the next section we explore some of the design factors that may influence interviewer effects.

5.3 DESIGN FACTORS THAT INFLUENCE INTERVIEWER EFFECTS

There is some research that attempts to relate specific design factors to interviewer effects with the aim of identifying the causes of interviewer error and methods for controlling it. As shown in Figure 5.3, the research has focused on four general areas of survey design: characteristics/behaviors of the interviewer, characteristics of the respondent, the questionnaire, and the general survey conditions or setting. In this section we summarize some findings from this literature and discuss their implications for survey fieldwork planning. As will be shown, research on the causes and influences of interviewer effects is quite sparse, and many factors that in theory are candidates for contributors to interviewer error have not been explored. In such cases we speculate on the potential contributions of such factors but offer no evidence of their importance.

5.3.1 Interviewer Effects and the Characteristics of Interviewers and Respondents

Demographic Characteristics

A number of characteristics and behaviors are thought to influence the magnitude of the interviewer effect. For example, an interviewer's age, race, gender, social class, and education level could have an influence on the responses to some questions with some respondents. For example, a mature interviewer may wield more influence with a child respondent than with respondents who are themselves mature. Similarly, an interviewer who is perceived as well educated, possibly holding a college degree, may be more influential with respondents having no formal education than interviewers who are perceived as being uneducated and unknowledgeable. One would imagine that this is particularly true for opinion and attitude questions for which the interviewer holds strong views. However, neither of these hypotheses has been verified in the literature.

The one interviewer characteristic that seems consistently to be important is that of race. Several authors have shown that racial matching of the interviewer and respondent yields a pattern of responses different from that obtained in the absence of matching (e.g., Williams, 1964; Schuman and Converse, 1971). There is some evidence, however, that this effect is important only when the subject matter of the survey is sensitive. The largest effects of race of interviewer are observed for questions dealing with racial issues (see, e.g., Hatchett and Schuman, 1975 or, more recently, Wilson and Olesen, 2002).

There is also some evidence in the literature that gender of the interviewer can make a difference in response patterns when the questions turn to gender roles. For example, Nealon (1983) examined gender effects in a survey of farm women and concluded that response patterns differed for male and female interviewers for questions on work involvement, decision making on the farm, financial arrangements, and satisfaction with farm programs. In other studies, females gave more feminist responses to males than to female interviewers on questions of women's rights and roles (Ballou and de Boca, 1980). There is also some evidence of larger interviewer variation for elderly respondents.

These results and other research suggest a pattern regarding the interaction of interviewer and respondent characteristics. When the subject matter of the survey turns to topics related to the characteristic, response effects can be expected to emerge. As an example, in a survey of teenagers, one might hypothesize that an elderly interviewer would elicit different responses on the topic of teenage sex attitudes than a more youthful interviewer. Similarly, for a survey regarding attitudes toward persons of wealth, an interviewer dressed expensively and driving an expensive car might obtain very different responses than an interviewer appearing to be of much more modest means. Therefore, to the extent that staffing for the data collection will allow it, matching the visible characteristics of the interviewer and respondent (age, race, gender, socioeconomic status) would seem to be a good strategy in most interview situations since the available evidence suggests that responses are usually more accurate with this strategy.

Interviewer Expectations

A number of researchers have suggested that the expectations that interviewers hold regarding the probable answers of respondents and whether respondents will be willing or able to answer certain questions have a profound effect on survey response. Concern by the interviewer that the respondent may react negatively to a question may cause the interviewer to rephrase the question or simply skip it. A study by Singer and Kohnke-Aguire (1979) tested this hypothesis and did not find a strong relationship between interviewer expectations regarding question difficulty and respondent behavior. Similar studies have also found only weak evidence of an interviewer expectations effect. However, not much is known about the complex relationships between interviewer expectations, interviewer behavior, and the question subject matter.

For example, interviewer expectations of the respondent's answers is a proposed cause of inconsistent probing behavior. Based on preconceived notions of the respondent's circumstances and status on the various measurements in a survey, the interviewer may decide when to probe or give clarification of the questionnaire concepts. For example, if the topic of the survey is crime victimizations, the interviewer may be more inclined to probe a response of "no" to a question about thefts in a high-crime area than he or she would in a more affluent area. This type of inconsistent probing behavior, which is purely at the discretion of the interviewer, is believed to be an important cause of interviewer variance. However, so far there has been no systematic study of this effect.

There is some evidence that more experienced interviewers obtain higher unit response rates than do less experienced interviewers. However, for item nonresponse, the opposite might be true. Stevens and Bailar (1976) compared item nonresponse rates in the CPS for experienced and inexperienced interviewers and found that missing data rates for inexperienced interviewers were somewhat lower for some questions, such as income questions. This may be the result of experienced interviewers having an expectation that respondents react negatively to such questions and either skip the question or settle too quickly for a refusal. Less experienced interviewers may be more naive regarding respondent reactions to these questions and exert more effort to obtain a response.

Interviewer Effects on Nonresponse

There is clear evidence in the survey methods literature that interviewers can have a considerable influence on the response rate for a survey—not only cooperation rates but contact rates as well. Groves and Couper (1998) provide a comprehensive review of the literature regarding the influences of interviewers on nonresponse in household surveys. Their analysis supports the following conclusions:

- Interviewers with greater experience tend to achieve higher rates of cooperation than do those with less experience.
- Interviewers who are confident about their ability to obtain cooperation tend to achieve higher cooperation rates.
- Standardization of the introduction to the survey (particularly for telephone interviews) is ineffective at obtaining cooperation. Introductions that are unscripted and adaptive to the householder's objections or concerns tend to be more effective.

There are plausible and fairly obvious explanations for these findings. Experienced interviewers may become more skillful in the art of persuasion and making effective use of their time. Confident interviewers may be more effective and skillful in applying the six psychological principles of survey participation studied in Chapter 3: reciprocation, consistency, social validation, authority, scarcity, and liking, particularly the latter principle. Flexible introductions allow interviewers to tailor the request to participate in the survey to particular concerns and reservations of the individual. This flexibility, as we learned in Chapter 3, can prevent initial reluctance to participate from becoming hardened resistance and refusal. Since interviewers vary considerably in experience, confidence, and social adroitness, interviewer variance for unit nonresponse is usually quite high in surveys.

Regarding item nonresponse in surveys, Bailar et al. (1977) found evidence of interviewer variation in item nonresponse for the 1970 U.S. Census of Population and Housing. As noted previously, differences in item nonresponse by interviewer experience levels were observed in other studies. Part of this effect is due to skip errors or "slips," which is reduced by experience with the instrument. Part of it is also interviewer reluctance to ask certain questions that typically elicit negative responses from respondents, a behavior that is learned by experienced interviewers. There is also some evidence from Swedish surveys that interviewers will sometimes skip one or more questions if they feel the questions are "out of place" or "out of context" at the point they are supposed to be asked in the interview. Although they intend to come back to them later in the interview, they often forget.

Computer-assisted interviewing (CAI) methods such as CATI and CAPI have reduced the former error since branching from one question to the next is now automated. The same is true to some extent for the latter effect, since it is usually not possible to proceed to the next question in a CAI instrument until an entry has been made for the current question. However, interviewers can still avoid asking the question if they are willing to enter a false entry. Unlike item nonresponse, this interviewer behavior is not likely to be detected. In Section 5.4 we discuss some methods for detecting interviewer falsification.

5.3.2 Interviewer Effects and the Design of the Questionnaire

Prior to the 1920s, interviewer variability was associated primarily with observational studies such as crop yield studies, housing valuations, and the like. Rice (1929) reported an interviewer effect in a study of destitute men involving two interviewers with very different opinions about societal problems. One interviewer, a prohibitionist, tended to find respondents who attributed their condition to alcohol. The other, a socialist, tended to find destitutes who blamed the social and economic conditions of the time. Since then, researchers have found more evidence that the behavior, mannerisms, appearance, vocalizations, and so on, of the interviewer all have the potential to influence the respondent in ways that vary from interviewer to interviewer. The decisions the interviewer makes regarding whether or how to ask a question, whether to probe for clarification, whether to provide feedback, how to interpret the respondent's response, and so on, can alter the data in ways that vary from one interviewer to another. The rapport the interviewer has with the respondent and the interviewer–respondent interaction can also influence answers. This realization led to the widespread use of standardized instruments and interviewing procedures. The fundamental purpose of standardization is to limit the ability of interviewer attitudes, beliefs, behaviors, and mannerisms to affect survey response by attempting to homogenize the respondent–interviewer interaction. However, there is considerable evidence in the literature which suggests that the primary cause of interviewer variability is not the interviewer per se but the questionnaire design, survey questions, and associated instructions.

Evidence from the literature suggests that low values of ρ_{int} can be expected for demographic characteristics and other items which are clearly and unambiguously stated requiring factual information that is well defined and easily accessed by respondents. Higher ρ_{int} values are associated with attitudinal or opinion questions, open-ended questions, sensitive or emotional questions, and difficult items that require interviewer clarification and probing. However, findings in the literature are somewhat inconsistent and suggest that these general guidelines tend to oversimplify a much more complex set of interactions between the questionnaire and the interviewer. Perhaps the literature on conversational interviewing can shed more light on this issue.

Schober and Conrad (1997) suggest that one problem with standardized interviewing is respondent situations often do not "map" easily into the definitions specified in the question. When this happens, the standardized interviewer may be pressured into deviating from the script of the interview and offering unscripted clarification and assistance to help the respondent respond to the question. Some interviewers do this well, others not so well, and the majority of standardized interviewers may not do it at all, choosing instead to follow the standardized approach of allowing respondents to resolve these difficulties on their own.

In one experiment (Schober and Conrad, 1997), 41 respondents were given scenarios (or vignettes) describing various situations and then interviewed about the scenarios using both standardized and conversational interviewing approaches. For half of the scenarios, the mapping of the situations described in the scenarios and the operational definitions of terms used in the question (which were not a priori known by the respondents) were quite straightforward. In the other half, the mappings were much more complicated.

For example, respondents were shown a floor plan and asked "How many half-bathrooms are there in this house?" When a room had two fixtures (a toilet and a sink), respondents had no difficulty recognizing it as a half-bathroom with either interviewing approach. However, when the room had only one fixture, the number of correct responses differed markedly between the two approaches. Across the 12 questions investigated in the study selected from ongoing U.S. government surveys, accuracy for both standardized interviewing and conversational interviewing was nearly perfect when mapping of the scenarios to the legal definitions was straightforward—97 percent and 98%, respectively. For complicated mappings, the standardized interview accuracy dropped to only 28% accurate compared to 87% for the conversational interview approach.

These results suggest that interviewer variance may indicate variation in the way interviewers handle those situations where mappings to operational definitions are complicated. Although interviewer variance increases the variance of an estimator, interviewing approaches that achieve greater accuracy in the responses could have an offsetting positive effect on the bias component of the mean squared error. Still, we should not accept procedures that we know will cause considerable interviewer variance with the hope that they will achieve lower response bias. Rather, better questionnaire designs, interviewing approaches, and training regimens are needed to control both bias and variance. In many cases it may not be possible to anticipate the many complexities that interviewers encounter in applying the survey definitions. In those situations, a more flexible interviewing approach that can aid the respondent in applying the operational definitions to his or her particular situation may be the best approach.

5.3.3 Interviewer Effects and the Survey Setting

There are a number of other factors related to the survey design that potentially could affect the magnitude of the interviewer effect. These include the mode of interview; the methods for hiring, training, supervising, and monitoring the interviewer performance; protocols for interviewing (e.g., standardized, conversational, or hybrid interviewing approaches); and other factors that characterize the conditions under which the interview takes place, such as privacy and respondent attentiveness to the survey task, that are referred to generally as the survey setting. In this section we summarize what is known regarding the influence of these factors on interviewer error.

Previously, we discussed the potential for interviewing protocols to affect interviewer variance. Also, our previous discussions dealt with the potential effects of telephone and face-to-face interviewing on interviewer error. The effects of CAI on interviewer error were treated to some extent in our discussion of interviewer effects on item nonresponse. In that discussion it was noted that the computer has virtually eliminated skip errors in interviewing and thus interviewer variability due to skip errors. However, concerns remain about the recording accuracy of CAI interviewers and whether random error due to miskeying responses may be a problem. To the extent that interviewer keying and computer proficiency vary across the interviewers, recording error could also be a source of interviewer variance.

Several studies have shown that data entry (keying) for both CATI and CAPI interviewers is not an important source of error in surveys. In fact, evidence provided by Lepkowski et al. (1998) suggests that recording accuracy is higher for CAI than for paper-and-pencil interviews (PAPI). For example, the number of keying errors made by CAPI interviewers was far fewer than the number of transcription errors made by PAPI interviewers. Tourangeau and Smith (1998) report smaller ρ_{int} values for CAPI interviewers than PAPI interviewers for measuring a number of sensitive questions. It appears that computerization reduces the interviewer effect, perhaps by enforcing a certain level of standardization on the interviewers. However, there have not been studies to our knowledge to compare the interviewer variance for CAPI and PAPI personal interviews.

A paper by Couper (1996) explored the effect of the change from paper-and-pencil interviewing to CAPI in the CPS on the setting of the interview, and the effect of the change on data quality and costs. He found that the change to CAPI had a large effect on the number of doorstep interviews (i.e., those conducted with the interviewer standing outdoors) conducted in the CPS. He found that CAPI interviews were much more likely to take place inside the respondent's home with the interviewer seated at a table near an electrical outlet where the laptop could be supported and powered by household electricity. As an example, whereas 58% of the CPS paper-and-pencil interviews are conducted in the respondent's home, more than 75% of the CAPI interviews are conducted in this setting.

Doorstep interviews are associated with poorer data quality. For example, they tend to be shorter, suggesting that respondents and interviewers may be more rushed at the doorstep than when seated comfortably inside a home. This can result in more top-of-the-head responses and satisficing (Krosnick, 1991), to get the interview over. When the interviewers are holding a laptop at the doorstep, they may be more concerned about the performance of the computer (battery giving out, fatigue from holding the machine, difficulty keying or seeing the screen) than the respondent's or their own performance.

Table 5.4   Effect of Length of Training on Interviewer Performance Measures[a]

Interviewer Behavior	Length of Training Program			
	½ Day	2 Days	5 Days	10 Days
Reading questions as worded	30	83	72	84
Probing closed questions	48	67	72	80
Probing open questions	16	44	52	69
Recording answers to closed questions	88	88	89	93
Recording answers to open questions[b]	55	80	67	83
Nonbiasing interpersonal behavior	66	95	85	90

Source: Fowler and Mangione (1990), Table 7.4.

[a] Entries are the percent of interviews rated as either excellent or satisfactory.

[b] n.s., all others $p < 0.01$.

However, a greater need to be in the home and seated during the interview can also have drawbacks. Respondents may refuse to allow the interviewer entry into the home, which may result in either a unit nonresponse or a "break-off" of the interview. For example, Couper notes that refusals for CAPI interviews were slightly higher than for paper-and-pencil. It is not clear, however, how much of this increase is due to the interview setting. In addition, interviews conducted in the home tend to be longer and, as a result, cost the survey organization more in interviewer labor charges.

Couper concludes that it is difficult to say whether the setting changes induced by CAPI improved data quality since that would depend on the trade-off between possibly lower response error and higher nonresponse error. Still, his analysis suggests the importance of considering the effects of changes in technology on the survey data collection process and data quality.

An important study by Fowler and Mangione (1985) found mixed results regarding the effect of training length on ρ_{int} values. Four lengths of training were tested: ½-, 2-, 5-, and 10-day training periods. The highest values of ρ_{int} were estimated for interviewers with the shortest and longest training periods. One plausible reason for the finding in the latter group is overconfidence as a result of training. Although the interviewers were being trained in the standardized interviewing approach, the longer training may have actually worked counter to standardization. As we noted in Section 5.1, for situations that require complex mapping of the respondent's circumstances to the operational definitions, the more highly trained interviewers may have been more apt to intervene in the response task and help the respondents arrive at a response.

Table 5.4 reports some additional findings from the Fowler–Mangione study. The adverse effects of inadequate training are obvious from the table: Interviewers are ill-prepared to handle some of the basic functions of interviewing, such as reading the questions as worded, probing, recording answers, and maintaining a neutral interpersonal behavior. All of these functions can be improved with additional training. Probing behavior especially benefits from longer training, perhaps as a result of interviewers acquiring greater knowledge regarding the research objectives underlying the questions.

Additional findings from their research may be summarized as follows:

- Interviewers with minimal training can handle the interpersonal aspects of enlisting cooperation and relating to respondents as well as more highly trained interviewers. The authors suggest that this may be partly due to the fact that these skills are enhanced more by experience than through training.
- Interviewers with more training were more task oriented; those with less training were more interpersonally oriented.
- Too much training can be counterproductive; however, the definition of "too much" varies from survey to survey.

5.3.4   Practical Implications for Fieldwork

All three components of the survey interview—interviewer, respondent, and questionnaire—determine what happens in the interview and the resulting data. However, the way the interviewer delivers the survey questions to the respondent and what is ultimately recorded on the questionnaire are purely in the hands of the interviewer. The traditional idea that the interviewers should adhere strictly to the questionnaire scripts assumes that the questionnaire and training guidelines cover all possible interviewing situations. However, it is not realistic to assume that all possible special cases that interviewers encounter can be anticipated. Even if they could be planned, the training required to ensure that all interviewers handle the myriad of situations in the same prescribed manner is not feasible from a cost perspective. Because of this, training guidelines should give interviewers principles of behavior (e.g., read the question exactly as worded and then probe in a neutral and nondirective manner) and let them then apply the principles in various situations as they arise.

The risk of this approach is that the interviewer error will increase as a result of variation in the way the interviewers present the questionnaire concepts and interact with the survey respondents. However, a number of strategies can be adopted to protect against these adverse consequences. Here is a list of suggested actions.

- Recruit interviewers who display excellent interpersonal skills and powers of persuasion, have good organizational skills, and whose visible characteristics are well matched to the characteristics of the sample members in their assignments, given the topics to be surveyed.
- Hire experienced interviewers whenever possible. When less experienced interviewers are used, the training should provide lessons and

practice sessions on adapting and tailoring the interview approach to the many situations they may encounter. This is particularly true in gaining cooperation from respondents. (For discussions of interviewer training methods, see Groves and McGonagle, 2001, and Campanelli et al., 1997a.)

- Interviewers with high confidence levels and a positive attitude toward the study are preferred over interviewers who display a lack of confidence and feel the study will not succeed. In some cases, these attitudes can be manipulated during training and on-the-job feedback.
- Training for the interviewers should cover the concepts and objectives behind the survey questions as well as the mechanics of the interview. Probing techniques should be covered in some depth and the length of training should be sufficient for the interviewers to develop some proficiency with probing under various response situations.
- Interviewer performance should be closely monitored initially until the interviewer displays adequate proficiency with the surveying tasks. Then monitoring of performance can be less frequent. Both positive and negative feedback on the monitoring results should be given to the interviewer as soon as possible.

In the next section we discuss methods for monitoring and evaluating the interviewers and providing feedback on their interviewing performance.

5.4 EVALUATION OF INTERVIEWER PERFORMANCE

In any survey involving interviewers, some kind of evaluation of interviewer performance during fieldwork is essential. This is necessary not only to control the quality of the interviewing, but also to identify potential problems in other areas of the survey process: for example, the questionnaire or interviewer instructions. Interviewer performance evaluation involves collecting data and other information on various aspects of the interviewer's work, evaluating these data, and taking some action on the basis of the evaluation. Four key areas of performance are usually targeted in the evaluation as follows:

1. *Detection and prevention of falsified information.* Falsification of data by interviewers is a risk for which every survey manager should prepare and seek methods to detect or deter. *Falsification* is also referred to as *fabrication, cheating, table-topping,* and *curbstoning* in the survey methods literature. (The latter term is based on the image of an interviewer sitting on the curb outside a sample housing unit falsifying the questionnaire.) This type of error may involve the fabrication of entire interviews, the deliberate skipping of some questions, or falsifying other information, such as whether a housing unit is vacant or occupied. The reasons some interviewers will resort to falsification are many. Perhaps the pressures to achieve a high response rate or to expe-

dite the fieldwork are too great for the survey. Interviewers may have personal problems which prevent them from working their assignments adequately and may resort to cheating as a way of getting the work done. Interviewers who are afraid to enter rough neighborhoods may resort to making up the interviews. If interviewer cheating is widespread, it can invalidate the results of the entire survey enterprise. For that reason, interviewer falsification should not be tolerated under any circumstances (see also Chapter 6).

2. *Compliance with the rules and guidelines for interviewing set forth in training.* An important contributor to interviewer variance is variation in the adherence by interviewers to the instructions and guidelines for administering the questionnaire. Some interviewers may closely follow the instructions; others do not. As a consequence, the quality of the data will vary by interviewer. As we saw in the earlier discussion, this may be particularly problematic when the respondent's situation does not easily conform to the operational definitions in the questionnaire. If interviewer noncompliance is widespread, the instructions and guidelines may be at fault rather than the interviewers. Poorly designed procedures are a burden to interviewers and respondents alike and the interviewers may attempt to "repair" the procedures as they go. If this is the case, the procedures should be revised.

3. *Performance on noninterview tasks such as administrative activities.* Interviewers perform many tasks outside the interview in preparing to interview, recording some results of the interview, gathering information from the field to help in sampling, and so on. Although these tasks may not have a direct bearing on the accuracy of the survey results, they can affect survey data quality indirectly and therefore should have some type of evaluation with feedback to the interviewers. For example, telephone interviewers must record the outcome of each call (busy, ring-no-answer, appointment, refusal, etc.). Errors in this process may provide biased process statistics such as response rates, the success rates of call attempts, ineligibility rates, and the like. Field interviewers must plan their routes to minimize costs and increase the effectiveness of their home visits. Inefficiencies or errors in this process can increase costs and even reduce response rates.

4. *Identification of problems in the interviewer–questionnaire interface.* Whether paper and pencil or electronic, one objective of survey questionnaire design is to provide an ergonomic interviewer-questionnaire interface. Usability in the context of the survey instrument refers to the ease with which interviewers interact with a PAPI, CATI, CAPI or other questionnaire. If the commands are complex, confusing, and awkward to use, interview quality will suffer.

A number of methods are available to the survey manager for evaluating these areas of interviewer performance. In Table 5.5 these methods are listed along with the areas of performance that are targeted by the methods.

One common design feature of all interviewer performance evaluation methods is their emphasis on less experienced or new interviewers. For

Table 5.5   Some Methods for Evaluating Interviewer Performance

| Evaluation Methods | Key Performance Area[a] | | | |
	Falsification	Interview Performance	Noninterview Tasks	Usability
Reinterview	●	·		
Verification recontact	●			
Observation		●	●	·
Audio-recording (with or without behavior coding)	● (without behavior coding)	● (with behavior coding)		·
Monitoring	●	●		·
Review of questionnaires	·	·	●	●
Performance and production measures	·	●	●	
Keystroke/trace file analysis	·			●
Mock interviews/ tests of knowledge or practice		●		●

[a] A large bullet (●) indicates a primary objective of the method, and a small bullet (·) indicates a secondary objective or that the method provides an indirect measure of performance. Note that some of the table entries have not yet been defined; however, these will be discussed later in the chapter.

example, at the initiation of a new survey, interviewers are evaluated more frequently than in the later stages of the survey. New hires may be evaluated more frequently than well-established interviewers. However, it is common to evaluate all interviewers at some level regardless of their length of experience or demonstrated skills and competence. There is some evidence of a curvilinear relationship between interviewer performance and experience. For example, item nonresponse and nonverbatim question delivery can be higher among more experienced interviewers. For this reason, the evaluation and improvement of interviewer performance should be a continuing process.

Another common method designed to improve interviewer performance is feedback to the interviewers either orally or using a printed feedback form. For feedback to be effective, it should be timely and contain positive messages about performance as well as areas where improvements are needed.

5.4.1   Reinterview Surveys and Verification Recontacts

Reinterview surveys and verification recontacts are a widely used method for evaluating face-to-face interviewer performance. In the context of quality evaluation, a reinterview survey involves selecting a relatively small, random sample of survey respondents, recontacting them within a short period of time after the original interview, and interviewing them again on some of the same topics that were included in the original interview. Reinterview surveys are also an important method for evaluating the nonsampling errors in surveys since, depending on how they are designed, they can provide information on nonsampling error components, interviewer performance, or both. In Chapter 8 we discuss some uses of reinterview surveys for estimating nonsampling error components. Here we discuss their use for interviewer performance evaluation.

The U.S. Census Bureau has evaluated interviewer performance using reinterviews since the 1950s. For example, in the Current Population Survey, interviewers are randomly selected each month, and approximately one-third of their assignments are reinterviewed. Part of this sample, about 75%, is used for estimating reliability; however, the other 25% is designed to evaluate interviewer performance. For this small sample, the primary objective is to detect and deter interviewer falsification (curbstoning). As Biemer and Stokes (1989) showed, the probability of detecting falsification with this type of design is very small. Rather, the real value of this approach is deterrence since interviewers who know that their assignments are being inspected for falsification are less likely to falsify.

To save costs, the reinterview is typically conducted by telephone, if available, and face-to-face as a last resort. Attempts to use a mailed questionnaire for reinterview have not been successful, due to the very low response rate. Properly designed, reinterview surveys are an effective means of identifying interviewers who have fabricated an entire interview. In addition to reasking some of the questions from the original interview, the respondent might also be asked directly if an interviewer conducted the interview. For example, the reinterviewer might ask:

- If an interviewer called or visited them during the interview period
- If so, the length of time the call or visit
- Whether specific questions from the questionnaire were asked
- Whether the incentive was offered and paid (if an incentive is used in the survey)

To some extent, the reinterview can also be used to detect other types of interviewer errors, such as errors in the respondent selection procedures, incidents of accepting proxy reports when self-reports are mandated, inappropriate use of telephone interviewing, and misclassifying occupied housing units as vacant. Further, when combined with reconciliation, reinterviews can provide information on interviewer performance within an interview.

Reconciliation refers to the process of comparing the reinterview responses to the survey questions with the original responses and resolving any discrepancies with the respondents. Reconciliation can be an effective means of identifying interviewers who skip certain questions, fail to probe when necessary and appropriate, and convey the wrong information to the respondent. An important shortcoming of the approach is that it seldom provides conclusive evidence of interview performance problems, as often the respondent's account is denied by the interviewer. Forsman and Schreiner (1991) provide a comprehensive review of alternative reinterview survey designs for both interviewer performance evaluation and the estimation of nonsampling error components.

A related approach that is widely used by survey organizations is the verification recontact. Like reinterview surveys, verification interviews are conducted primarily by telephone for a small subsample of interviews, say 10 to 20% of the main sample. However, the verification may not include reasking any questions from the original interview since the purpose of the contact is just to verify that the interview was conducted with the appropriate respondent using the appropriate methods. For example, the verification interview may ask: "Did an interviewer from our organization conduct an interview with you in your home last week?" This contact can also be used to gather information on the respondent's perception of the interviewer's demeanor during the interview (courtesy, flexibility, knowledge of the survey, appearance, etc.).

Some survey designs, such as panel surveys and recurring surveys, require that respondents are recontacted periodically for additional interviews. For example, for the CPS, respondents are contacted in four consecutive months, followed by no contacts for eight consecutive months followed by four additional consecutive, monthly contacts. Other U.S. panel surveys, such as the Survey of Income and Program Participation and the National Crime Victimization Survey, require less frequent contacts. One concern regarding the use of reinterviews for assessing interviewer performance is that the additional respondent burden brought on by the reinterview contact could cause some respondents to refuse to participate later during the regular panel survey contacts. However, studies conducted at the U.S. Census Bureau suggest that the effects of reinterviews on panel survey response rates are small or nonexistent.

5.4.2 Audio Recordings, Monitoring, and Other Observations

A number of methods are available for observing the behaviors of interviewers during a live interview performance. For face-to-face interviewing, an interviewer's supervisor may accompany the interviewer on his or her rounds to observe how the interviewer carries out various tasks, particularly interviewing. For centralized telephone interviews, unobtrusive call monitoring is typically used. Trained monitors, who may be supervisory personnel or interviewers who have a dual role as monitors, listen to all or parts of a sample of telephone interviews and note positive and negative attributes of interviewer performance. Interviews may also be recorded using tape recorders or laptop computers with computer audio-recorded interviewing (CARI) technology installed. Since these recordings can be reviewed repeatedly by different observers, they allow more detailed and reliable information on interviewer performance than observations or call monitoring.

Supervisory observations of face-to-face interviewers provide a wider range of information on a particular interviewer's performance than that of any other method. The supervisor obtains information not only on the interviewer's approach to interviewing, but also on the interviewer's organizational skills and ability to plan and schedule contacts and many other aspects of the job. One drawback of the approach is that the presence of the supervisor during the interview may inhibit the interviewer so that the observed behavior is atypical of the interviewer's usual behavior. For example, the interviewer may be on his or her "best behavior" in the supervisor's presence. In most cases, staged "good behavior" is not easy to sustain over a long period of time, and eventually, very poorly performing interviewers will be detected. Nevertheless, supervisory observations are not ideal for detecting deliberate violations of the interview guidelines.

Call monitoring of telephone interviewers is a routine part of survey operations in most centralized telephone facilities. There are essentially three dimensions of a call monitoring system: the selection of interviewers and interviews to monitor, the information to be recorded during the monitoring session, and the approach to providing monitoring results to interviewers and instrument designers in the form of constructive feedback. The selection of interviewers should be random or in some other manner that is unpredictable by the interviewers. Sampling rates may vary depending on the skill and experience of the interviewer, the budget for the survey, and the quality objectives of the monitoring operation. A typical goal of continuing monitoring is a minimum of 10% of the interviews, and each interviewer should be monitored each week (see Couper et al., 1992).

In many telephone centers, the monitors provide feedback to the interviewers immediately following monitoring. Often, the comments are subjective and unsystematic, varying with the interest and orientation of the observer. Behavior coding schemes have been developed that provide more objective and systematic feedback. Studying interviewer and respondent behaviors often reveals problems that interviewers or respondents have with survey questions, signaling deficiencies in wording or design of the survey questions. Studying the behavior patterns may indicate problem questions to the survey manager and may suggest how to improve them.

For evaluating interviewer performance, the focus of behavior coding is typically on four behaviors that affect interviewer variation: question delivery, probing behavior, feedback to respondents, and pace or clarity. Figure 5.4 presents an example of a coding scheme used in one telephone center that was adapted from Mathiowetz and Cannell (1980). This simple scheme was used for coding live interviews, where coding is simultaneous with the interviewing. With live coding, the reliability of the coders' judgments can be a problem. To

QUESTION DELIVERY
11 Reads question as worded
12 Minor wording changes
13 Major wording changes
14 Fails to read question

FEEDBACK
31 Delivers feedback appropriately
32 Delivers feedback inappropriately
33 Fails to deliver feedback

PROBING BEHAVIOR
21 Probes appropriately
22 Probes inappropriately
23 Fails to repeat

PACE/TIMING
41 Reads too fast or too slow
42 Timing between items too fast
43 Timing between items too slow

Figure 5.4   Simple coding scheme for telephone call monitoring.

Table 5.6   Use of Behavior Coding Results for Interviewer and Question Performance[a]

	Interviewer Type (Rustemeyer, 1977)			Question Type (Mathiowetz and Cannell, 1980)	
	Experienced	End of training	New	Open	Closed
Question Delivery					
Exactly as written	66.9	66.4	66.9	95.8	95.4
With minor changes	22.5	17.9	19.9	19.9	3.7
With major changes	5.2	3.6	3.9	0.5	0.4
Not read	3.3	8.9	6.0	1.8	0.5
Probing Behavior					
Correct probe	80.7	86.0	80.3	79.2	85.6
Incorrect probe	19.4	14.1	19.6	20.8	14.4

Source: Groves (1989), Table 8.3.

[a] Entries are the percentages by type based on all coded interview behaviors of each type.

compensate for this, coding procedures tend to be more simplistic. More elaborate schemes have been devised for coding from tape-recorded interviews by adding items that capture problems in the respondent's understanding of the question. For example, coders may record whether respondents asked to have the question repeated, requested clarification, refused to answer, and so on. See Cannell and Oksenberg (1988) for some useful guidelines for designing a behavior coding system for telephone surveys.

As an illustration of the use of behavior coding for evaluating the interview process, we provide Table 5.6. Results from two studies are summarized in this table. The research conducted by Rustemeyer (1977) is based on mock interviews with over 200 interviews and focuses on three groups of interviewers: "experienced," defined as interviewers having at least three months of field experience; "end of training," defined as interviewers who had completed only two or three assignments; and "new," defined as interviewers fresh out of training. There are essentially no differences among the three groups in the percentage of questions that are read exactly as worded. The biggest difference occurs for questions that are skipped. Here experienced interviewers are best followed by new interviewers. This result suggests that skip errors are largely a function of inexperience with the navigational features of the paper-and-pencil questionnaire. Note that experienced interviewers have the highest percentage of major wording changes. This could suggest questionnaire wording difficulties in the questionnaire. Experienced interviewers may be more aware of these and attempt to resolve the difficulties by altering the wording.

The Rustemeyer study also captured information on probing behavior. These are verbal interchanges between the interviewer and the respondent which aim to clarify the response and the meaning of the question so that the appropriate response category can be assigned. Correct probes are neutral and attempt to clarify the response task without leading the respondent to a specific response. An incorrect probe is one that is leading or directive. For this behavior, the relationship between experience and correctness is curvilinear, with the best results occurring for the middle group. Behavior coding can also be used to look at interviewer performance as a function of question type, as in the Mathiowetz and Cannell (1980) study. These results are consistent with our earlier remarks regarding the risk of interviewer variance for open-ended questions.

Among six studies mentioned in Groves (1989), the variation in interviewer behavior by survey is quite pronounced. For example, the proportion of questions delivered exactly as worded ranges between approximately 57 and 96%. In addition, there is considerable variation within the same survey across the interviewers. In their study of telephone interviewers, Cannell and Oksenberg (1988) found that unacceptable question delivery was at least seven times more likely among the worst interviewers than among the best.

5.4.3 Other Methods of Evaluating Interviewer Performance

A number of other techniques are commonly used in surveys for evaluating interviewer performance. One method that is used for PAPI surveys is to review the completed questionnaires to identify problems with improperly completed forms. For PAPI and CAPI surveys, production and performance measures such as nonresponse rates, uncodable information, the distribution of work completed by time (e.g., for time of day, day of the week, week of the field period, etc.), and other statistics on the survey process can be quite informative regarding the quality of the interviewing process. In addition, they are inexpensive and easily conducted. Some cases of interviewer falsification have been detected in the CPS by selecting for evaluation reinterview interviewers who turn in a high proportion of unemployed persons in an area known to have low unemployment or who have reported much lower than expected travel mileage and higher than expected vacancy rates.

With the widespread use of CAI questionnaires, the traditional measures of PAPI interviewer performance are no longer sufficient in a CAPI environment. The review of forms for legibility, completeness, answers within range, and so on, is unnecessary using CAI since many of these functions are now automated. Now, keystroke and trace file analysis has taken the place of the inspection of paper questionnaires in surveys.

Keystroke files are a record of every keystroke that an interviewer enters as he or she moves through the CAI survey instrument during the interview. *Trace files* capture only those functions that are executed by the system. Both files contain a large volume of unstructured data that must be analyzed using statistical methods to identify patterns in the trace files that reveal performance problems. However, by tabulating various codes and keystrokes by interviewer, interview, or questions, interviewers and questionnaires can be evaluated.

Couper et al. (1997) present a preliminary evaluation of keystroke file analysis as a tool for assessing interviewer performance using CAPI. Their work highlights the use of this tool as a means of identifying problems in the human–machine interface, or what has become known as the *usability* of CAI questionnaires. Keystroke data from the U.S. Study on Asset and Health Dynamics of the Oldest Old (AHEAD), a national study of adults aged 70 and older, were analyzed in two ways. One involved the detailed review and coding of each keystroke file in a fashion similar to behavior coding. The other approach simply aggregated selected keystroke behaviors across the interviewers and interviews.

Their analysis detected a number of inefficiencies and errors that interviewers were making that would be difficult to identify without keystroke analysis. As an example, they found that 86% of interviewers used an erroneous function key at least once and 58% of these were due to interviewers hitting by mistake a key adjacent to the desired function key. This result led to improvements in the selection of keys to perform various common functions that interviewers perform. It is not yet known whether using keystroke files, time stamps, and detailed call record data can be used to improve interviewer performance, but they can function as process data for improving the interview process and the questionnaire development process.

CHAPTER 6

Data Collection Modes and Associated Errors

The *mode of data collection* refers to the medium that is used in a survey for contacting sample members and obtaining their responses to survey questions. Today, a number of modes are in use that can be classified in terms of three dimensions: degree of contact with the respondent, degree of data collector–interviewer involvement, and degree of computer assistance. A presentation of the various modes of data collection using this classification scheme is shown in Table 6.1. Note the number of computer-assisted interviewing (CAI) modes in the table. Use of the computer for collecting survey data continues to increase as new technologies are developed for communicating and interfacing with respondents in their homes, at work, and during travel (Nicholls et al., 1997). Some of these are discussed in detail in this section.

There are essentially three principal modes of data collection: face-to-face (or personal visit) surveys, where the interviewer and respondent are physically present during the interview; telephone surveys, where the interview is conducted by an interviewer over the telephone; and mail surveys, where the questionnaire is mailed to the sample members and returned by mail. Face-to-face and telephone surveys are referred to as interviewer-administered modes, whereas mail surveys are self-administered. Other modes of data collection we discuss include the use of administrative registers and direct observation.

In this chapter these modes are described in more detail and their main features are examined. There is a choice when it comes to data collection mode. With each mode there are relative advantages and disadvantages. Mode selection is often complex and related to the goals of the survey, mode characteristics, various design issues, and the methodological and financial resources available. In this chapter we present these advantages and disadvantages for a number of modes and describe how the choice of mode affects data quality.

188

Table 6.1 Data Collection Modes as a Function of Data Collector Involvement, Respondent Contact, and Degree of Computer Assistance[a]

	High Data Collector Involvement		Low Data Collector Involvement	
	Paper	Computer	Paper	Computer
Direct contact with respondent	Face-to-face (PAPI)	CAPI	Diary	CASI, ACASI
Indirect contact with respondent	Telephone (PAPI)	CATI	Mail, fax, e-mail	TDE, e-mail, Web, DBM, EMS, VRE
No contact with respondent	Direct observation	CADE	Administrative records	EDI

[a] ACASI, audio CASI; CADE, computer-assisted data entry; CAPI, computer-assisted personal interviewing; CASI, computer-assisted self-interviewing; CATI, computer-assisted telephone interviewing; DBM, disk by mail; EDI, electronic data interchange; EMS, electronic mail survey; PAPI, paper-and-pencil interviewing; T-ACASI, telephone ACASI; TDE, touch-tone data entry; VRE, voice recognition entry.

6.1 MODES OF DATA COLLECTION

In this section we review a number of different data collection modes listed in Table 6.1. Each mode is described and comments are made regarding what is generally known about its characteristics.

6.1.1 Face-to-Face Interviewing

Face-to-face interviewing is the oldest mode of interview since it does not rely on modern communication technologies. Because it provides for the maximum degree of communication and interaction between the interviewer and respondent, face-to-face interviewing is often associated with good data quality and is viewed by many survey researchers as the preferred mode of data collection for most survey topics. This view has been challenged in recent decades mostly because of the measurement errors associated with sensitive topics. Indeed, the list of topics that can be adversely influenced by the interviewer's presence is increasing as further research on interviewer effects is conducted. Let us consider some of the advantages and disadvantages of the mode.

Advantages and Disadvantages
Face-to-face interviewing is usually associated with high budget/cost surveys since it requires the interviewer to visit or meet with the respondent in a home, a workplace, or a public place. Travel is usually a high-cost component of this

mode. In some cases interviewers may have to travel long distances, completing only one interview in a day's time. In general, face-to-face interview is usually more costly than other modes of data collection.

Face-to-face interviews are known to generate social desirability bias for some types of questions. As discussed in Chapter 4, social desirability bias is a phenomenon that may occur with sensitive questions. For some topics, there may be a tendency in face-to-face surveys for the respondents to be more concerned about how they are viewed by the interviewer than in providing accurate answers. This phenomenon manifests itself in various ways. Survey topics involving socially stigmatized behaviors such as consumption of alcohol or drug use or the number of sexual partners are typically underreported in face-to-face interviews. For attitude questions, the respondent might choose not to disclose extreme attitudes regarding political preferences or racial issues in the interview. There is also evidence in the literature that respondents tend to want to be perceived as being knowledgeable and up-to-date regarding current events in society when in fact that is not the case.

However, social desirability bias is not the only source of measurement error of concern in interviewer-assisted surveys. As noted in Chapter 5, interviewers themselves are an important error source. Interviewers have a tendency to affect respondents in ways that differ among interviewers. Each interviewer has his or her own set of behaviors, work procedures, question-delivery and question-wording techniques, probing techniques, strategies about when to probe, pace, and methods for recording responses to open-ended questions. Because of these differing interviewing styles, the interviewers as a collective contribute correlated interviewer error to the total survey error. As noted in Chapter 5, this error component can increase the standard errors of the estimates considerably and is not normally reflected in the variance estimates calculated in sample surveys.

One approach for minimizing interviewer effects is that of standardized interviewing procedures. The idea is to train interviewers so that they all perform in nearly the same way and therefore eliminate the potential for interviewer variance. However, as mentioned in Chapter 5, there is a controversy regarding the standardized versus conversational and other styles of interviewing. These issues suggest that the use of interviewers in surveys and methods for reducing their effects are still being researched actively in the field of survey methodology.

Despite these concerns about interviewer variance and social desirability bias, face-to-face interviewing is still considered the most flexible mode available, for a number of reasons. The mode allows for long, complex interviews that may last for an hour or more. Since the interviewer is in the presence of the respondent, he or she can use many tactics to gain cooperation, can make direct observations during the interview that can be recorded for later analysis, can ensure that the respondent's answers are not affected by the presence of others, and can provide probes for more complete and accurate answers when necessary.

As mentioned, face-to-face interviewers can administer visual aids and response cards for sensitive questions and can build rapport and confidence in their interaction with the respondent. Perhaps the most valuable feature is the fact that response rates can usually be kept at relatively high levels. The potential for face-to-face interviewers to gain cooperation, both initially and, if necessary, through refusal conversion is unmatched by modes with no face-to-face contact.

Another advantage of face-to-face interviewing is its high coverage of the general population for sample designs, such as area probability sampling. Area sampling entails sampling from an area frame such as a map or a photograph, and face-to-face interviewing is usually suitable for collecting data on those units that are sampled in the ultimate step (see Chapters 3 and 9). Despite the possibility for interviewers to control the presence of others during the interview, contamination by others might be a problem, both in the respondent's home and in other places, such as at work or in a public place. The presence of others can impede the respondent's ability to be completely open and honest in answering certain types of questions.

Perhaps the most serious practical problem with face-to-face interviews is the cost associated with its use. Since interviewers must travel to the units, workloads are much smaller than in telephone interviews. Thus, it usually requires more time and personnel resources per completed case than do, say, telephone surveys. In addition, the administration of training activities and observations of interviewer performance is time-consuming and costly.

Finally, an important concern associated with face-to-face interviewing is that the interviewer might falsify the interview. This behavior is referred to in the field as fabrication, cheating, curbstoning or table-topping (see Chapter 5). This behavior is known to all survey organizations and can be very difficult to detect. The interviewer may falsify an entire interview or simply skip questions during an interview to save time and respondent burden and then later fill in the skipped questions by making up data. A study of this phenomenon is found in Biemer and Stokes (1989). The behavior can be discovered by means of reinterviews or verification interviews and analyses of response distributions (see Chapter 8). One reason for interviewers to falsify interviews might be the difficulty in accessing certain population segments: for instance, high-crime areas or in apartment buildings protected by security staff.

Computer-Assisted Personal Interviewing
With the advent of systems for computer-assisted interviewing (CAI), the computerized variant of face-to-face interviewing termed CAPI (computer-assisted personal interviewing) is being used more and more. The interview is conducted by means of a program stored in a laptop computer. The program is such that survey questions appear on the screen in the order prescribed for each respondent. The interviewer asks the questions and enters the answers provided by the respondent.

Computer-assisted interviewing was first used in centralized telephone interview settings, referred to as computer-assisted telephone interviewing (CATI); see Section 6.1.2. In centralized CATI facilities, interviewers are housed together in large rooms, referred to as *telephone facilities* or *calling centers*, with up to 100 interviewing stations or more. Cases are typically assigned to interviewers from a centralized database on demand. When the interviewer completes a call, the call outcome is entered into the database. If the complete interview is not conducted, the case is made available for assignment to another interviewer to complete at another time. Usually, some fraction (e.g., 5 to 10%) of all interviews are monitored by a supervisor for quality control purposes.

In theory it would seem that using a computer program to control the interview would eliminate some errors. For example, if the questionnaire contains skips, the computer program can choose the correct route based on previous answers. Also, if the questionnaire for a household survey contains questions concerning more than one household member, the name of the person being asked about in the question can be filled in by the computer so that this information automatically becomes part of the question. Thus, the computer software automatically "personalizes" the questionnaire using "fills," which replace the standard generic language common in paper questionnaires with wording adapted for the particular respondent being interviewed. Questions concerning all vehicles owned by a household, events that are being investigated in some detail, and questions concerning all workplaces of a business can also be handled this way, thereby avoiding confusion on the part of both interviewers and respondents.

Certain data processing activities can also take place during the interview, such as online editing and coding (see Chapter 7). As discussed by Groves and Tortora (1998), these theoretical and logical advantages associated with CAI are not always supported by data from methodological studies in the sense that differences are always statistically significant. Nevertheless the studies usually show clear reductions in indicators of measurement error.

Effects on Data Quality

The literature on CAPI's effects on data quality is very sparse. There are some studies comparing face-to-face paper-and-pencil interviewing (PAPI) with CAPI regarding such process variables as skip error rates, item nonresponse, out-of-range responses, and interviewer variance, but the differences reported are not substantial. Instead, the evidence suggests that survey organizations and survey sponsors opt for CAPI for reasons other than data quality. Savings in costs and time are important, and such gains have been accomplished by some organizations. Other organizations, however, report that CAPI actually costs more than PAPI, but comparing the two modes on the basis of costs alone may not be relevant since CAPI is much more feature-laden than is PAPI. For example, CAPI provides greater opportunities to conduct interviews of much greater complexity than PAPI, to use dependent interviewing (where responses obtained on earlier data collections can be part of the question wording), and to use advanced interviewer–respondent interaction in, for instance, online editing. Therefore, even if CAPI costs somewhat more than PAPI, the value added to the interview as a result of these features may be well worth the increased costs.

Converting from a PAPI to a CAPI Environment

For survey organizations, there are costs associated with the transition from a noncomputerized environment to a computerized one. These are startup costs for hardware, programming, and interviewer training. In some cases, CAPI programming errors are discovered after the interviewing has begun. It seems that many organizations have continuing problems with debugging CAPI applications and may occasionally experience a loss of cases due to errors.

Thus, there are a number of reasons to make the transfer from PAPI to CAPI, but CAPI cannot be used for all surveys and survey organizations. Sometimes the startup cost can be high even for a large organization, and moving from one CAPI system to another can be extremely burdensome. For organizations that are small, even commercially available CAPI systems might not yet be economically feasible.

For organizations that already have a CAPI system in place, it might not always be worthwhile to use the system for one-time surveys. Neither may it be feasible if the survey has to be carried out very quickly. There might not be enough time to do the technical work and train the interviewers. So most large organizations must be prepared to conduct both PAPI and CAPI, depending on the circumstances.

Face-to-face interviewing is expensive but highly flexible, provides high response rates, but may generate social desirability bias and interviewer variance.

6.1.2 Telephone Interviewing

Telephone interviewing has not always been accepted as a data collection methodology for social and economic research. In the 1936 U.S. presidential election, the results from a telephone survey based on telephone directory listings predicted a landslide victory for Landon over Roosevelt, which was published widely. When Roosevelt won, the telephone methodology was faulted for the erroneous prediction (Katz and Cantril, 1937). Actually, at that time, only 35% of the households in the United States had telephones, and the telephone population was disproportionately Republican, which explains the preference of telephone households for Landon.

Unfortunately, it took 40 years for telephone surveys to begin to receive acceptance among U.S. social scientists as a legitimate mode of interview. Two reasons for the increased interest in the mode are the lower cost of data collection than for face-to-face surveys and the increased coverage of the population by telephone both in the United States and worldwide. For example, Groves and Kahn (1979) provide a detailed examination of the cost and error components for an RDD and area frame face-to-face survey. They show that RDD telephone surveys can provide data quality which is comparable to face-to-face, but at a much lower cost. However, today, use of telephone surveys is again being challenged as a sole mode of data collection. The primary reason for this is the trend toward lower and lower response rates in surveys. However, the telephone is still used for many surveys, particularly in combination with other modes, called *mixed-mode surveys*.

Below, we discuss some of the advantages and disadvantages of this mode of data collection as a stand-alone mode as well as in combination with other modes of data collection.

Advantages and Disadvantages

Since they both use interviewers for communicating with the respondent, it is not surprising to find that the characteristics of face-to-face and telephone interviewing are very similar. Both modes have the potential to create interviewer variance and social desirability bias in the survey results. However, the literature suggests that these effects are somewhat less in the telephone mode (see Chapters 4 and 5).

Telephone interviewing is somewhat less flexible than face-to-face interviewing in that many things that are possible in face-to-face interviewing simply cannot be done over the telephone. For instance, visual aids cannot be used unless they are mailed to the respondent before the interview, which is a very impractical procedure. Also, questions delivered over the telephone cannot be too complicated and should not contain more than six but preferably fewer response alternatives.

Complexities such as asking the respondent to go through records and make calculations are extremely difficult to perform in a telephone interview setting. The social desirability bias might be smaller than in face-to-face interviewing because of the anonymity of the interviewer, but this very anonymity is not advantageous for developing rapport and persuading the respondent to participate or perform his or her best during the interview. As a result, there is a tendency for more acquiescence and extreme responses (choosing the endpoints of scales or choosing the first or last of response categories provided) over the telephone and less considered responses in general.

For example, the number of "don't know" and "no answer" responses tend to be larger over the telephone than in face-to-face surveys, and the faster pace leads to shorter replies. Also, interviewer variance might be larger than in face-to-face interviewing, due to the fact that telephone interviewers are usually assigned larger workloads than what is feasible in face-to-face interviewing. Typically, response rates are lower in telephone surveys than in face-to-face surveys of comparable type and size.

Telephone surveys tend to be considerably shorter than face-to-face interviews. Although there are examples of successful telephone surveys having average interview durations of one hour, these are quite rare for surveys of the general population. Most telephone interview designers try to limit the interview time to 30 minutes or less. Telephone surveys can be set up more quickly and are less expensive than face-to-face surveys.

Centralized telephone interviewing offers special advantages, since general administration, supervision, feedback, and training are more easily accomplished in such a setting. In some countries use of the telephone is not possible as the main data collection mode because of low rates of telephone coverage (referred to as *penetration rates*). The major problems associated with the telephone mode are that not all sample members have telephones, that answering machines and caller-ID equipment can prevent access to the sample unit, and that the questionnaire cannot be too long or utilize features such as visual aids and many response alternatives.

Computer-Assisted Telephone Interviewing

Today, telephone interviews are often conducted using CAI technology. In fact, CATI was the first computer-assisted data collection methodology to enter the scene about three decades ago and has now become standard practice in many organizations. It no longer belongs in the category of new technology. The benefits and drawbacks of CATI are very similar to those of CAPI, discussed in Section 6.1.1. There is also generally wide-spread agreement that both CATI and CAPI do not add appreciably to measurement errors simply by virtue of the computerization of the instrument (Couper et al., 1998). The key advantage of CATI seems to be the ability to avoid erroneous skips that plague complex paper questionnaires, and thus item nonresponse rates for CATI are generally lower than for PAPI.

TELEPHONE INTERVIEWING COMPARED TO FACE-TO-FACE INTERVIEWING

Advantages	Disadvantages
• Less costly	• Less flexible than face-to-face surveys
• Interviewer variance is usually smaller than for face-to-face surveys	• No ability to use a visual medium of communication
• Social desirability bias usually smaller than in face-to-face surveys	• Interviews must be shorter than in face-to-face surveys
	• No coverage of nontelephone units

6.1.3   Mail Surveys

In a mail survey a questionnaire is mailed to each sample member who is asked to complete the questionnaire and then send it back to the survey organization. Since no interaction with a data collector is involved, the mode is referred to as *self-administered*. As such, it is imperative that the questionnaire be self-explanatory to the respondent (i.e., questions and instructions must be easily understood by the respondent since there is no help available from the outside). To a much greater extent than for interviewer-assisted modes, the quality of the resulting data hinges on the quality of the questionnaire design.

In addition, as discussed in detail in Chapter 3, the response rate to mail surveys can vary tremendously by survey organization, owing, to a great extent, to the skill and knowledge of the survey staff. The care that is needed to implement mail surveys successfully is quite evident from the discussion in that chapter. In this section we describe some of the advantages and disadvantages of mail surveys compared to other modes of data collection.

Advantages and Disadvantages

During the last 20 years the mail survey has gone through a revival. As the cost of the interviewer-assisted modes has increased considerably, mail surveys have become an attractive option. Mail surveys are quite inexpensive to implement, which makes them the preferred mode for low-budget surveys. In addition, mail surveys may have some quality advantages over interviewer-assisted modes, due to the reduced risk of social desirability bias associated with self-administration. Since the visual communication medium is available for this mode, it has advantages over telephone surveys for situations where providing the respondent with a visual aid is necessary: for example, the use of maps to indicate various sites visited on a recent trip to a national park or for questions that contain more than, say, six response alternatives.

Since the respondent has possession of the questionnaire, mail surveys also provide time for the respondent to give thoughtful answers and look up records. Finally, mail surveys do not generate the kind of variance contribution from interviewers, and errors generated by question order or response alternative order are reduced (Dillman, 2000).

However, mail surveys are known to generate lower response rates than do interviewer-assisted modes. Therefore, it is usually combined with follow-ups conducted by interviewers (CATI or CAPI mixed-mode surveys). A few decades ago the mail survey was considered inferior to interviews because of the low response rates and the fact that the survey organization has little or no control over the response process. For example, it is not possible to ensure that the intended sample person completes the questionnaire or that he or she collaborates with others as appropriate and as intended in the survey design. When the questionnaire is not returned, it is not known whether this is because the questionnaire never reached the sample person or if it did reach the sample person but he or she failed to return the questionnaire. In some cases, the sample person may not even be eligible to complete the questionnaire, yet no information is available for the survey staff to make that determination. This is problematic for both the computation of response rates and adjusting the estimates for nonresponse. There is also a greater risk of considerable item nonresponse rates associated with the mode.

For household surveys, mail surveys present problems when the sample person is to be selected at random from within the household. Although schemes such as the last birthday method (Chapter 9) have been tried, there is concern that these methods do not produce random samples and should not be used in rigorous scientific research. Therefore, mail surveys are not recommended in situations where the questionnaire cannot be mailed directly to a specific, named respondent.

Mail surveys also require a long field period to obtain acceptable response rates. Usually, eight weeks or more is required from mailing the first mailing to the final return. Also, the questionnaire has to be simple in the sense that the respondent understands the concepts and can navigate through the questionnaire without problems. Of course, respondents must be on a literacy level that allows them to read the questions and the instructions. In the United States a general rule is used that the reading level of the questionnaire should be at a fifth-grade level for mail questionnaires used in government surveys of the general population.

Recent Research

A considerable part of the revival in the use of mail surveys can be attributed to the work of Don Dillman at Washington State University in Pullman, Washington, U.S.A. (see, e.g., Dillman, 1978, 2000). Dillman has developed efficient strategies for mail survey data collection, consisting of step-by-step operations that, if followed, will result in response rates that are larger than for other documented strategies. Dillman's strategies use social exchange theory as a basis and were described in Chapter 3. He has also started to develop a theory for self-administered modes based on theories of graphic language, cognition, and visual perception combined with the theories used by Cialdini (1984) and Groves and Couper (1998) to increase cooperation in interview surveys. Some of these social and psychological influences (reciprocation, commitment and consistency, social proof, authority, scarcity, and liking), discussed in Chapter 3, can also be used in the design and conduct of mail surveys.

Thus, there are two sets of principles at play. One concerns development of the questionnaire in such a way that it becomes self-contained, so that the organization of the information in the questionnaire and the navigational guide offered to the respondent are such that completion of the questionnaire is ensured. Viewed from this perspective, it appears that these features aid the respondent in much the same way as interviewers guide respondents in interviewer-assisted modes.

The second set of principles involves implementation of the data collection in a prescribed, step-by-step fashion. This involves the use of advance letters, multiple follow-ups, real stamps on the return envelopes, personalized mailings, progressively urgent appeals to the respondent, and other strategies for increasing response rates. These principles are described in Chapter 3 as well as in much more detail in Jenkins and Dillman (1997), Dillman (2000), and Redline and Dillman (2002). The principles have been and will continue to be revised as new information is gained through experimental studies.

A *mail survey* does not generate interviewer variance and is very suitable for collecting some types of sensitive information. Although inexpensive, a mail survey takes time to carry out, often produces relatively low response rates, and the risk for item nonresponse is considerable.

6.1.4   Diary Surveys

Diary surveys are used for the purpose of collecting information on events retrospectively. The structured questionnaire is replaced by a diary where respondents enter information about frequent events, such as household purchases, food intake, daily travel, television viewing, and use of time. To avoid recall errors, respondents are asked to record information soon after the events have occurred. Often, this means that respondents should record information on a daily basis or even more frequently. Thus, successful completion of the diary requires that respondents take a very active role in recording information.

Diary surveys differ from mail surveys because interviewers are usually required to contact a respondent at least twice. At the first contact, the interviewer delivers the diary, gains the respondent's cooperation, and explains the data recording procedures. At the second contact, the interviewer collects the diary, and if it is not completed, assists the respondent in its completion. Because of the need for a high level of commitment from the respondent to maintain the diary accurately, diary reporting periods are fairly short, typically varying in length from one day to two weeks.

For example, when the purpose of the survey is to collect data on household expenditures, the respondent is asked to record the specific purchase as soon as possible after its completion, together with the price of the merchandise. Such detailed information cannot be collected via interviews because the respondent will not remember some purchases, particularly small ones, for more than a few days. The interviewer is a critical component of the method since for some types of purchases, interviewers can assist the respondent in providing more accurate information than the respondent alone can be expected to provide.

Particularly for large purchases, such as cars or boats, the risk of *telescoping* effects is high (see Chapter 4). Such purchases are salient events to many respondents, and there is evidence that such events have a tendency to be telescoped forward in the reporting period. Some respondents may disregard the payment of bills as purchases, and consequently, in some expenditure surveys, credit card payments are considered out of scope. Instead, the respondent is asked to record each single purchase at the time it is made.

Despite efforts to stimulate cooperation by interviewer contacts, incentives, reminders, and so on, there is a clear tendency for respondents gradually to lose interest in the recording task, as evidenced by fewer purchases reported at the end of the collection period than at the beginning (Silberstein and Scott, 1991).

There is also a clear risk of conditioning on the part of respondents. As a result of participation in the survey, the respondent might change his or her purchasing behavior. For instance, when the respondent notices the amounts spent on specific goods, he or she might change the purchase pattern temporarily. Soon after the data collection, the purchase pattern is usually back to normal, resulting in biased estimates.

For many items, recall errors are such that not all purchases are recorded, resulting in underestimates of purchases. It is very difficult for respondents to think constantly about keeping receipts from eating out or buying gas. Thus underreporting is a big problem in these areas. These recall errors increase with the length of the data collection period.

In some designs proxy reporting is a problem. If there is one diary respondent per household, underreporting will occur. Personal diaries, of course, eliminate the effects of proxy reporting. Social desirability bias is smaller in diary surveys than in interviews but can still be a problem, especially if there is just one diary per household. Typically, purchases of alcohol and tobacco are underreported in expenditure surveys. Also, as in all paper-and-pencil self-administered modes, the literacy and education levels are important prerequisites to obtaining good-quality data.

There is evidence that the design of the diary can be an important error source. There are essentially two designs. One consists of more-or-less blank pages, and the other is more structured, with preprinted headings or sections. Comparisons between the two have provided inconclusive results (Tucker, 1992). Recently, two issues for the design of diary surveys have been discussed extensively in the literature. One issue concerns the optimal length of the recording period. Years ago, four-week diaries were used for purchases in household expenditure surveys. However, in recent years, this practice has been abandoned because of the extensive respondent burden involved and the fact that the completeness of responses declines during such a long period. As a result, shorter collection periods for diary surveys are common today.

Another issue concerns the types of events that can be studied using diaries. A recent trend is to limit the scope of the survey to avoid the heavy respondent burden involved. There are examples of expenditure surveys that collect data on food items only. However, this trend is not present in all countries. For instance, one design used in Holland is such that it consists of two diary components. The first is a two-week complete diary and the second is a one-year recordkeeping of expenses above a certain amount.

A specific feature associated with diary surveys is that the placement of diaries follows a prescribed pattern. Typically, the sample is dispersed in time throughout a whole year to accommodate for the collection of all kinds of purchases or all kinds of activities. If surveys like these were restricted to parts of the year, the occurrence of some events would be heavily underestimated, due to seasonal patterns. Therefore, data collection is conducted by dividing the sample into subsamples that start their recording at prespecified dates. It is important that there is an even distribution of the sample, even by a prespecified day of the week. Practical reasons usually prevent exact adherence to the placement plan. Most organizations therefore allow minor deviations from the plan. One must bear in mind, however, that letting a sample member start the recording at a date other than the one prescribed by the design will result in measurement errors due to "nonresponse in time" even though actual nonresponse is avoided by such a decision (Lyberg, 1991).

6.1.5   Other Self-Administered Modes

There are a number of computerized versions of self-administered modes that have helped enhance the usefulness of self-administration (see Ramos et al., 1998). *Electronic data interchange* (EDI) is a technique to obtain economic and other data directly from sampled businesses' own computer-stored records. Although still in its infancy, the long-term goal is for the survey organizations to collect data from these records, thereby reducing the need for traditional surveys in some fields. Obviously, there are a number of problems to solve before widespread use of EDI can be anticipated. There is a need for convincing businesses to use standard messaging formats that can be used to transfer statistical data to the collecting agencies. Experience shows that standardization can be difficult to accomplish since not only businesses are involved but also developers of accounting software. But the very idea of linking a business's database to the collecting organization is an appealing one and shows great promise since businesses always are interested in less paperwork (Keller, 1994).

In *disk by mail* (DBM), a diskette is sent to the respondent via postal mail. The diskette contains a self-administered questionnaire and respondents install and fill out the questionnaire on their own computers. The responses are saved on disk and mailed back to the survey organization. DBM works almost like computerized interviewing since the DBM program controls the question flow, provides instructions, and performs edit checks. Clearly, the use of DBM has the potential of saving costs and reducing errors, and if viewed as an alternative to the interviewer-assisted mode, no interviewer variance is involved.

Electronic mail survey (EMS) is a mode where computerized self-administered questionnaires are transmitted to respondents and returned via electronic mail or delivered via the Internet or the World Wide Web (WWW). DBM and EMS have many elements in common. The basic difference is that

Web surveys have current or potential features that DBM and electronic mail do not have. With the graphic and multimedia capabilities of the WWW, the design choices are almost unlimited, and it is important for survey mode researchers to intensify their work on developing guidelines for Web survey design (see Couper et al., 2001). There are currently two lines of development. One is led by large survey organizations and is characterized by a careful step-by-step development that appreciates the fact that not everybody has access to or wants to use the computer to answer survey questions. Therefore, DBM and EMS are usually offered as alternatives to other collection modes by large survey organizations. Another line of development is led by organizations that might not even have survey work as a business. Examples of the latter are newspapers and TV networks conducting investigations called, for example, "Today's Web question," resulting in an abundance of "surveys" based on self-selection.

DBM and EMS have a number of limitations, which have been summarized by Ramos et al. (1998):

1. There is no readily available software for electronic questionnaires that suit survey organizations using capabilities found in most CAPI and CATI programs, such as skips, edit checks, and randomization of response alternatives. The software has to be developed in-house.
2. The questionnaire design capabilities are limited, which restricts the scope of applications.
3. E-mail is based on nonstandardized software, with no standardization in sight.
4. Respondent access to modems and communication software is limited.
5. E-mail addresses often change.
6. There are many issues associated with confidentiality and security. Confidentiality must be assured and guarantees must be provided that installment of the communication package will not lead to virus attacks.
7. Coverage error is a concern, due to limited respondent access to properly equipped PCs and limited modem and Internet access.

Despite the limitations, there are reasons to expect measurement error to be reduced by DBM and EMS compared to traditional mail surveys. These reasons are related to those behind CATI and CAPI improving traditional interviewing. DBM and EMS provide a more controlled collection environment. Controlled routing and embedded edits decrease measurement error and item nonresponse. There is also evidence that answers seem more thoughtful in DBM and EMS applications.

Touch-tone data entry (TDE) is a collection mode in which the respondent calls a computer and responds to questions asked by the computer. The answering process is such that the respondent enters data by using the keypad of a touch-tone telephone. After entering the response, the computer asks the respondent to verify the answers provided.

TDE is a very common communication technique in today's society. It is used for ordering, transactions, and routing of telephone calls. At this time, TDE has only been used in surveys that are quite short (say, 5 to 10 minutes) and data entry is simple. It is particularly suited for situations where the information requested is either numeric or can be transformed to a numeric code. TDE is an alternative to a mail questionnaire or an interview and the development has been triggered by concerns associated with costs. Also, TDE allows respondents to choose the time for interview since the service is open 24 hours a day, seven days a week.

Another advantage is that the collection time per respondent is shortened in many TDE applications compared to manual modes. For the collecting agency, it is advantageous to eliminate interviewer involvement, not only because of costs but also for practical reasons associated with the 24-hour access for respondents. The obvious disadvantage is the limited scope of application. Since TDE works best for data collection situations involving short, numeric, and repetitive entries, there may be few surveys where it can be used successfully.

Voice recognition entry (VRE) is a variant of TDE. Instead of using a keypad to enter a response, the respondent speaks the answers into the telephone and the computer verifies it by repeating the response. This technology has become fairly common for very simple exchanges of information in a number of situations in society. Some examples of VRE applications include banking by phone, automated order entry, automated call routing, and voice mail. VRE is more difficult to apply since spoken answers are more easily misinterpreted or not recognized by the computer. On the other hand, the technology works for any kind of telephone, and respondents seem to think it is simpler than TDE (see Clayton and Werking, 1998).

As summarized by Dillman (2000), TDE and VRE are limited technologies. The respondents lose the overview they have in mail questionnaires, and they do not get the help that is provided by interviewers in interview surveys. The design of TDE and VRE is complicated since all situations must be anticipated in the script, and appropriate messages must be formulated for these situations. On the positive side, all stimuli are delivered in a completely uniform way, which is not possible with "live" interviews.

Within the last decade, a new mode of data collection and its variants has gained popularity, particularly in surveys dealing with sensitive topics. This mode is *computer-assisted self-interviewing* (CASI). As discussed earlier in this chapter, interviewer-assisted modes are generally not recommended when the survey deals with sensitive topics and other topics prone to social desirability bias. CASI is a computer technology that combines the benefits of face-to-face interviewing, self-administration, and computer assistance. With CASI, part of the interview (e.g., questions that are not sensitive and could benefit from interviewer assistance) can still be conducted by the interviewer. However, when sensitive topics arise, the computer can literally be turned around to take over the interview and guide the respondent through the sensitive questions.

Recall that one early low-tech solution to this problem described in Chapters 3 and 4 is the *randomized response* (RR) *technique* (Warner, 1965). With the advent of computer-assisted collection methods, the RR method has been replaced by techniques in which a laptop computer is essentially turned over to respondents for part or all of an interview. In CASI (sometimes referred to as *text-only CASI*), respondents view the questions on a computer screen and enter their answers using a computer keypad. This form of CASI requires respondents to have a certain literacy level to be able to read the questions. This requirement is eliminated with *audio CASI* (or ACASI). Using ACASI, respondents can both listen to questions using headphones and read the questions on the screen and then enter their answers by pressing labeled keys. Both (text-only) CASI and ACASI require that the interviewer deliver the instrument on the CAPI laptop computer and then wait until the respondent has finished the task of answering the questions to retrieve the computer. A natural and cost-effective alternative is to implement computerized self-interviewing in a telephone setting. *Telephone audio CASI* (T-ACASI) is almost identical to TDE and can be either through respondent call-in or through interviewer call-out. The primary difference is that with T-ACASI, an interviewer initiates the call and then turns the interview over to the computer and the respondent for the remainder of the interview. TDE usually does not involve an interviewer. Calls are either made by the respondent to a computer or by the computer to a respondent.

The CASI variants originated from the need for good measurements of sensitive attributes. Viewed from that perspective, they seem to be successful since they generate increases in reporting (see Tourangeau and Smith, 1998; Turner et al., 1998). The CASI development differs fundamentally from CATI and CAPI. The latter technologies changed the interviewer role so that it became more standardized. CASI changes the interviewer–respondent interaction completely and the interviewer assumes a more administrative role, while the collection format is still an "interview" with all advantages associated with that mode compared to the mail format. As discussed by Turner et al. (1998) and Tourangeau and Smith (1998), audio-CASI seems to reduce underreporting of certain sexual behaviors as measured not only by face-to-face interviews but also by self-administered paper questionnaires.

6.1.6 Administrative Records

For some data collection objectives, the use of administrative records is a cost-effective and high-quality alternative to surveys. *Administrative records* are records that contain information used in making decisions or determinations or for taking actions affecting individual subjects of the records. Most nations keep records to be able to take actions regarding licensing, taxing, regulating, paying, and so on. It is, of course, very tempting to use such data for statistical purposes, and there are indeed a number of such register-based collections in countries with a tradition of record-keeping. In some countries the use of administrative records for statistical purposes has a long history. As mentioned in Chapter 1, early censuses had administrative purposes, and some vital record systems in Europe are more than 300 years old. Recent advances in computer technology and record linkage methodology have opened up new possibilities to use records not only for primary collection purposes but also as auxiliary information in some estimation processes.

Thus, the situation usually is as follows: In some countries authorities and organizations build administrative records about a group of persons, businesses, or other entities. The purpose is to take action, if necessary, regarding individual entities. In a taxation register, all individuals and businesses are recorded and each record contains information regarding a number of variables, such as occupation, income, age, type of business, number of employees, address, number of work sites, total revenue, and so on. The organization in charge of the register needs information about the subjects so that the correct action can be taken regarding each subject, such as calculating the correct amount of tax for each person or business. Obviously, with a record like this it is possible to move from an individual-related action situation to one where one is interested in characteristics and attributes of groups of individual entities (i.e., statistics based on register or record data). For instance, it would be possible to provide tables on occupation distribution or via record linkage match survey data with occupation register data.

In principle, there is no difference between error structures found in administrative data compared to data collected via other modes. There is a tendency toward an unjustified reliance on the quality of administrative data among producers of statistics. In some applications concerning the evaluation of survey data, administrative record data are used as the gold standard. Sometimes this is justified since the administrative data have been collected under completely different circumstances than the corresponding survey data. But on other occasions, the data quality of the administrative record might be far from a gold standard.

The major concerns with data quality in administrative records include the following:

- In most cases there has been no influence on administrative data generation, updating, and coding on the part of the survey researcher interested in using the administrative data. For example, occupation descriptions provided by taxpayers are usually taken for granted by the record keeper since that variable is of marginal interest to the record keeper or is used just to screen certain segments of the population for greater scrutiny without the ambition to cover the segment entirely.

- Those in charge of administrative records usually do not have a statistical background, resulting in processes that are, at least for some variables, imperfectly collected, recorded, and controlled.

- Usually, there are conceptual differences between the statistical application and the administrative use.
- The quality of the administrative data is seldom assessed.

The major advantages associated with the use of administrative data are:

- The records and the data already exist and are inexpensive to use.
- In many applications there is no need for contacts with register units and therefore there is no respondent burden.

The use of administrative data is discussed further in Zanutto and Zaslavsky (2002) and Hidiroglou et al. (1995b).

6.1.6 Direct Observation

Direct observation as a mode of data collection is the recording of measurements and events using an observer's senses (vision, hearing, taste) or physical measurement devices. In this mode the observer assumes the role of the respondent. Direct observation is not uncommon in survey research. Applications include estimates of crop yield, counting number of drivers not using their seat belts, measuring TV viewing by electronic devices, measuring pollution via mechanical instruments, and assessing land use by interpreting aerial photos. Direct observation also has a long tradition in anthropology and psychology. Some of that tradition has been transferred to recent cognitive survey research where respondent and interviewer behavior is studied via observations.

The error structures in direct observation are of two kinds. When observers use their senses, the error pattern that emerges is very similar to that generated by interviewers. Observers may misperceive the information to be recorded, they exhibit variability within and between themselves, and their behavior may change over time. The same types of control programs used for interviewers can be used for observers. The second kind of error structure emerges when mechanical devices are used for measurement. Studies using these devices must have a program for validation and instrument calibration to ensure that systematic errors do not occur (see Fecso, 1991).

6.2 DECISION REGARDING MODE

In every survey, a decision must be made as to which mode of data collection should be used. A number of factors enter into this decision process, including the desired level of data quality, the survey budget, the content of the survey instrument (question type, number of response alternatives, number of questions, complexity, and the need for visual aids), the time available for data collection, the type of population that is being studied, and the information available on the sampling frame for this population. It is also necessary to view the data collection together with other steps in the survey process, such as access to the sample units for data collection and data processing alternatives available.

Often, however, the mode decision is clear because many alternatives are deemed unrealistic from a financial point of view or are otherwise not practical for the particular study. Examples of situations with only one mode alternative are the collection of data on expenditures of different kinds and the collection of data on crop yields. In the former case the literature uniformly suggests the use of the diary method because to keep track of purchases, it is necessary for the respondent to make notes of purchases in a continuing fashion to avoid major problems with recall effects. With the diary method it is possible for respondents to record purchases soon after they have been made. Any other known data collection method is unsuitable because respondents cannot possibly recall all purchases made during a specific time period unless they concern goods such as houses, cars, washing machines, and other investmentlike purchases. In the latter case with crop yields, the only possible collection methods are eye estimates made by observers or harvesting crops on sampled plots.

The mode decision is more complex when several alternatives exist. Some decision factors are relatively easy to assess. For instance, the cost can usually be predicted relatively easily, and so can the nonresponse rates given that a set of relevant design factors has been established, such as sampling method and the system for nonresponse follow-up. When measurement errors are crucial to the designer, the decision becomes more difficult.

There are two ways of deciding which data collection mode is best in terms of minimizing the measurement error (for reviews and additional references, see Lyberg and Kasprzyk, 1991; de Leeuw and Collins, 1997). One method involves the application of a set of general principles or "rules of thumb" associated with each mode. For example, if strict sample-size requirements are placed on the survey, it might be difficult to use a mail or a face-to-face survey, since they cannot be adjusted easily for over- or underestimates of response rates. When the survey budget is low, a mail survey might be the only option. Complicated questionnaires with branches and skips do not favor mail surveys. If social desirability bias is a concern, interviewer-assisted modes could be problematic. Questions with many response alternatives do not function well in an interview setting, but in face-to-face interviewing there is the ability to use flashcards (i.e., cards that show the response alternatives to the respondent, thereby creating a "self-administered situation"). If a high response rate is crucial, a face-to-face interview is to be preferred. If a high-degree of geographic dispersion is necessary (e.g., samples with very little geographic clustering of the sample units), face-to-face interviews are usually too expensive, due to travel costs. Interviewer control is easier to carry out in a centralized telephone interview setting than in a decentralized telephone interviewer or a face-to-face interviewer organization. The list of general rules of thumb goes on and on.

Typically, the decision is made on the basis of the priorities of the researcher as to how he or she values the various advantages and disadvantages of each option. Prior research on mode comparisons are usually quite helpful in making these choices. If there is very little information in the literature regarding which mode of data collection is preferred for the particular set of phenomena to be studied, it may be necessary to conduct a study to address that issue.

In designing a mode comparison study, there is a choice as to what type of mode effect should be measured: a pure mode effect or a mode system effect. A *pure mode effect* is essentially a measurement bias that is specifically attributable to the mode. To measure the pure mode effect accurately, one would have to compare the results of two collections where the only difference between the collections is the mode (i.e., the same questions are asked in the two collections, the same characteristics are measured, the same population is studied during a specific reference period, etc.). The resulting difference between the response distributions generated by the two collections would then constitute an estimate of the mode difference (Biemer, 1988; Groves, 1989).

However, in practice things are more complex. An estimate of the kind described must rely on a laboratory experiment or a field test whose realism is doubtful (see Chapter 10). Very rarely do we have a situation where it is possible, feasible, or desirable to keep all design factors intact and just vary the mode. This may be possible only in very simple and unrealistic applications, which may have little relationship with the complex realities of large-scale data collection. Further, the survey practitioner's main interest usually lies not just with the pure mode effect, but also with the combined effects of mode and other design factors. The following example illustrates the difficulty and undesirability of experimenting pure mode effects in practical situations.

Example 6.2.1　Suppose that we wish to compare telephone and face-to-face interviewing. Interviewer skills are probably different in a face-to-face interviewing organization than in a more-or-less closely monitored centralized telephone interviewing organization. The sample population for a telephone survey might differ from that for a face-to-face survey, simply because not all persons can be contacted via the telephone, resulting in coverage differences. A face-to-face questionnaire that uses visual aids shown by the interviewer cannot easily be transformed into one that can be used in a telephone interview. Thus, if a survey practitioner considers converting from, say, a face-to-face interview to a telephone interview, the mode is not the only thing that changes in the conversions. To conduct a fair test of telephone interviewing, design factors such as interviewers, questionnaire, the field period, performance monitoring and supervision, and so on, should be optimized for the telephone mode. This suggests that many other design factors other than the mode of interview should vary between the two modes.

One way out of this dilemma associated with mode comparisons is, instead, to measure *a mode system effect*: to focus on comparing whole systems of data collection rather than trying to isolate the effect of the mode. A mode system effect is easier and more realistic to estimate than a pure mode effect. By *system* we mean an entire data collection process designed around a specific mode. A system might include design factors such as interviewer hiring, interviewer training, interviewer supervision, questionnaire contents, number of callback attempts, refusal conversion strategies, sampling system, and sampling frame coverage. The difference between two systems can be measured along a number of dimensions, such as the response distribution, the unit nonresponse rate, the item nonresponse rate, the time needed to accomplish a given response rate, and the cost of the data collection. Some of these dimensions provide obvious comparative measures since a high response rate is better than a low response rate, a low cost is better than a high cost, and a short production time is better than a longer one. When it comes to the distribution of responses for certain items (i.e., the dimension that reflects measurement bias), we cannot assess a system difference unless we have some gold standard with which to compare the two outcomes. For some characteristics it is possible to declare one response distribution outcome as superior to the other. For example, sometimes we assume that more frequent reporting of certain behaviors or events is a sign of less recall error or less social desirability bias. In those cases "more is considered better," but one can never be completely sure unless an evaluation study is performed (Moore, 1988; Silberstein and Scott, 1991).

In general, data collection systems do not consist of one mode only, since mixed-mode surveys are the norm these days. For example, a common situation is to start with a relatively inexpensive mode, such as mail, and after a number of follow-up attempts, use telephone interviewing for the mail nonrespondents. In some surveys, face-to-face interviews have also been introduced to follow up mail nonrespondents that do not have telephones. There are also examples of surveys that begin with an interviewer-assisted mode and then a mail survey for follow-up for the mail nonrespondents. Some respondents prefer being sent a mail questionnaire rather than being interviewed (Carrol et al., 1986).

Even though the issue of a pure mode effect might seem of academic interest only, it quickly becomes important when we consider the situation above. Thus the fact that most surveys are forced to utilize more than one mode directly from the start makes the mode effect an important design consideration. The first step is to choose a *main* mode that can best accommodate the survey situation, where all important design factors are taken into account. The main mode is used to its maximum potential with certain constraints, such as a prespecified number of callbacks or reminders; then another mode is used for the principal purpose of increasing the response rate. This is a necessary and most practical approach but is a strategy of compromise. If a particular main mode is chosen because it is deemed appropriate for the survey measurement situation at hand, a switch to another mode should result in data of lesser quality. It is therefore important to make sure that the size of the effect of using complementary modes is as small as possible.

In some surveys the mode effects are small because the same questionnaire can be used across all modes. Most problems occur when mail is combined with an interviewer-assisted mode. Unless the mail questionnaire is so simple that it can readily be used in interviewing, we are in trouble. A mail questionnaire with many response alternatives is not suitable in a telephone interview. If there are, say, eight response alternatives for a specific question, it is very difficult for the respondent in a telephone follow-up to remember them all and distinguish between them. It is also difficult and sometimes awkward for the interviewer to read all response alternatives. In these cases there is a risk that the respondent uses a response strategy that favors the first few response alternatives or that the respondent interrupts the interviewer as soon as a reasonably fitting alternative has been read by the interviewer and that the interviewer skips reading the remaining alternatives to achieve a smooth interview.

If a mail questionnaire has been chosen because some questions are very sensitive, any conversion to an interview-assisted mode may be risky. The conversion could result in larger measurement errors, and that loss in response quality must be weighed against any gains that may result in response rates. When a specific mode has been chosen for its measurement characteristics, a switch to another mode to, say, increase response rates can be very difficult to address in the survey planning stage; often, ad hoc strategies with minimal control are adopted to handle this trade-off situation. It is easier to plan for questionnaire format issues. Questions in a mail questionnaire can sometimes be constructed so that they fit both interview and self-administered situations. Unfortunately, this often means that the measurement potential is lowered for all modes involved (see Chapter 4).

A conversion from interviewing by telephone to face-to-face interviewing is usually less problematic. A face-to-face interviewer usually has more options to convert refusals and more opportunities to guide and assist respondents than does a telephone interviewer. Also, a face-to-face interviewer can provide visual aids, something that is usually not possible with telephone interviewing.

We have discussed the situation with a mixed-mode strategy where one mode is combined with one or two other modes to achieve a higher response rate. There are also other instances where mixed-mode strategies are appropriate. One application occurs when one mode is used to screen a population for a subpopulation with some rare attribute that we are interested in, and a second mode is used to measure that attribute and related ones for the subpopulation. For example, a telephone survey may be used to identify businesses in the United States that export goods to Sweden. The resulting subpopulation is then surveyed by means of a mail questionnaire or a face-to-face interview.

A third application of a mixed-mode strategy is rather common in panel surveys, where the purpose of the strategy is to reduce costs. Panel studies are surveys that return to the same sample units at periodic intervals to collect new measurements to measure changes over time at the respondent level. Each new round of interviewing constitutes a "wave." In the Current Population Survey conducted in the United States, a mixture of face-to-face and telephone interviews is used (U.S. Department of Labor and U.S. Bureau of the Census, 2000). Mixed-mode strategies in panel surveys try to take advantage of the best features of each mode. Typically, face-to-face interviewing is used in the first wave of the panel survey, which improves on the otherwise incomplete coverage associated with nontelephone units while establishing a relationship with the respondent to maintain cooperation and response in future rounds of the panel (Kalton et al., 1989). Telephone interviewing (possibly decentralized, that is, telephone interviewing from interviewers' homes) is used in subsequent waves unless the respondent prefers the face-to-face mode or does not have access to a telephone.

During the last couple of decades, data collection methods in surveys have undergone a technological transfer. To a large extent, data are now collected by computer-assisted means. The telephone is still a dominating mode, but self-administered modes have seen a revival for several reasons. First, it has become apparent that self-administration is the best mode when it comes to collecting certain types of sensitive data, because social desirability bias is small. Second, it turns out that self-administered modes can utilize computer assistance, and third, compared to interviews, self-administration is inexpensive.

In many countries, however, the literacy level is so low that self-administered modes are impossible to use. Also, telephone coverage rates are still very low in many countries, including developing countries, where the survey researcher must rely on face-to-face interviewing when surveying human populations. In some environments there is only one option of data collection mode, face-to-face interview, no matter what the survey topic is. Thus the survey world looks very different depending on the country and the population to be surveyed. In Table 6.2 we have provided examples of factors that dictate the choice of main mode.

6.3   SOME EXAMPLES OF MODE EFFECTS

We have seen that a study aiming at estimating the pure mode effect is both difficult to design and not necessarily relevant. Instead, mode comparison studies are usually designed to compare whole systems of data collection (i.e., the outcomes of each system with all its components are assessed regarding a set of relevant evaluation criteria). Examples of such criteria are unit and item nonresponse rates, completeness of reporting, similarity of response

Table 6.2   Factors to Consider When Determining the Optimal Main Data Collection Mode

Factor	Implication for Mode Choice
Concepts to be measured	If a visual medium is required, a telephone survey can be ruled out. Complex concepts usually benefit from interviewer assistance.
Target population to be surveyed	Can the nontelephone population be ignored? If so, consider the telephone mode. Literacy level: Mail modes require literacy rates at or above the national average.
Contact information available on frame	If a name and address, mail or face-to-face interview should be considered. If a telephone number, consider a telephone survey.
Saliency of the topic	If much persuasion is needed to obtain adequate response rates, mail surveys must be ruled out.
Speed of completion	If needed very quickly, telephone is best. If needed in weeks, a mail survey may be feasible.
Scope and size of the sample	If a local survey, all three main modes may be feasible. For a national survey, cost may be the reigning factor that suggests mail or telephone survey.
Sample dispersion	Maximum dispersion suggests a mail or telephone survey. In face-to-face surveys, some clustering is almost always needed.
Frame coverage of target population	If only poor coverage frames are available, use a face-to-face survey, RDD, or mixed-mode. Good coverage implies flexibility regarding mode choice. Availability of an address permits advance letters and prepaid incentives.
Nonresponse	Interview modes usually generate higher response rates than self-administered. Ability to persuade reluctant sample units depends on richness of media (e.g., in mail surveys, motivation is limited to written materials). Nonresponse is confounded with coverage problems in mail and telephone modes. Mail questionnaires might be regarded as junk mail and thrown away by sample unit. Mail questionnaires cannot be completed by parts of human populations, due to literacy problems.
Interviewer	Interviewer can generate response errors, such as social desirability bias. Interviewer-assisted mode is not good for collecting sensitive information. Interviewer necessary for visual aids and probing. Centralized telephone interviewing reduces costs and errors compared to noncentralized interviewing. Telephone interviewers can have larger workloads due to no travel burden.
Respondent	There is some evidence that respondents prefer self-administered surveys. Self-administered modes are suitable for collecting sensitive information. If the response task is difficult, interviewer assistance is necessary.
Instrument	Mail questionnaires must be relatively simple but are suitable when questions contain many response alternatives. Complex instruments call for the interview mode. Mixed-mode must use questionnaires that can be used in all modes.
Cost	Everything else may be secondary if mail is the only mode that can be afforded. The telephone interview mode is less expensive than the face-to-face mode.

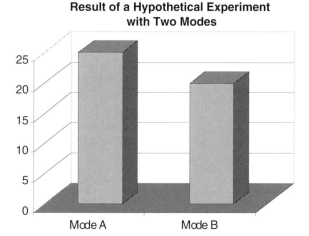

Figure 6.1   Which mode is better? The results of a mode comparison study may be inconclusive. In this hypothetical example, the two modes yield significantly different estimates of the parameter, but this information alone is inadequate to determine the better mode (i.e., the mode that yields an estimate closer to the true parameter value). However, if information is available on the direction of the bias for the characteristic under study, the better mode can be identified. For example, if we know that the bias tends to be negative (i.e., estimates tend to underestimate the parameter), mode A is preferred, since it yields a higher estimate. If the bias tends to be positive, mode B is preferred.

distributions, presence of social desirability responses, length of responses to open-ended questions, length of collection period, and cost of the collection.

De Leeuw and van der Zouwen (1988) have conducted a meta-analysis of 28 research studies comparing face-to-face and telephone interviewing and summarizing the results as if one large study had been conducted. The largest effect was obtained for length of responses to open-ended questions and then in favor of face-to-face interviewing. There also seems to be a tendency for less social desirability bias over the telephone. The large mode effects occur for sensitive variables. For general topics and if no visual communication is needed, telephone interviewing produced about the same quality as face-to-face interviewing.

Any comparison study of two mode systems, A and B, ultimately produces two estimates: one as a result of mode system A and another as a result of mode system B. Even if we have a set of evaluation criteria, it is difficult to assess which mode produces better data quality. Is it the one that produces the smallest measurement error, highest response rate, or that does well on a number of other evaluation criteria? The answer depends on what the researcher believes to be important. Quality components are often in conflict with each other, and even though we are very concerned about data quality, we cannot ignore such factors as cost and the time it takes to produce survey results (see Figure 6.1).

Table 6.3   Mode Effects in a Women's Health Study

Experimental Group	Mean Reported Sex Partners			Percent Admitting Illicit Drug Use
	Past Year	Past 5 Years	Lifetime	
Self-administered questions (SA)	1.72	3.88	6.54	40.9
Conventional SA	1.56	3.37	6.88	42.5
Computer-assisted SA	1.89	4.40	6.25	39.3
Interviewer-administered questions (IA)	1.44	2.82	5.43	40.7
Conventional IA	1.56	2.86	4.58	39.3
Computer-assisted IA	1.36	2.79	6.27	42.2

Source: Adapted from Tourangeau and Smith (1998) and Tourangeau et al. 1997.

Table 6.4   Examples of Estimates of Sensitive Attributes by Mode

A. Percent of U.S. Women Who Have Had One or More Abortions

Original Report in Interviewer-Administered Questionnaire	Subsequent Reports in ACASI		
	No Abortions	1 Abortion	2+ Abortions
No Abortions	95.4	3.5	1.0
1 Abortion	1.8	92.5	5.8
2 Abortions	0.4	2.2	97.4

B. Percentages of U.S. Males Ages 15–19 Reporting Male–Male Sexual Contacts

Mode	Reporting Male–Male Sexual Contacts (%)
Self-administered PAPI	1.1
ACASI	4.7

C. Estimates of Prevalence of Sensitive Behaviors in the U.S. Population

Mode	Having Limited Sexual Experience (No Sex in Past 6 Months) (%)	Most Recent Sexual Relationship Lasted Less Than 6 Months (%)
CAPI	1.5	5.8
T-ACASI	8.0	21.3

Source: Adapted from Turner et al. (1996, 1998).

Perhaps the main problem with mode studies is that we seldom know much about the measurement bias. One way to get information is to compare estimates A and B with a gold standard, as discussed in Section 6.1.6. Such an evaluation study is quite possible to conduct, but the result might be unreliable since evaluation studies are expensive and therefore can be conducted on a limited scale only. An alternative route is to assume a likely direction of the measurement bias. In general, sensitive items are underreported and there is overwhelming evidence that modes that eliminate social desirability bias produce more reporting.

For an illustration of this phenomenon, see Tables 6.3 and 6.4. Here it is assumed that more reporting is better reporting, and for sensitive variables this seems to be a general truth. But no rule is without exceptions. The perception of what might constitute a sensitive variable might very well vary between age and gender groups. For instance, there is evidence that men in their teens, when asked about sexual experiences, tend to overreport. There are other variables considered sensitive where the reporting pattern might vary between groups. Income is such a variable, where the social desirability bias is positive, and as a result, members of some income groups might overreport their income. Thus, to be sure about measurement biases, one has to estimate them by means of evaluation studies.

Examples of other studies include Cannell et al. (1987), who found that question order effects can vary between telephone and face-to-face interviewing. Also, reports of health events were consistently higher in telephone interviews than in face-to-face interviews. Substantive differences between the modes were, however, small. Sykes and Collins (1988) conducted three experimental studies comparing face-to-face and telephone interviewing in Britain. Among other things, they found that answers to open-ended questions tended to be shorter in the telephone mode, which could partially explain the faster pace in the telephone mode. The issue of mode is both complicated and not so complicated. It is complicated in the sense that a pure mode effect is very difficult to assess. It is not so complicated in the sense that in practice the choice of mode is often straightforward, due to design constraints such as cost or topic. Often, modes have to be combined to accomplish acceptable response rates and cost-efficiency, but for some topics and questions, some modes are simply not suitable.

CHAPTER 7

Data Processing: Errors and Their Control

Data processing is a set of activities aimed at converting the survey data from its raw state as the output from data collection to a cleaned and corrected state that be can used in analysis, presentation, and dissemination. During this process, the data may be changed by a number of operations which are intended to improve their accuracy. The data may be checked, compared, corrected, keyed or scanned, coded, tabulated, and so on, until the survey manager is satisfied that the results are "fit for use." The sequence of data processing steps range from the simple (e.g., data keying) to the complex, involving editing, imputation, weighting, and so on. Operations such as data entry, editing, and coding can be expensive, time consuming, and costly. Increasingly, therefore, data processing operations are becoming automated. Recent technological innovations have greatly affected the way these operations are performed by reducing the need for a large manual workforce. Technology has also allowed greater integration of data processing with other survey processes. For example, some data processing steps can be accomplished during the data collection phase, thereby reducing costs and total production time while improving data accuracy.

Data processing operations may be quite prone to human error when performed manually. By reducing reliance on manual labor, automation reduces the types of errors in the data caused by manual processing, but it may also introduce other types of errors that are specific to the technology used. Therefore, it is important that data processing steps include various measures to control errors introduced by machines as well as humans.

The errors that may be introduced in the data processing stage tend to be neglected in some surveys. In general, knowledge about data processing and its associated errors is very limited in survey organizations. Operations are sometimes run without any particular quality control efforts, and the effects of errors on the overall accuracy as measured by the MSE are often unknown except perhaps for national data series of great importance. In this chapter we

Table 7.1 Data Processing Steps for PAPI Questionnaires

Process Step	Description
1. PAPI questionnaire	Questionnaires are collected in the field and work units are formed.
2. Scan edit	Entries are inspected to avoid data entry problems caused by stray marks, ambiguity, etc.
3. Data entry	Questionnaire data are registered via keying, scanning, or other optical sensing.
4. Editing	Logical editing is performed on registered data. Corrective measures such as imputation are part of the editing.
5. Coding	Open-ended responses are coded.
6. File preparation	Data are weighted and checked. Files are prepared for public and/or client use.
7. Output file	Output files are analyzed, documented, and delivered to the client.

describe the main data processing operations, their error structures, and the measures that can be taken to control and reduce these errors.

7.1 OVERVIEW OF DATA PROCESSING STEPS

The data processing steps vary depending on mode of data collection for the survey and technology available to assist in data processing. Table 7.1 shows the data processing steps for PAPI questionnaires. Similar steps are used for CAI questionnaires except that the data entry step can be omitted. The steps in the process for PAPI consist of the following:

1. *The PAPI questionnaire* is used to collect information for the variables under study. Some questions corresponding to these variables are closed-ended, requiring the interviewer (or respondent) to check a box representing a response alternative. For example, marital status may be coded as "1 = single," "2 = married," "3 = divorced," "4 = widow/widower," and "5 = never married." If the question is unanswered, a "98 = blank" may be coded or "99 = refused" if the respondent refused to answer in an interviewer-assisted mode. For open-ended questions, a free-format response is written in a blank field on the questionnaire. For example, a response to the question "What is your occupation?" may be "I am a flight attendant for SAS" or "plumber."

2. *Scan edit* is a preparatory operation involving several steps. First, as questionnaires are received by the survey organization, they are coded as received into the receipt control system and inspected for obvious problems, such as blank pages or missing data for key items that must be completed for questionnaires to be usable. For interviewer-assisted modes, the question-

naires that are rejected as being incomplete by this process may be sent back to the interviewers for completion. For mail modes, the questionnaire might be routed to a telephone follow-up process for completion. In cases where there is no nonresponse follow-up, the questionnaires may be passed on to data entry but ultimately may be coded as nonrespondents due to incomplete data. After this step, the questionnaires are often divided into small batches called *work units* to facilitate subsequent process steps. Finally, these work units receive a final inspection to ensure that they are ready to proceed to the next stage of processing. The entire scan edit proceeds very quickly since additional inspections are built into the process downstream.

3. *Data entry* is the process step where the questionnaire data are converted into computer-readable form. Data can be entered manually using keying equipment or automatically by using scanning or optical recognition devices. Sometimes this process step is called *data capture*.

4. *Editing* is a set of methodologies verifying that data-captured responses are plausible and, if not, correcting the data. Editing rules can be developed for each variable or for combinations of variables. The editing rules specify likely variable values or likely values of combinations of variables, often as acceptable value intervals. Typically, the editing reveals obvious errors and responses highly suspicious of being erroneous. Measures are taken to correct some of them. Sometimes, missing values are inserted or recorded values are corrected after recontacting interviewers or respondents. In other cases, values are inserted or changed by means of deducing the correct value based on other information on the record. Most editing is performed automatically by specially designed computer software.

5. *Coding* is a procedure for classifying open-ended responses into predefined categories that are identified by numeric or alphanumeric code numbers. For example, there may be thousands of different responses to the open-ended question "What is your occupation?" To be able to use this information in subsequent analysis, each response is assigned one of a much smaller number (say, 400 to 500) of code numbers which identify the specific occupation category for the response. For the occupation categories to be consistent across surveys by different organizations, a standard occupation classification (SOC) system is used. A typical SOC code book may contain several hundred occupation titles and/or descriptions with a three-digit code number corresponding to each. In most classification standards, the first digit represents a broad or main category, and the second and third digits represent increasingly detailed categories. Thus, for the response "flight attendant," a coder consults the SOC code book and looks up the code number for flight attendant. Suppose that the code number is 712; the "7" might correspond to the main category "restaurant service worker," 71 might correspond to "waiter and waitresses," and 712 to "flight attendant." In *automated coding*, a computer program assigns these code numbers to the majority of cases while the cases that are too difficult to code accurately by computer are coded manually.

6. In the *file preparation* step, a number of activities are involved. First, weights have to be computed for the sample units (see Chapter 9). Often, the weights are developed in three steps: First, a base weight is constructed, which is the inverse of the probability of selection for a sample member; second, the base weight is adjusted to compensate for nonresponse error; and third, further adjustments of the weights might be performed to adjust for frame coverage error, depending on the availability of external information. These postsurvey adjustments are intended to achieve additional improvements in the accuracy of the estimate. File preparation may also involve the suppression of data for disclosure avoidance and confidentiality purposes, and missing data or faulty data may be *imputed* (i.e., the faulty or missing value is replaced by a value that is predicted from a statistical model built on other responses in the data file). The file preparation step results in an analysis file that serves as the output file, and the output file is used to produce statistics and analyses.

7. The *data analysis* step may be executed by the producer of the survey data, the client or survey sponsor, or other data users. There may also be combinations of these approaches. The producer can perform one set of analyses, and others using the analysis file can perform various secondary analyses. It is important that the output file be documented properly. One common set of documentation is the provision of a generic data structure, sometimes referred to as a *data record description*. This documents what variables are on the file, all the code numbers associated with each variable (e.g., for the marital status variable, 1 = "single," 2 = "married," . . . , 98 = blank, 99 = refused).

The new technologies have opened up many possibilities to integrate the data processing steps (see Couper and Nicholls, 1998). Therefore, the sequence of steps might be very different from the one described above for some surveys. For example, it is possible to integrate data capture and coding into one step; similarly, data capture and editing can be integrated with coding. It is also possible to integrate editing and coding with data collection through the use of CAI technology. The advantage of this type of integration is that inconsistencies in the data or insufficient information for coding can be resolved with the respondent immediately. This reduces follow-up costs and may result in better information from the respondent. Many other possibilities for combing the various data processing steps are possible. The goal of integration is to increase the efficiency of the operations while improving data quality.

As is clear from the earlier discussion, the data can be modified extensively during data processing and therefore can be an important source of nonsampling error. The editing is intended to improve data quality, but it still misses many errors and can even introduce new ones. Automation can reduce some of the errors made by manual processing but might introduce new errors. For example, in optical recognition data capture operations, the recognition errors are not distributed uniformly across digits and other characters, which can introduce systematic errors (i.e., biases).

7.2   NATURE OF DATA PROCESSING ERROR

The literature on the data processing error and control is quite small relative to that on the measurement error (especially, respondent errors and questionnaire effects) and nonresponse. This is unfortunate since some processing steps, such as coding, can be very error-prone, particularly coding of complex concepts such as occupation, industry, field of study (in education surveys), and various types of medical coding. For instance, coding error rates or coding disagreement rates can reach 20% levels for some variables, especially if the coding staff is not well trained.

Despite its potential to influence survey results, data processing error is considered by many survey methodologists as relatively uninteresting. Perhaps this is because, unlike nonresponse and questionnaire design, cognitive models and sociological theories are not directly applicable to data processing operations. This may explain the dearth of literature on data processing topics. Although there is ample evidence of the importance of data processing error in survey work, the associated error structures are essentially unknown and unexplored.

In this chapter we discuss the many ways in which data processing operations can contribute to both systematic and variable errors in the survey data. We also examine how automation tends to generate systematic errors, while manual operations can generate both systematic and variable errors. For example, the errors made by coders and editors resemble those made by interviewers, at least theoretically. Like interviewers, editors, coders, and other data processing operators can generate systematic errors that bias all the units in their individual work assignment in the same way. For interviewers, we call this type of systematic error correlated interviewer error and developed the concept of interviewer variance in Chapter 5. Similarly, coders, editors, and other operators can give rise to variance components due to correlated error: coder variance, editor variance, and so on. For example, a particular coder may misunderstand the instructions for coding technicians and engineers in that many engineers are coded erroneously as technicians. We can measure the effect of these systematic errors on the estimates by interpenetrating coder assignments. If coder assignments are interpenetrated so that all coders have approximately the same number of technicians and engineers in their assignments, this particular coder would tend to have a higher proportion of technicians than the other coders. Thus, as we saw in Chapter 5, this coder-level systematic error can dramatically increase the variance of the estimate of the number of technicians in the population. Furthermore, this increase in variance is not fully reflected in the estimates of standard errors, so the standard errors of the occupation estimates may be understated. The sizes of the correlated variances and biases have to be estimated by means of specially designed experiments. This is seldom done in practice. A more efficient approach is to adopt methods such as better supervision, remedial training, and other continuous quality improvement techniques that reduce or eliminate coder variance.

Data processing operations have traditionally accounted for a very large portion of the total survey budget. As we will see later, in some surveys the editing alone consumes up to 40% of the entire survey budget (U.S. Federal Committee on Statistical Methodology, 1990). However, the data processing share of the total resources for surveys has been reduced considerably by new innovations in CASIC technology. In addition, the total survey error associated with data collection and data processing has been affected by these changes. Some reasons for these reduced costs and potentially reduced total error have been documented in a number of articles, including Lyberg and Kasprzyk (1997), Bethlehem and van de Pol (1998), Speizer and Buckley (1998), Biemer and Caspar (1994), and Lyberg et al. (1998). The following describes some of these developments:

- New developments in data processing automation have decreased the need for large manual operations. Data processing in a manual environment is very labor intensive and time consuming. As the data processing workload in a processing center peaks, many untrained, temporary staff may need to be hired to meet the schedule. For much of the coding, editing, and data capture previously done by human labor, computers have reduced our dependency on human resources. In addition, computers perform these tasks in a standardized, consistent manner, which eliminates the types of variable errors human operators make (e.g., operator variance). As we shall see, however, the computer programs may also introduce errors, so whether or how much automation improves data quality is not obvious.

- As discussed in Chapter 6, developments in CASIC technology, both new and old, have created new opportunities to merge data collection activities and data processing operations. We foresee greater and greater integration of data collection and data processing systems in the future as well as opportunities to optimize the entire survey system.

- The manual parts of the data processing system resemble an assembly line common in many other production environments (e.g., automobile manufacturing). Therefore, it was natural for survey designers and survey managers to apply schemes for quality assurance, such as acceptance sampling, used routinely in quality control in industrial applications. Acceptance sampling guarantees a prespecified average outgoing quality level (AOQL) for an operation. For example, if we believe that the maximum defective rate for some product is 1%, an acceptance sampling scheme can be devised to guarantee this level of outgoing quality.

Acceptance sampling was originally developed by Dodge and Romig (1944) for industrial settings. In the 1960s and 1970s, these methods were extended for use in many manual survey operations. Such methods have been referred to in the survey literature as the *administrative applications of quality control methods*. Acceptance sampling can be used to ensure that the error in a process is controlled without large fluctuations in quality and that the pre-specified quality levels are achieved. However, these methods have been criticized by Deming (1986) and others, primarily as a result of the absence of a feedback loop in the approach. That is, although errors in an operation are identified, there is no way to improve continually an operation since the causes of the errors are not identified and eliminated.

It should be noted, however, that the main objective of acceptance sampling is to find the defects in a product. Typically, the method works by first creating batches or lots from the survey materials (e.g., questionnaires). A sample is drawn from each lot and checked for conformance to a specified standard. The lot is then classified as either meeting or not meeting the quality specifications. If the lot is found to contain too many defects on the basis of the sample check, it is rejected. At that point, the lot might be reworked and the operator that produced the rejected lot may undergo a greater level of quality control screening. Typically, the amount of rework is rather stable unless there is an element of continuous improvement associated with the application of acceptance sampling, in which case it should decrease over time.

Statistics Canada and other organizations have used acceptance sampling for many processing applications over the years but have incorporated a feedback loop to the operators for the purpose of generating quality improvements. These organizations continue to use acceptance sampling in this way and have found it to be preferred to other quality control methods (e.g., continuous quality improvement) in operations experiencing high staff turnover rates (see Mudryk et al., 1996, 2001a) Acceptance sampling has some shortcomings as a means of continually lowering the error rate for the outgoing product over time. Continuous quality improvement (CQI) methods emphasize improvement in the underlying process rather than screening the product. Clearly, there is room for both methods, as we discuss in Section 7.7.

- As mentioned previously, for manual data processing, the operators generate errors that statistically resemble those made by interviewers. Coders, editors, and keyers tend to affect the values that are recorded for elements that have been assigned to them in uniform ways. They generate correlated errors, and these errors inflate the total survey error in ways that are unknown unless special experiments are performed. There are reasons to believe that in some surveys these variance contributions can be substantial. In addition to correlated errors, manual operations may also generate variable errors and systematic errors that are common to all operators. There have been very few studies measuring the total effect on survey data as a result of data processing errors, due primarily to declining survey budgets and lack of attention to the problem. Two exceptions are Jabine and Tepping (1973) and U.S. Bureau of the Census (1974). We believe the goal is not necessarily quality evaluation but quality improvement, which will obviate the need for extensive quality evaluations of these error sources.

In the next few sections we provide some results regarding the error in data processing operations and some methods for reducing and controlling the errors.

Data processing is a neglected error source in survey research. Coders, keyers, editors, and other operators contribute correlated error to the mean squared error, similar to the contributions made by interviewers. The increased use of technology eliminates correlated error but may generate new errors.

7.3   DATA CAPTURE ERRORS

7.3.1   Data Keying

Data capture is the phase of the survey where information recorded on the form or questionnaire is converted to a computer-readable format. The information may be captured by keying, mark character recognition (MCR), intelligent character recognition (ICR), and even by voice recognition entry (VRE). MCR detects the presence or absence of a mark on a scanned image, but not the shape of the mark, so it is not appropriate for hand- or machine-printed characters. It is commonly used on forms where the respondents fill in small circles, called *bubbles*, to indicate their responses to questions. ICR turns images of hand-printed characters (not cursive) into machine-readable characters. Optical character recognition (OCR) converts machine-generated characters (e.g., bar codes) into computer-readable characters. VRE is a method for converting voice patterns into machine-readable characters. Recently, new forms of data capture technology have evolved, including facsimile transmission, electronic data interchange (EDI), and e-mail transmission through the Internet.

Data keying is a tedious and labor-intensive task that is prone to error unless controlled properly. Keying can be avoided for interviewer-assisted modes by using CATI, CAPI, and other CAI technologies. For mail surveys, keying can be avoided by using specially designed paper questionnaires that are first scanned and then the scanned images are converted to computer-readable characters using OCR and ICR technologies. Next, we discuss some advantages and disadvantages of these options.

In the data processing literature, three definitions of a keying error rate are prevalent:

1. $\dfrac{\text{no. of fields (or variable values) with at least one character keyed erroneously}}{\text{total no. of fields}}$

2. $\dfrac{\text{no. of characters keyed erroneously}}{\text{total no. of characters keyed}}$

3. $\dfrac{\text{no. of records with at least one keying error}}{\text{total no. of records}}$

Definition 1 measures the proportion of fields that are in error. For example, a response to a question on income may be considered as a field. If 100 persons respond to this question, the denominator of this measure is 100, while the numerator is the number of these fields that were keyed in error. Definition 2 measures the proportion of all keyed characters that were keyed in error. Therefore, if 10,000 characters were keyed and 200 were keyed in error, the keying error rate is 200/10,000 = 0.02 or 2%. Finally, the last definition measures the proportion of records that have at least one keying error. Other definitions are possible, as demonstrated by the following example.

Example 7.3.1   Suppose that a response to a question on income is 17,400. If the value keyed differs from 17,400, we say that a keying error has occurred. The following three entries are examples of keying errors: 1740, 17,500, or 17,599. In the first case, the keyer failed to key the final "0"; in the second case, one digit was miskeyed; and in the third case, three digits were miskeyed. Which of the three errors should be considered most serious? It is obvious that the first case is more serious than the other two. Do the definitions above reflect these errors appropriately?

In most keying operations, independent rekey verification is the primary method of quality control for keying. These methods involve independently rekeying either all or a sample of a keyer's work and then comparing the newly keyed entries with the original entries. Discrepancies between these two keyings can be checked by computer and adjudicated. Modern data entry systems will do these checks "on the fly" during the second keying, allowing the second keyer, rather than a third person, to serve as the adjudicator. In small operations (e.g., a few hundred questionnaires), keying quality control can be implemented quite simply using dependent review of the keyed entries against the hardcopy source documents (questionnaires). However, this method is likely to miss a considerable number of keying errors. Later in the chapter we discuss the effectiveness of dependent verification compared with independent verification.

The concern over a few, very serious keying errors in the survey data set and the fact that keying is a relatively inexpensive operation has motivated the extensive use of keying quality control activities in many organizations. Quite commonly, 100% independent rekey verification is used despite the fact that compared to some other survey operations, the keying error rate is quite small, particularly for closed-ended question responses. For instance, the error levels in the U.S. Fourth Follow-up Survey of the National Longitudinal Study, based on a quality control sample of questionnaires, was about 1.6% prior to quality control verification and correction (Henderson and Allen, 1981). Similarly, error levels in the 1988 U.S. Survey of Income and Program Participation were in the neighborhood of 0.1% (Jabine et al., 1990). In a study designed to determine the quality of keying of long-form questionnaires in the 1990 U.S. Census, keyers committed keystroke mistakes (or omissions) in 0.62% of the fields in the initial stages of production keying (U.S. Bureau of the Census, 1993). These rates relate to errors in entries of individual variable values (definition 1) and apply to operations and operators specializing in keying.

Computer technology has initiated new concerns about the number of errors made by CAPI interviewers who are viewed as essentially unskilled data keyers. Dielman and Couper (1995), however, report a keying error rate of 0.095% in a study focused on keying errors in a CAPI environment, thereby providing some evidence that keying errors are not a significant problem in CAPI surveys. In a Swedish study where error rates were observed in terms of percent of records, the average error rate per record was 1.2% by definition 3 (see Lyberg et al., 1977).

The concern over keying errors is that very large errors are possible, as Example 7.3.1 illustrated. Such very large errors can usually be detected in the editing step but it is not wise to rely on editing to control the quality in keying. This suggests that some type of quality control is needed for keying.

In fact, the studies on error rates cited above were all conducted under a system of quality control, including editing procedures that can detect and correct some types of keying errors. Had there been no quality control, the error rates would probably have been higher since keyers produce better quality results when they know their work is being verified than when these checks are removed. This is particularly true when keyer performance evaluations are based on both productivity and accuracy.

The error rate for any keying process depends on a number of factors, such as keyer experience, the variation in error rates among keyers, the keyer turnover rates, the amount and type of quality control used for the keying process, the legibility of the data to be keyed, the ease with which keyers can identify what fields should be keyed, and the amount and quality of the scan editing process needed to prepare forms for keying.

Keying error rates are usually small, but the error effects can be considerable. Comparisons of keying error rates across surveys and organizations are often difficult to do, due to varying definitions of error rates.

7.3.2 Scanning

The U.S. Census Bureau pioneered in the use of FOSDIC (Film Optical Sensing Device for Input to Computers), which is a process similar to OMR except the marks are sensed from a microfilmed copy of an appropriately designed questionnaire rather than a scanned copy. FOSDIC uses a beam of light to identify whether the response bubbles have been marked. This process has an average error rate of about 0.02% (Brooks and Bailar, 1978).

The U.S. National Center for Education Statistics (NCES) used OMR for the High School and Beyond Study (Jones et al., 1986). OMR is very effective for multiple-choice questions; however, it is not able to sense handwriting, as noted above. ICR has been suggested as a solution to the data capture problem when the data are handwritten responses. The U.S. Census Bureau has sponsored several conferences on this topic to stimulate new research in ICR. The general findings from these conferences suggest that the feasibility of ICR depends on the application since some applications require considerably less sophistication than others. However, accuracy of ICR systems for capturing written responses has improved dramatically over the last few years, to the point where "machine performance in reading words and phrases may now be good enough to decrease the cost and time needed to carry out a census without decreasing the accuracy of the results." (Geist et al., 1994).

Two types of errors can occur with ICR, substitution and rejection. A *substitution error* occurs when the ICR software misinterprets a character. A *rejection error* occurs when the ICR software cannot interpret a character and therefore rejects it. Rejected characters must be corrected manually and then re-entered into the system. Consequently, reject errors are expensive to handle but contribute no error to the data entry process as long as they are handled properly.

Substitution error rates are usually small. In a study conducted for the 1970 Swedish Census, ICR was used to read handwritten responses, and only 0.14% of the digits were substituted. However, this rate should not be generalized for other surveys, since, as mentioned previously, the accuracy of ICR depends on the application. In this application, the process of writing digits was highly standardized. Staff having special training in writing ICR digits wrote the digits. Obviously, this is a situation very different from one where the general population is asked to enter data on income and other variables for ICR processing.

Tozer and Jaensch (1994) report that the substitution rate in the Retail Census conducted by the Australian Bureau of Statistics was about 1% on a digit basis. Similar levels have been reported by the U.K. Employment Department (Thomas, 1994). In the 1991 Canadian Census of Agriculture, substitution levels for alphabetic and alphanumeric characters were 2 and 5%, respectively (Vezina, 1994).

Even though the substitution error rate is small in many applications, the resulting error can be quite problematic and serious. Statistics Canada's experience (Mudryk et al., 2002) is that the substitution error can be systematic in nature, thereby causing severe problems. For instance, consider a scanner in production that has a smudge on the optical window. This scanner will probably produce many systematic substitution errors for as long as this problem exists. There are in fact many other problems that can cause similar systematic substitutions, such as low luminance of the scanner light bulb, dirt in various components, and paper transport problems.

This suggests that quality control should include frequent equipment recalibration and configuration to ensure that this substitution error is kept to a minimum. Substitution errors can also be a function of the type of data that are being read or interpreted. For example, bar codes and tick boxes have a much lower inherent substitution rate than numeric data, which in turn is also lower than alphabetic or alphanumeric data. As well, the condition of the incoming documents is also a factor, since in general, documents prepared in a controlled environment have a lower substitution error rate than do those prepared in an uncontrolled environment.

Recently, Statistics Sweden and some other agencies have started scanning entire survey forms and using ICR to convert the information to computer-readable format. Thus, no data entry in the usual sense is necessary. Further, some functions of the editing process can also be accomplished during the data capture process. Documents can be stored and retrieved during editing and correction. Although this research is still in its infancy, Blom and Lyberg (1998) report that reject rates range from 7 to 35% on a position level in initial scanning studies performed at Statistics Sweden.

Finally, new developments have been made in the application of ICR to survey data. As described by Mudryk et al. (2001a), ICR has been combined with automated mark and image recognition, supplemented with manual capture, by operators who "key from image" using a "heads-up data capture technique." This means that the ICR technology does not reject a character, but in fact always offer an interpretation of it, with a corresponding confidence level for the field. A field having an acceptable ICR confidence level is deemed recognizable by the ICR software and the ICR interpretation is accepted. Fields not having an acceptable confidence level are sent to a "key from image" operation, where operators read the optically captured image and manually enter the text of the image. This data entry is simple and no paper handling is required, which reduces costs considerably. Furthermore, these images can then be used for paperless editing purposes, which further reduces costs.

There are two types of error associated with intelligent character recognition: *substitution* and *rejection*.

7.4 POST–DATA CAPTURE EDITING

7.4.1 Definition of Terms

There are a number of definitions of editing. For example, according to the U.S. Federal Committee of Statistical Methodology (1990) *editing* is defined as "procedures designed and used for detecting erroneous and/or questionable survey data (survey response data or identification type data) with the goal of correcting (manually and/or via electronic means) as much erroneous data (not necessarily all of the questioned data) as possible, usually prior to data imputation and summary procedures." The International Work Session on Statistical Editing has endorsed the Economic Commission for Europe (ECE) definition (1995), which simply states: "Editing is the activity aimed at detecting and correcting errors in data." A third definition (Granquist and Kovar, 1997) extends the latter somewhat: "Editing is the identification and, if necessary, correction of errors and outliers in individual data used for statistics production."

Note that none of the definitions state that all errors be corrected or even identified. Therefore, we need to elaborate the goals of editing. They are:

1. *To provide information about data quality.* A visual inspection or *scan edit* of survey questionnaires as they come in from the field can reveal much about how the interviewers (for PAPI interview modes) are performing their roles and where respondents may be struggling with their roles as providers of accurate information. Important information on data collection problems can often be gleaned from interviewer notes and comments, or there may be evidence that some interviewers are not completing their forms properly. For example, they may forget to enter some observational data or may not skip all questions that should be skipped. These problems may indicate that the interviewer needs to be retrained in these areas.

Following this scan editing procedure the data are keyed or otherwise converted to computer-readable format, where they can be checked further. Analysis can be performed to determine the extent of missing data values, out-of-range values, and implausible values. This analysis can give us additional information on data quality and potential problems with the questionnaire or the interviewing process. At this stage, data corrections may be entered to improve the data. Sometimes this can be done accurately just by using other information that is provided on the questionnaire. However, in many cases these corrections may require recontacts with the respondent or the interviewer to fill in or clarify one or more critical pieces of data.

In CAI applications some of the editing can take place at the time of data collection and be performed by interviewers. The CAI software can be applied so that the interviewer is notified about inconsistencies or implausible values during the course of the interview and the interviewer can ask the respondent for clarification. In PAPI applications it is possible to let the interviewer carry out some of the editing before the questionnaire is sent off to the agency. Although it is still quite common for special staff to edit the forms manually, in large surveys editing is almost always done, at least in part, automatically, using special software.

2. *To provide information about future survey improvements.* The lessons that we learned from inspecting questionnaires and reviewing data through computer analysis can often suggest improvements to the data collection procedures that will eliminate these types of errors for future surveys. For this

reason it is important that survey designers try to understand the root causes of errors in data so that problems in data collection or data processing can be corrected.

3. *To simply "clean up" the data so that further processing is possible.* This has been discussed previously as the primary motivation for editing. Indeed, when cleaning up the data, it seems natural that we try to identify and correct as many errors as possible since it is presumed that each error we can eliminate brings the data closer to the quality ideal. However, there are reasons to believe that, in general, the resources that are spent on this activity are much larger than the value of the improvements that can be reaped from the activity. That is, cost–error optimization suggests that a large share of resources typically devoted to editing could be better spent improving other components of the survey process.

An overview of editing methods is given in United Nations (1994b).

7.4.2   Basic Concepts Used in Editing

Data editing consists of a set of rules that are applied to the survey data. Some examples of editing rules are: A value should always appear in specific positions of the file; some values of a variable are not allowed; some combinations of values are not allowed; the value of a sum should be identical to the sum of its components; and a specific value is allowed only if it is included in a predefined interval. We call these edits *deterministic edits*, which, if violated, point to errors with certainty. In the editing process, edit rules are applied and error messages are printed or displayed on a computer screen for manual intervention. Of course, software that can both detect and correct errors according to these specifications can be developed. For a small survey it might not be efficient to use such software, but in large surveys it is essential. The process for correcting the data based on the editing rules may use only the data that are available on the data record or may involve data collected via recontacts with respondents or interviewers.

Fatal or critical edits are supposed to detect erroneous variable values that must be corrected for the data record to be usable. Examples of errors that may be considered as fatal are:

- Identification number errors (e.g., a business ID number is invalid)
- Item nonresponse for key variables
- Invalid values on key items (e.g., out-of-range values for age)
- Values that are inconsistent (e.g., birth mother's age is less than the child's age)
- Defined relationships between variables are not satisfied (e.g., net after-tax income is less than gross income before taxes)
- Values that are extreme or unreasonable

Query edits are supposed to identify suspicious values. Suspicious values are values that may be in error, but that determination cannot be made without further investigation. For example, a farm operator may value his or her land at 10 times the value of the land that surrounds the farm. This could be a keying error in which an extra "0" was appended to the farmer's report. However, it may also be the actual estimate provided by the farm operator. A check of the original survey form should reveal which it is.

Query edits should be pursued only if they have the potential to affect the survey estimates in a noticeable way. Since it is not clear whether an error has occurred, each query edit has to be investigated in more depth. Such investigations can be time consuming and costly. It is therefore important that the rules for query edits be designed so that only errors that have a substantial probability of affecting the estimates be identified and pursued. These editing rules might also be termed *stochastic edits* (as opposed to deterministic edits), since there is uncertainty as to whether they identify actual errors in the data.

Editing performed at the record or questionnaire level is called *microediting*. Recently, there has been a growing interest in *macroediting*, edits that are performed on aggregate data (e.g., means or totals) or other checks that are applied to the entire body of records. If an aggregate value is deemed suspicious on the basis of a macroediting rule, the individual records that comprise the aggregate are inspected to determine whether the cause of the anomaly in the aggregate can be attributed to one or a few erroneous records. However, if an aggregate is not suspicious, the individual records comprising the aggregate pass the edit. Of course, even though an aggregate clears the edit, there may still be errors in the individual record values comprising the aggregate. However, since these errors have no discernible effects on the estimates (as evidenced by the aggregate value), they do not have to be investigated and corrected.

Thus, editing can be divided into two phases: an identification phase (deterministic or stochastic) followed by a correction or adjustment phase. The latter phase may involve recontacting interviewers and respondents or may simply involve suppressing the erroneous variable or replacing it with a more suitable deduced or imputed value. Thus, imputation can be an important remedy for edit failures as well as item nonresponse. Imputation applied in editing is discussed in Pierzchala (1990, 1995), Fellegi and Holt (1976), and Legault and Roumelis (1992).

In summary, data can be edited at different stages of the survey process, as follows:

- *Editing during the interview*
  The interviewer alone or in collaboration with the respondent conducts edit checks as specified in the interviewer instructions. Consistency checks, range checks, variable relationships, and missing values can be built into the CAI software so that these edits are performed as the interview progresses. However, the number of real-time edit queries should

be restricted to the most important edit checks in order not to unduly prolong or interrupt the interview. In addition, there is evidence (obvious revisions on the form and the like) that some amount of self-editing is conducted by mail survey respondents. The questionnaire designers should encourage such initiatives through proper instructions.

- *Editing prior to data capture*
  This type of editing is usually manual, and takes place after questionnaires have been delivered to the survey organization. The editing at this stage is usually quite general, involving only classifying questionnaires into accept, reject, needs action, and so on, with regard to further processing. For example, a questionnaire may be rejected and not processed further if it is mostly blank or contains obvious errors.

- *Editing during data capture*
  Data capture may involve keying, ICR, OMR, or other scanning technologies. During this phase it is possible to use editing software in an interactive process. The editing can be done on a variable basis or on a record basis. Typically, the software stops the process when edits fail and data capture can continue only after the operator has taken certain measures. These measures include a specific acceptance on the operator's part of the value identified, or a change made by the operator, or a flagging that serves as a reminder for future action.

- *Editing after data capture*
  The majority of editing occurs at this stage, which involves a system of critical and query (micro-) edits, macroediting, and deductive and model-based imputations. Often, this editing is automated with limited manual intervention.

- *Output editing*
  This is the final editing, which focuses on the values as they are presented to the users. Aggregates in table cells or other statistical estimates are checked to see if the values are reasonable. One common method is to compare the results with those of earlier rounds if the survey is continuing or with external sources if the survey is a one-time event. Suspicious aggregates are treated with macroediting procedures.

From time to time survey organizations publish tables and other outputs containing large errors that are discovered too late, and worse, by the users. Large errors in the published results may escape the notice of all involved, due to the lack of information needed to identify the errors or because the error is such that a comparison with previous estimates is not a relevant method for all variables and surveys. The methods for output editing appear not to have been developed sufficiently, although there are graphical systems in use that seem reliable (see Houston and Bruce, 1993). It is important that data collection organizations have policies and action plans in place to deal with the unfortunate situation where large errors have been published. Such policies and plans should address customer concerns, establish and analyze the causes of the errors, and make sure that the entire organization learns from the mistakes.

7.4.3 Editing in Practice

Editing is an essential stage of the survey process, but the problems created by editing can be quite serious. Editing is costly and time consuming, despite all the new technology. As mentioned, in some surveys, editing alone can consume 40% of the entire survey budget (U.S. Federal Committee on Statistical Methodology, 1990). Extensive editing may delay release of the survey data, thus reducing its relevance to data users. The consensus of the survey community seems to be that the amount of editing should be based on cost–error optimization strategies. There are many alternative options for reducing survey error, and the amount of resources to devote to editing should be balanced against these other options, especially since there is evidence that many editing systems do not improve data quality appreciably. Extensive editing that does not achieve noticeable quality improvements is termed *overediting* by Granquist and Kovar (1997). They note that one explanation for overediting is that it is much simpler to deal with post–data collection problems than most other errors, such as nonresponse and measurement errors that occur during the data collection process. Survey producers who overedit the data are guilty of not operating by the total survey error optimization principles espoused in this book. They prefer to fix errors after the fact rather than adhering to Deming's idea of building quality into every stage of the process (Chapter 1).

A key lesson from contemporary research on editing is that not all errors in the data should be fixed, but rather, one should try to implement *selective editing*. Various studies show that selective editing can result in considerable savings in time and money without any degradation in the accuracy of the final estimates. In selective editing, less than 100% of all suspicious entries are reviewed. Instead, query edits are selected based on the importance of the sampling unit, the importance of the variable under study, the severity of the error, and the cost of investigating the suspicious entry. The importance of a sampling unit would usually be based on the size of the unit or the unit's weight in the estimation process. For an agricultural survey, larger farms would be given higher priority in the editing process than smaller farms, which contribute much less to estimates of production and other agricultural statistics. Units that have large weights have a larger potential for contributing to the mean squared error. Suspicious entries for large weight units would be given priority over units with smaller weights.

The cost of investigating the suspicious entries may include more than just labor and computer costs. Recontacting respondents to correct an entry increases respondent burden since each contact costs the respondent time and effort, not to mention the potential stress of being told that entries on the form are in error or are suspicious. This may reduce response rates for later waves of panel surveys or future implementations of a periodic survey. For example, in large-scale establishment and agricultural surveys, very large units are almost always in the sample, since their probabilities of selection make them *certainty units* (Chapter 9). Thus, contacts with these units to correct a data item should be minimized to avoid future reluctance to participate, or even nonresponse. These concerns also need to be taken into account by the editing process designers.

Another recommendation is to collect data on the editing process itself, referred to as *process data*, which are important in monitoring the quality and efficiency of the editing operation. The assumption is that if the process data indicate that the process is operating as designed, outgoing data quality should also be acceptable. Monitoring an operation for conformance with prespecified standards involves measurement and monitoring of the key process statistics for the operation. Table 7.2 provides examples of some key process statistics suggested by Granquist et al. (1997) that can be used for this purpose.

The specific variable estimates (6)–(9) in Table 7.2 should be ordered by size. All estimates (1)–(9) should be used in a continuing fashion to identify where the process is deficient and changes should be made. Such process changes might include omitting inefficient edits, concentration on the vital few that contribute most to the quality of survey estimates, or making changes to the questionnaire.

Recent literature on editing suggests that there are many alternatives to traditional exhaustive microediting. We have discussed various forms of selective editing, macroediting, and improvements through the measurement and analysis of key editing process variables. As pointed out by Bethlehem and van de Pol (1998), managers who prepare data sets for release may need to be convinced that these alternatives are not harmful to quality since they often represent a reduction in the amount of editing normally conducted for surveys. Historically, exhaustive editing has been viewed as something absolutely essential for survey quality; however, current thinking is that what matters is total survey data quality rather than perfection for any component of the survey process.

In Granquist and Kovar (1997) a number of examples are provided illustrating the fact that overediting is quite common. The concerns are that editing is very expensive and consumes a large part of many survey budgets and that the effect on estimates is often questionable. They argue that too many minor changes are being made at a significant expense. The minor changes have very small effects on the estimates. Hedlin's (1993) study of the Swedish Annual Survey of Manufacturing reports that 50% of the changes made during the editing process resulted in a less than 1% change in the final estimate. This result is consistent with general experiences in Australia, Canada, and the United States that it is a relatively small number of important changes that have a large effect on the estimate.

Table 7.2   Examples of General Key Process Ratio Statistics for Monitoring Editing Performance

| Process Variable | Definition of Ratio | | Purpose |
	Numerator	Denominator	
1. Edit failure rate	No. of objects with edit failures	No. of objects edited	An estimate of the amount of verification
2. Recontact rate	No. of recontacts	No. of objects edited	An estimate of the number of recontacts
3. Output editing rate	No. of objects verified in output editing	No. of objects verified	To check whether the editing process is well balanced between microediting and output editing
4. Correction rate	No. of objects corrected	No. of objects edited	An estimate of the effect of the editing process (a small value of variable 4 compared to a value of variable 1 indicates that the editing process is inefficient)
5. Recontact productivity	No. of objects recontacted with at least one imputed variable	No. of objects recontacted	An estimate of the efficiency of recontacts
6. Edit failure rate by variable	No. of objects with edit failures on variable X	No. of objects edited on variable X	An estimate of the amount of verification per variable
7. Edit failure rate by edit check	No. of objects with edit failures caused by edit check K	No. of objects edited with edit check K	An estimate of the amount of verification that each edit check has caused
8. Editing change rate by variable	No. of objects with changes on variable X	No. of objects with a value for variable X	An estimate of the effects of the editing process per variable
9. Edit success rate by variable	No. of objects with changes on variable X	No. of objects with edit failures on variable X	An estimate of how successfully the edits identify errors per variable

Source: Adapted from Granquist et al. (1997).

There is also a risk that editing can add errors, for example, as a result of illogical or erroneous editing rules. Granquist and Kovar provide the following startling example: "Overedited survey data will often lead analysts to rediscover the editor's models, generally with an undue degree of confidence. For example, it was not until a demographer 'discovered' that wives are on average two years younger than their husbands that the edit rule which performed this exact imputation was removed from the Canadian Census system!"

Editing is an essential part of the survey design, but it should focus on those changes that have a considerable effect on the survey estimates. The only way to identify the most efficient edit rules is to conduct experiments that compare edited and unedited data. Thus, resources put on editing must be balanced against other needs (see Chapter 10).

Editing can consume a large fraction of a total survey budget. To maximize the benefits of editing, results from the editing phase should be used to improve the survey processes; for example, to improve the questionnaire, interviewer training, and other survey processes that may cause edit failures. There is also a need to improve the editing process itself by using methods that more effectively detect potential errors. Macroediting and selective editing are examples of such methods.

7.5 CODING

7.5.1 Coding Error

Coding may not be necessary for all surveys, but for many surveys it is a very important operation as well as a potentially damaging source of error. Coding is a classification process in which raw survey data, often in the form of responses to open-ended questions, are assigned code numbers or categories that are suitable for estimation, tabulation, and analysis. Coding can be conducted manually by an operator or coder or automatically by specially designed coding software. Sometimes a combination is preferable where the computer codes the easiest situations and the coder codes the remaining ones.

The coding operation has three basic input components, as shown in Figure 7.1:

1. *Response.* Each element in a sample or a population is to be coded with respect to a specific variable by means of descriptions (answers to questions related to the variable) that are usually verbal.

2. *Nomenclature.* There exists a prespecified set of code numbers for this variable. This set consists of numbers (e.g., 00 to 99) corresponding to specific categories of the variable. A description of the category is associated with each code number. The set of code numbers is sometimes referred to as the coding *standard* or *nomenclature*.

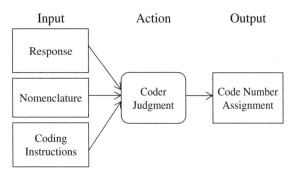

Figure 7.1 Generic coding process.

3. *Coding instructions.* There are coding instructions or rules that relate 1 and 2. These rules may be in the form of instructions to the coder for associating the key components of the open-ended response with the code numbers in the nomenclature.

Based on this input the coder makes a judgment and assigns a code number for the element.

The problems that can occur during the coding process are many, and these are not always realized by survey organizations. Some of these are:

- Most coding is susceptible to errors. The errors occur because the coding rules are not always properly applied, and the coding rules themselves are deficient. For example, nomenclatures cannot be constructed so that all possible open-ended responses are covered, and even highly skilled coders will often disagree about the proper code number to assign. For some operations, coder disagreement rates may be as high as 20%.
- Developing a quality coding operation is difficult since coding can be a highly subjective activity. Sometimes the open-ended responses are not adequate to assign a code number unambiguously, so coders have to use their judgment and read "between the lines" to code the response. Coding skills may require much time to develop, but unfortunately, staff turnover in these operations can be high.
- Coding operations can be quite large for major surveys and are difficult to manage. Controlling the error in such operations is challenging.

Coding operations can take a number of forms. Coding can be conducted manually at a number of different sites (referred to as *manual decentralized coding*) or manually in a single site (*manual centralized coding*). It can also be automated with manual coding of residual cases, or the coding can be manual computer-assisted. Typical forms of decentralized coding include coding performed by interviewers in connection with data collection, or coding performed by respondents. Centralized coding is coding performed within the survey organization by staff who are more or less specialized. *Automated coding* means that a computer program attempts to match the response descriptions to computer-stored nomenclature descriptions (dictionary) and then assigns code numbers when matches are acceptable according to specified matching criteria. Residual cases (i.e., cases that cannot be matched acceptably by the computer) are diverted to a manual coding operation. In computer-assisted coding, coders interact with the coding software. The software may present a list of code numbers that match well to the response description to the coder. The coder selects the appropriate code number from this list, or the software may code on demand when coders ask the software for assistance when they are in doubt about the correct code number.

Variables that typically require coding include industry, occupation, academic field of study, place of work, and home purchases. For these variables the nomenclature might contain hundreds of code numbers or categories. We assume that a true code number exists for each element and for each variable under study. *A coding error occurs if an element is assigned a code number other than the correct one.* This seemingly simple definition needs further elaboration.

First, it is often difficult to determine the correct code number, due to ambiguities in the response or even the characteristics of the element to be coded. The basic assumption is that each element belongs to one and only one category. In practice there are difficult situations where a specific description is such that it can be assigned different code numbers, depending on interpretation. An example of this is the response "telephone operator" to an occupation question. The nomenclature for occupation has, say, two possible categories for telephone operators: one for operators who work in telephone companies and the other for operators who work in other types of companies. Without auxiliary information this element cannot be coded on finer levels of the nomenclature.

Second, even if the response is detailed, problems might arise in assigning true code numbers. Studies show that the variation between coding experts can be considerable, and as a consequence, true code numbers have to be defined via an operational rule. One such operational rule is the majority rule. Let us say that three or more coding experts independently code a response. The true code number is then defined as the one assigned by a majority of the experts. In cases where a majority has not been reached, additional rules must be used.

Third, a response on, say, occupation may be coded correctly but still be the wrong occupation for the person. For example, an airline pilot filling out a mail questionnaire may answer the question "What is your occupation?" with "flight attendant" for some reason. The coder assigns the code 712, which may be the correct code number for this occupation. However, there is still an error here which should be attributed to measurement error on the part of the pilot rather than coding error. This emphasizes that part of the error for occupation coding is respondent error and part is coding error. Improvements in the coding operation may have little or no effect on respondent errors.

Most standards or nomenclatures are built systematically using a system of numbers. The first digit represents a main categorization; the second represents a more detailed categorization within the main, and so on. Since this system defines a series of levels, and since many statistical outputs (e.g., tables) use the higher coding levels more often than the detailed ones, an error that occurs on a high level, is considered more serious than one that occurs on a lower level. The main point is that as soon as an error occurs on a specific level, all subsequent levels are erroneously coded as well. For instance, an error in the first position of the code number affects the presentation of results on any level. On the other hand, coding errors that occur on lower levels only will not affect the outputs as long as these lower levels are not part of the outputs.

The coding error rates can be calculated in various ways. Error rates can be calculated for a specific survey variable, for a specific code number or code number level (digit position), and for individual coders. These error rates measure the gross errors since they measure the total number of misclassifications. However, misclassifications can be offsetting when elements are erroneously classified in or out of a particular code number. In that case, net errors, which may be much smaller than gross errors, may be more relevant. Typically, coding errors affecting tables are net errors. The following two examples provide some motivation for attempting to reduce coding error rates and improve coded data.

Example 7.5.1 Suppose that we want to investigate people with specific occupations for a health study. We have at our disposal a census file that contains occupations for all individuals. Our interest is restricted to miners, stonecutters, and house painters in our study of, say, pulmonary diseases. By means of the code numbers associated with these three occupations, it is possible to screen the appropriate subpopulation. When investigating the screened subpopulation, we detect some people that do not belong to the subpopulation because due to coding errors, they have been assigned erroneously one of the three categories we are interested in. The removal of these people is merely a financial and administrative problem. Much worse is the fact that an unknown number of miners, stonecutters, and house painters are hidden under false code numbers and we have no chance finding them.

Example 7.5.2 This example concerns labor force surveys and the estimation of parameters for gross changes (i.e., the total number of switches between different categories of occupation and industry). Some studies (Lyberg, 1981) have shown that relatively few changes between industry categories and between occupation categories are real changes. Most changes are due to coding errors. The publishing of such results would create an exaggerated picture of the mobility on the labor market. The solution to these problems is, of course, to try to minimize the coding error rates by various measures.

Numerous studies show that the coding error frequencies can be substantial (i.e., the gross errors can be large). For instance, in the 1965 Swedish Census of Population, the error rate for the variable industry varied between 8.2 and 14.5% at various points in the coding process. The higher rates were observed during early stages of the coding operation, and the lower rates were observed during later stages when coders were more experienced. In the 1970 Swedish Census the estimated error rate for the variable occupation was 13.5% and for industry was 9.9%, and for more simple one-digit variables such as relationship to head of household and number of hours worked, the error rates varied between 3.7 and 11.5%. In the 1975 Swedish Census the estimated error rate for occupation was 7.8% and for industry it was 3.5%. All one-digit variables had estimated error rates between 0.5 and 1.0%. The vastly improved error rates obtained in 1975 are explained by new procedures for error control that we will discuss later.

In the evaluation of the 1970 U.S. Census of Population, the estimated error rates for industry and occupation were 9.1 and 13.3%, respectively (U.S. Bureau of the Census, 1974). In a pretest for the 1980 U.S. Census, coders with experience from the U.S. Current Population Survey had estimated error rates for industry and occupation of 6.9 and 8.1%, respectively (U.S. Bureau of the Census, 1977).

During the last decades it has been difficult to find similar evaluation studies since most large organizations now use a mixture of automated, semi-automated, and manual approaches. Organizations that use manual coding tend not to conduct evaluation studies of their coding operations. There are exceptions, of course. One is an RTI study from 1991 described in Biemer and Caspar (1994), where estimated error rates for industry and occupation coding were 17 and 21%, respectively. Another is reported in Campanelli et al. (1997b) describing studies on the quality of occupational coding in the U.K. In their studies, estimates of correlated coder variance were obtained. The values for ρ_{coders} (see Chapter 5 for methodological details) were very small and design effects varied between 1 and 1.79, depending on which major occupational group they concerned. A vast majority of the design effects were between 1 and 1.13, showing that the variance inflation factor due to correlated coder variance was relatively modest given workload sizes between 300 and 400. Despite the lack of studies during recent years, the problem picture is clear. If coding is left uncontrolled, error rates are large, which in turn might lead to increased survey error and flawed analyses.

7.5.2   Controlling Manual Coding Error

There are basically two different methodologies available for controlling manual coding: dependent verification and independent verification. In *depen-*

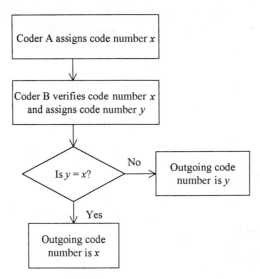

Figure 7.2   Dependent verification of coding.

dent verification production coder A codes an element. The code number assigned by A is then reviewed by verifier B. B inspects the code number and decides if it is correct. If it is considered correct, it remains unchanged; otherwise, B changes the code number to one that B thinks is the correct one. This scheme is represented in Figure 7.2.

Dependent verification is very inefficient. A common rule of thumb is that only about 50% of errors are corrected. Some studies show that the error reduction rate can be even smaller. The cognitive mechanism that creates this low change rate is that the verifier's judgment is strongly influenced by the code number already assigned. The tendency is that only obvious, unequivocal errors are corrected with this method. Less obvious errors tend to remain unchanged because the verifier often may reason that since the original code number is not absolutely wrong, it should not be changed. In other words, there is a tendency to defer to the original coder's judgment.

In *independent verification* the basis for such doubts on the part of the verifier is removed (i.e., the verifier does not have access to the originally assigned code number). With independent verification, the first coder, say coder A, assigns a code number to an element denoted by x_A. The same element is coded again by a second coder, say coder B, who assigns the code number x_B. Since the two coders work independently, neither is aware of the code number assigned by the other. The two code numbers are compared and the following decision rule is applied to determine the final, outgoing code number:

- If $x_A = x_B$, then x_A is the outgoing code number.
- If $x_A \neq x_B$, the element goes to a third coder, say coder C, who assigns the code number x_C.

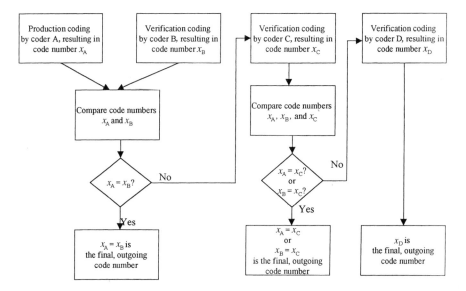

Figure 7.3 Two-way independent verification with adjudication.

- If $x_A = x_C$, then x_A is the outgoing code number.
- If $x_B = x_C$, then x_B is the outgoing code number.
- If x_A, x_B, and x_C are all different the element is coded by a fourth coder, say coder D, who decides (adjudicates) the final decision, x_D, as the outgoing code number.

This system is called *two-way independent verification with adjudication*. It is depicted in Figure 7.3. A variation of this approach used by RTI and others is to use only three coders and, if necessary, decide to assign x_C as the outgoing code number. Another variation is the three-way independent verification, which has three independent coders code every element and decide on the final code number using a majority code number rule and a fourth coder adjudication if all three initial coders disagree. However, Lyberg (1981) has shown that two-way independent verification with adjudication produces the same result and costs less than three-way verification with adjudication.

Is independent verification better than dependent verification? The basic assumption for independent verification schemes is that the outgoing code number generated by the decision rule is the correct one since the probability of two or three independently assigned code numbers being all in agreement and erroneous is quite small. However, it is important that the coders generating the comparison code numbers have approximately equal coding skills. Otherwise, the coders may seldom agree, and thus too many cases are sent to the adjudicator, thereby increasing adjudicator workload and potentially adjudicator error. In addition, it is possible for two poor coders to overrule one good coder simply because the former do not know the instructions properly. Let us look at the following example from mortality medical coding (Harris, 1974).

Example 7.5.3   The coding instruction states the following: "If diagnosis X is listed in the record, code number 111 should be assigned." An inattentive coder might stop here, assign code number 111 if diagnosis X is present, and move on to the next element. This coder would miss the second part of this instruction, which states: "However, if diagnosis Y is also listed, code number 112 should be assigned. Two inattentive coders that are paired with an attentive coder would result in an outgoing code number 111 and an error charged to the attentive coder.

Fortunately, situations such as this are rare in most coding operations, as various studies have shown. Although independent verification is more costly than dependent verification, it produces the correct code number much more often. The two-way independent system with adjudication is the least expensive of the independent schemes that we are aware of and it is used in many organizations around the world.

At Statistics Canada, independent verification is done within the framework of an acceptance sampling quality control system. In their system the verifiers are on different skill levels. The first-level verifier is more experienced than the production coder and the second-level verifier is more experienced than the first-level verifier. The first-level verifier independently codes a sample of the production coder's work. Disagreements are sent to the second-level verifier for resolution according to the majority rule. As the two verifiers are always more experienced than the production coder, the situation described above with two poor coders overruling a good one is unlikely ever to occur.

There are two methods for verification of coding, *dependent* and *independent verification*.

7.5.3   Automated Coding

The problems with coding manifested by large coding error rates, large quality variation in general, and high costs make it natural to consider using the computer to automate at least parts of the coding process. As we have seen, the computer is used extensively in editing, and computer coding can be viewed as an extension of these efforts. The first automation efforts date back to 1963 when the U.S. Census Bureau launched an automated system for its geographic coding. But the challenge lies in developing systems for the most error-prone and costly variables such as industry and occupation (I&O), education, and purchases relying on open-ended free verbal descriptions. Free open-ended verbal descriptions also appear in "other—please specify" response alternatives, which are common in many surveys. The basic features of any automated coding system are:

1. There should be a computer-stored dictionary or database comprising words or parts of words with associated code numbers. This is the equivalence to the nomenclature used in manual coding.
2. Responses are entered on-line or via some other medium such as scanning or keying.
3. Responses are matched with dictionary descriptions, and based on that matching and an accompanying decision rule, code numbers are assigned or responses are left uncoded.
4. By collecting and analyzing key process data the coding process is evaluated and improved continuously.

There are different levels of automation in systems used. One line of development is manual coding combined with computer-assisted coding (CAC), where the coders can ask the CAC system for help when they are not sure about which code number to assign. For instance, they can key the first three letters of a response and the CAC system provides a number of suggestions that the coder can choose from. Sometimes, the CAC software presupposes that responses are entered in standardized ways so that the software can be used to its full potential. As shown in Bushnell (1996), CAC can increase correlated coder variance since the system gives coders some individual freedom regarding the extent to which assistance is used. This potential problem is not considered an important drawback, however.

A second line of development is automated coding in batch mode. Here all responses to be coded are fed into the computer and are either assigned a code number or are left uncoded by the software. The uncoded (residual) cases are then sent to manual coding, which might or might not be combined with a CAC system. In automated coding, the *database* or *code dictionary* replaces the nomenclature descriptions used in manual coding. Construction of the dictionary can be based on the contents of coding manuals, but experience shows that using empirical patterns of responses is much more efficient. The actual compilation of the dictionary can be done manually or by means of a computer program.

There are two types of *matching* in automated coding: exact and inexact matching. In *exact matching* a response entered must be identical with a dictionary entry for a code number to be assigned. One might think that such a simple algorithm must result in large portions of responses not being assigned a code number, but it depends on the application and language. In Sweden, coding of consumer purchases and occupation data often have simple one-word structures and are suitable for exact matching. Coding degrees (i.e., the proportion of responses that are coded by the automated system) in such applications have been in the range 60 to 80% (Lyberg and Dean, 1992).

Allowing inexact matching can increase the coding degree but will usually increase the error rate. By *inexact matching* we mean that a response is considered a match if it is sufficiently similar to one of the dictionary entries. Simple types of rules for inexact matching include ignoring the word order of input descriptions, successive truncation of input descriptions, identification of highly informative words or phrases that are associated with a certain code number, and assigning heuristic weights to words where the weights are proportional to each word's information content. The importance of having sophisticated matching algorithms is somewhat overrated, however. Because the distributions of code numbers are often quite skewed (i.e., some code numbers occur with much higher frequency than other code numbers), concentrating on relatively few categories and common descriptions associated with these categories usually is very effective. Again, we see the power of the Pareto principle!

There are a number of key process variables that need to be studied during the automated coding process. Here are some examples:

- Coding degree (i.e., the proportion of responses coded automatically).
- Changes in coding degree after dictionary updates.
- Coding degree by category for manual versus automated coding.
- Cost.
- Coding error rate by coding mode (i.e., manual, CAC, automated), category, and dictionary update.
- CAC data on how often the system is consulted by the coder.

All these variables are easily measured except for error rates where there is need for a parallel expert coding on a sampling basis.

The goal of automated coding is to develop a dictionary that maximizes the coding degree, thereby reducing manual coding. At the same time, the dictionary should be such that the coding error for the automated part is very close to zero. The latter is accomplished by allowing only unambiguous entries in the dictionary. If ambiguous entries are allowed, a simultaneous assessment of coding degree and error rate becomes crucial. It is important first to achieve an acceptable error rate and then to increase the coding degree without increasing the error rate to unacceptable levels.

An ever-expanding dictionary is not always the best strategy for increasing the coding degree. For example, in a Swedish household expenditure survey the dictionary of purchases was updated 17 times during the year of data processing. The size of the dictionary increased from 1459 descriptions to 4230 descriptions during this process. After the third update the coding degree for the sample that was coded by that dictionary then containing 1760 descriptions was 67%. Despite a continuing addition of new descriptions, later versions of the dictionary never coded more than 73%. The occupation dictionary used in the 1980 Swedish census comprised 11,000 descriptions generating a coding degree of 68%. A doubling of the dictionary in the 1985 census gave virtually the same result. The explanation is the skewed distributions of code numbers assigned.

Most applications show cost savings compared to manual coding. Most large organizations use some kind of automation in their coding of difficult variables since it has become increasingly difficult to hire large pools of coders for temporary work in censuses and other large surveys. The U.S. Census Bureau has developed an expert system called the Automated Industry and Occupation Coding System (AIOCS). The AIOCS is designed to simulate manual coding by identifying informative words and less informative words, synonyms, misspellings, and abbreviations. When exact match occurs, code number assignment is straightforward. When matching is inexact, code numbers are assigned using probabilistic weights. The AIOCS has been able to code 50% of the cases with error rates around 10% (see Chen et al., 1993).

The U.S. Census Bureau has also developed another system called Parallel Automated Coding Expert (PACE). The PACE system uses data parallel computing techniques and is implemented on a massive parallel supercomputer. A large expert-coded database was used as a "training" device for the system, which uses an application of memory-based reasoning to identify nearest neighbors in the database. PACE considers in parallel all words provided in the responses. This input stream is then compared to the database and the system produces a set of possible near neighbors, and the final code number is based on a scoring algorithm. PACE has been able to code up to 63% of industry and occupation descriptions. System descriptions are found in Creecy et al. (1992) and Knaus (1987). Recent evaluations of automated coding in connection with the 2000 U.S. Census are found in Kirk et al. (2001) and Gillman (2000).

Statistics Canada has developed a system called Automated Coding by Text Recognition (ACTR). It uses word standardization techniques to match input text files to a reference file of phrases, the output resulting in a code number for the text. Like other systems it requires the user to provide a reference file of phrases or texts and their associated code numbers. A system description is found in Wenzowski (1996). The Australian Bureau of Statistics has developed the Australian Standard Classification of Occupations (ASCO) CAC system. This system is very efficient and probably superior to a fully automated system since it has many features that allow relatively untrained coders to perform in a way consistent with manual expert coders. Consistency with manual expert coding has exceeded 95%. The system has a user-friendly interface, providing potential matches and on-line help screens, shortened data entry descriptions, and fast searching and matching procedures. Other agencies using CAC include the U.S. National Center for Education Statistics, Statistics New Zealand, Statistics Netherlands, and the U.K. Office for National Statistics.

Automated coding can be performed in batch or with computer assistance. Verbal descriptions are matched with a computer-stored dictionary, and when acceptable matches occur, the software chooses or suggests a code number to be assigned. Any residual coding is performed manually. Error rates depend on the matching criteria used.

7.6 FILE PREPARATION

The last step in data processing is preparation of the data file. The file consists of individual data records, one for each sample unit. The file can be used by the statistical organization to produce estimates displayed in tables, diagrams, or other graphical means. The file can also be seen as part of a database from which users can compile their own estimates, for instance for specific subgroups. For the file to function properly, two things have to be done: (1) each responding sample unit should be assigned a weight, and (2) measures have to be taken which limit the risk of disclosing information on individual sample units.

7.6.1 Weighting

The principle behind any estimation procedure in a probability sample survey is that each sample unit represents many population units. When the sample is selected such that every member of the frame population has an equal chance of selection, sample means and proportions are good estimates of frame population means and proportions for the respondent stratum (see Chapter 3). However, when the probabilities of selection are not equal, a weight must be attached to each unit in order to obtain an estimate. The justification for this is discussed in some detail in Chapter 9. In addition, weighting can also compensate for nonresponse and frame noncoverage.

Usually, each unit in the sample is first assigned a base weight equal to the inverse of the selection probability. Then adjustment factors are applied to these weights to compensate for nonresponse and noncoverage. However, the final weight is the product of a number of factors, not just those mentioned. For example, when the sample is selected using multistage sampling, adjustment factors may be applied for different stage units (counties or districts, census tracts, etc.). These adjustments are intended to force conformance of the weighted sampling distributions for certain variables to external benchmark sources such as census population size projections.

The primary purpose of using these postsurvey adjustments of the base weights is to reduce the bias in the survey estimates caused by nonresponse and noncoverage. However, major adjustments to the base weights can be problematic since the adjustments indicate that some groups of the population are substantially underrepresented in the sample. This suggests that the responding sample is not representative of the target population. The adjustments serve to correct for nonrepresentativeness but are based on the strong assumption that the sample respondents within an adjustment group have characteristics similar to those that are missing within the group. Particularly in surveys where nonresponse and noncoverage are fairly extensive, post-survey adjustments can reduce the bias but are by no means a panacea for nonobservation errors.

When weighting adjustments are complex, there is a risk that they are not computed correctly and may even increase the mean squared error of an estimate. Although a brief introduction to weighting is provided in Chapter 9, the more technical issues in estimation and weighting are beyond the scope of this book. However, there is an extensive literature on weighting and the interested reader is referred to Horvitz and Thompson (1952), Elliott and Little (2000), Pfeffermann et al. (1998), and Hidiroglou et al. (1995a).

7.6.2 Disclosure Avoidance Issues

Virtually all national statistical institutes and many other survey organizations have policies regarding the release of macrodata and microdata to external users. *Macrodata* refers to files containing tabulations, counts, and frequencies. *Microdata* refers to files containing records that provide data about individual persons, households, establishments, or other units. The term *disclosure avoidance* refers to efforts to reduce the risk that a specific unit in the population is identified as a unit in the sample when such a disclosure could reveal information about the unit that is generally unknown. Thus, for any proposed release of tabulations or microdata, the acceptability of the level of risk of disclosure must be evaluated.

Quite often, there is a conflict between two competing needs. On the one hand, society needs detailed data on individuals, businesses, and organizations. On the other hand, respondents in samples of such populations must be assured that the information they provide is used in such a way that their confidentiality is protected. This means that macrodata released to the users should not be so detailed that individuals in the population can be identified and, further, that microdata files should be free from information on names, addresses, and other unique identifiers.

Since direct identifiers are easy to define, their removal from the released micodata files is straightforward. However, even when direct identifiers are removed, it may still be possible to *reidentify* (i.e., determine the identity of) a specific unit on the data file through analysis of the characteristics of the unit. This type of reidentification is sometimes referred to as *inadvertent direct disclosure* (i.d.d.). An i.d.d. occurs when two things happen:

1. An intruder (i.e., person trying to make a reidentification) recognizes an individual member of a population included in a macro- or microdata file.

2. The intruder learns something about that population member that he or she did not know from another source.

By combining information on a number of common variables such as geographic location, turnover, number of employees in business surveys and gender, age, and occupation in surveys of individuals, the risk of i.d.d.'s may be quite high unless steps are taken to limit the risk.

Example 7.6.1   Sweeney (2000) has experimented with 1990 U.S. Census summary data to determine how many individuals within geographically situated populations had combinations of demographic values that occurred infrequently. She found that combinations of few characteristics often combine in populations to uniquely or nearly uniquely identify some individuals. For example, she reports that 87% (216 million of 248 million) of the population in the United States had reported characteristics that probably made them unique based only on {5-*digit ZIP, gender, date of birth*}. About half of the U.S. population (132 million of 248 million, or 53%) are likely to be identified uniquely by only {*place, gender, date of birth*}, where place is basically the city, town, or municipality in which the person resides. Even at the county level, {*county, gender, date of birth*} are likely to uniquely identify 18% of the U.S. population. For this reason, publicly released data sets should not contain geographic identifiers at the county level or below, or date of birth, although age may be permissible.

Much of the research on confidentiality protection concerns the reduction of the risks of a unique identification of an individual unit. There are a number of methods developed to reduce these risks while preserving a balance between the needs of individual data providers. These methods, called *disclosure avoidance techniques*, should be applied in the preparation of the data file. Some of the techniques were mentioned in Chapter 3 since confidentiality assurance can be a means to enhance response rates.

Methods for Macrodata
The techniques available for disclosure avoidance for tabulations fall into three general classes: cell suppression, rolling-up the data, and disturbing the data. In *cell suppression*, a table value that has a high potential for disclosure is simply omitted and replaced by an asterisk or other symbol which indicates that the cell is omitted due to confidentiality concerns.

An example of *rolling up the data* is to combine the rows or columns of a table to form larger class intervals or new groupings of characteristics so that the number of cases comprising a cell exceeds the minimum cell size threshold. This may be a simpler solution than the suppression of individual items, but it tends to reduce the descriptive and analytical value of the table.

Disturbing the data refers to changing the figures of a tabulation in some systematic fashion, with the result that the figures are insufficiently exact to disclose information about individual units but are not distorted enough to impair the informative value of the table. Ordinary *rounding* is the simplest example. Figures in a table may, for example, be rounded to the nearest multiple of 5. However, there is a growing body of techniques for avoiding disclosure involving the introduction of random error into the figures to be published. By introducing "noise" into the file of microdata, the possibility of disclosure in any tabulations produced from the file is avoided. This method may simplify matters for the data producer, but it creates problems for the user (Dalenius, 1974).

Most large statistical organizations have rules in place that decide when tables have to be suppressed. For instance, there are rules that identify cells where the n largest contributors contribute more than $100k\%$ to the total cell value. The following example taken from Flygare and Block (2001) motivates the need for such rules.

Example 7.6.2   In a table showing total turnover for businesses in a specific industry, two companies dominate one table cell. The cell also contains values for three other companies that are very small compared to the two large ones. The table cell shows that the total turnover for these five companies is $3,295,000. The president of the second largest company knows that her own company's turnover is $921,000. She is confident that the three small companies together have at most half of her own company's turnover. She is now able to guess the total turnover of the largest company in the cell. If the three small companies contribute nothing to the cell's value, the largest company's total turnover is $3,295,000 - 921,000 = 2,374,000. If the three small companies contribute half of her own company's turnover, the largest company's total turnover is $3,295,000 - 921,000 - 921,000/2 = 1,913,000. Thus, if this particular table cell had been published, the president of the second largest company in that cell could have guessed that the total turnover of their greatest competitor is somewhere in the interval (1,913,000, 2,374,000).

Methods for Microdata

It has long been recognized that it is difficult to protect a microdata set from disclosure because of the possibility of matching to outside data sources (Bethlehem et al., 1990). Additionally, there are no accepted measures of disclosure risk for a microdata file, so there is no standard which can be applied to assure that protection is adequate.

To reduce the potential for disclosure, virtually all public-use microdata files (1) include data from only a sample of the population, (2) do not include obvious identifiers, (3) limit geographic detail, and (4) limit the number of variables on the file. Additional methods used to disguise high-visibility variables include:

- *Top coding* (or *bottom coding*). All values of a variable that exceed some maximum, say *M*, are replaced by *M*. For example, all incomes above 100,000 per year are replaced by 100,000. Bottom coding is similar except that values are truncated from the bottom.

- *Recoding into intervals.* The values of a variable are recoded into meaningful intervals, and instead of reporting the exact value of a variable, only the corresponding interval is reported. For example, the variable income might be recoded as <40,000, 40,000–100,000, and >100,000.

- *Adding or multiplying by random numbers (noise).* For each unit on the file, a random number is selected and added to the variable value.

- *Swapping or rank swapping* (also called *switching*). For a sample of the records, find a match in the database on a set of predetermined variables while swapping all other variables.

- *Blanking out selected variables and imputing for them* (also called *blank and impute*). For a few records from the microdata file, selected variables are blanked out and replaced by imputed values.

- *Aggregating across small groups of respondents and replacing one unit's reported value with the average* (also called *blurring*). There are many possible ways to implement blurring. Groups of records for averaging may be formed by matching on other variables or by sorting the variable of interest. The number of records in a group (whose data will be averaged) may be fixed or random. The average associated with a particular group may be assigned to all members of a group or to the "middle" member. It may be performed on more than one variable with different groupings for each variable.

In Citteur and Willenborg (1993) a review of disclosure avoidance techniques is provided.

When disclosure avoidance techniques are not satisfactory for research purposes, other options are available. For example, in the United States a researcher whose needs are not met by the "sanitized" public-use microdata file can usually pay the data producer to make special tabulations of the source file. This will provide the researcher with the same tables that he or she would have created had the full microdata file been accessible. In addition, restricted-use data files can be created which are made available only to serious researchers and under very strict guidelines. The files are essentially unmasked except perhaps by the deletion of direct identifiers. However, the researchers and agency representatives who want access to the file have to sign contracts where they swear not to breach confidentiality. Violations of the agreement may result in sanctions against the researcher or research institution or both, fines, or even imprisonment. In some cases, researchers can be named as temporary employees of the producing agency so that they can have free access to the data while working within the agency. Most other countries have similar procedures in place.

> *Disclosure avoidance techniques* are used to protect the confidentiality of data providers. One set of techniques can be applied on macrodata and another on microdata. It is essential that every survey organization adheres to the data protection rules that have been decided for each country.

The field of confidentiality protection and statistical disclosure methodology has generated considerable interest among statisticians. The literature on the topic is very extensive. Useful reviews are provided by Fienberg and Willenborg (1998) and Willenborg and De Waal (1996).

7.7 APPLICATIONS OF CONTINUOUS QUALITY IMPROVEMENT: THE CASE OF CODING

One way to decrease costs of verification is to administer the verification on a sampling basis using statistical quality control theory. The major *quality control methods* available are acceptance sampling, process control, and combinations of the two. A general reference is Ryan (2000). These are approaches that have been used for decades to ensure the quality of data processing operations. Typically, the items to be controlled by acceptance sampling are first divided into work units (lots), where a *work unit* is the amount of work that one operator (i.e., keyer, editor, coder, etc.) can do in a short time period, like a day or two. A sampling scheme is then implemented to select items from each work unit for inspection. The items are checked using a method such as independent verification. Any errors discovered are tallied and, in some cases, corrected. If the number of errors in the work unit is less than or equal to a prespecified acceptance number, the work unit is passed; otherwise, it fails and the entire work unit is reworked. At that point, the operator might be switched from sampling inspection to total inspection, where all items are verified until the error rate drops to an acceptable level. This inspection approach sometimes provides feedback to the operators to help them improve their performance. Typically, they are just given the results of the review. If they do not pass, they may also receive remedial training or instructions regarding what they need to do to avoid failing inspection in the future. A simplified scheme for acceptance sampling of a coder's work is shown in Figure 7.4.

Some statistical organizations, including Statistics Canada, have used acceptance sampling successfully to achieve good quality in survey data processing operations. However, some have questioned mass-inspection methods and the use of acceptance sampling (e.g., Deming, 1986; Biemer and Caspar, 1994). Some of the arguments against acceptance sampling include the following:

- Mass inspection is costly since a team of verifiers is needed to check the work and since it leads to reworking some work units.

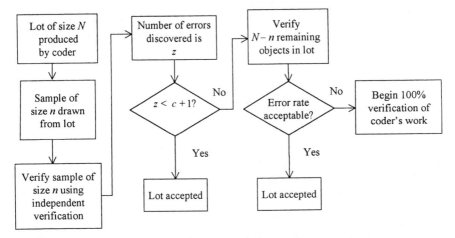

Figure 7.4 Acceptance sampling for samples of size n from lots of size N using acceptance number c.

- Coding studies have shown that dependent verification is often ineffective at identifying errors. Independent verification is much better, but if the error rate is small, 100% verification may be required to lower it further.
- Responsibility for improving quality is given to the verifiers rather than the production coders.
- Implicit in the inspection philosophy is the principle that operators are responsible for all the errors found by the verifiers. Indeed, that is why inspection provides feedback to the operators: to inform them that errors were found in their work units and that they need to do something to avoid errors in the future.

But the cause of errors may not be the operators but, rather, the way the operation is set up. Inspection seems to disregard this fact, which is primarily the reason that a number of quality experts have declared that quality cannot be achieved through inspection. If the procedures are ambiguous or the materials the operators have to work with are inherently deficient, inspection and verification will be limited in the degree of quality improvements they will achieve. For example, two coders who implement the instructions exactly could still arrive at different code numbers because of problems inherent in the coding process itself.

The literature on continuous quality improvement (CQI) distinguishes between two types of variation in the quality of some output from an operation: special cause and common cause. *Special cause variation* could arise because of errors made by individual coders, whereas *common cause variation* is due to the process itself. Initial quality improvement efforts should focus on

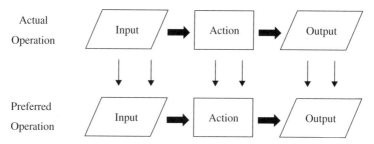

Figure 7.5 Continuous quality improvement. The inputs, actions, and outputs for an operation can be associated with preferred inputs, actions, and outputs. With each cycle of an operation, the actual operation is brought closer to the preferred operation by reducing the number of differences (nonconformities) between the two.

eliminating the special cause errors since they are usually responsible for most of the errors in a process. However, once the special cause errors are addressed, improvement efforts should focus on reducing the common cause errors. However, reducing the common cause errors will require changing the process in such a way that the error rate can be lowered. For example, it may require changing from manual to automated coding or even changing the nomenclature.

Quite often, common cause errors are mistaken for special cause errors. For example, operators are blamed for errors that are really system errors. For example, if all operators are making the same errors repeatedly, there is evidence that the cause of the errors is not the operator but what the operators are being asked to do. Thus providing feedback to operators based on observations of common cause variation does nothing to reduce these errors simply because the operator is not the problem. In fact, this type of feedback might demoralize the operators, who feel that the quality control system is unfair.

Given all these problems with acceptance sampling and similar quality control methods involving inspection, there is a need for alternative strategies for controlling quality in data processing operations. One method that we found does not have the drawbacks mentioned is CQI.

CQI is fundamentally different from inspection methods. It uses a team approach to improving quality based on the notion that data quality improvement is an iterative process. The basic idea of CQI is shown in Figure 7.5. This figure shows a typical operation consisting of some inputs, followed by the actions of the operators, which result in outputs from the process. There is the *actual process* that is currently being implemented and there is the *ideal* or *preferred process*. The latter process is one that is free of nonconformities of any kind (i.e., a process with no, or very few, errors). One way to define nonconformities in the current process is to compare its inputs, actions, and outputs with the preferred processes' inputs, actions, and outputs. Any differences between the actual and preferred processes are to be eliminated by the following five-step approach:

1. Conduct one cycle of an operation.
2. Identify the nonconformities in the operation based on the results from this cycle.
3. Discover the root causes of the nonconformities through a process that involves teams whose members represent all personnel with a potential to affect the quality of the operation.
4. Eliminate the root causes of the nonconformities by some types of activities or changes in the operation.
5. Return to step 1 to repeat this process for a new cycle of the operation.

Thus, CQI may be considered a process that continually removes noncomformities in an operation with each cycle of the operation process unit so that the operation comes closer and closer to the ideal operation. Since the ideal operation is one that has esssentially no deficiencies (usually an unattainable goal), CQI is continual and never-ending. Yet since each cycle brings new improvements, the operation will continue to improve over time. Theoretically, then, the error rate for an operation should be decreasing continually, but perhaps never reaching zero.

In actual experience, an operation may have many nonconformities and it is unrealistic to think that all of these can be identified and removed from the process in one cycle. Some problems may take a number of cycles to eliminate and the staff resources may be such that only a few problems can be addressed at each cycle. Therefore, a strategy is needed to determine which nonconformities to address at each cycle. A very useful tool for this purpose is the Pareto principle.

The *Pareto principle*, which we refer to several times in this book, is sometimes called the *80/20 rule*; that is, 80% of the problems in an operation arise from 20% of the nonconformities (or errors). The essential idea of the Pareto principle is that the nonconformities in an operation be ranked from most important to least important and that we start at the top of this list and work our way down, solving the most important problems first. The ordering is usually displayed in a Pareto chart (see Figure 7.6 for a simple example).

The Pareto principle suggests that the 20% of the activities operators perform in an operation are responsible for 80% of the errors made by operators in the operation. Of course, the 80/20 rule is an approximate one, but it is amazing how often the principle holds true for all sorts of processes. The CQI strategy focuses on root causes of errors rather than assuming that the operator is the sole cause. CQI views the operators as part of a team that includes supervisors and other workers who may be involved in the process. The goal of the team is to identify and eliminate the root causes of errors in the process wherever they can be traced. Sometimes the root cause may be the operators, who may not understand their jobs. But just as often the root causes can lie in the process itself or even "upstream," in the processes that precede the operation that is to be improved. As an example, a root cause of

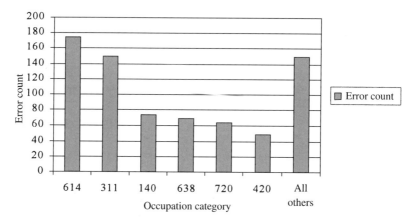

Figure 7.6   Pareto chart.

coding error could be the way interviewers collect the occupation and industry data. The coding quality team should be given the authority to recommend
changes in the data collection process so that further improvements can be
realized in the coding operation.

Biemer and Caspar (1994) describe an application of CQI for industry and
occupation coding that is used by RTI. Their approach can be summarized as
follows:

1. The coding operation is performed for a period of one week and a list
   is made of all the code numbers (categories) that were coded with at
   least one error.

2. A Pareto analysis is performed to identify the most frequently miscoded
   code numbers.

3. A coding quality team meets to discuss these code numbers and the
   causes of the errors in more detail.

4. Actions are taken to implement improvement measures selected by the
   quality team.

5. Effects of implementation are measured.

RTI used teams that were composed of coders, coding supervisors, and a
quality control specialist. The process studied was regular industry and occupation coding using two-way independent verification with adjudication in
cases where the production coder and the verifier disagreed. With this system
the adjudicator is considered an expert rather than just a third coder, so the
adjudicator is the final authority regarding the correct outgoing code number.
Any disagreement between the production code number and the adjudicator's
choice was considered a nonconformity or error.

On a weekly basis, the Pareto principle was applied for code numbers in error. Thus, the teams identified those code numbers or categories that were erroneously coded most frequently. On top of the list were five industry and five occupation categories. The teams met to discuss possible reasons that these categories were used erroneously so often. Many problems had their roots in the data collection phase. Responses to the industry and occupation questions often did not contain important information that the coders needed to distinguish between possible code numbers. The interviewers normally had no experience with coding, so they were not aware that this information was needed. Normally, it would not be a big change for them also to begin collecting this important information. It was rather a matter of developing better instructions and training interviewers in applying them. Therefore, new instructions were sent to the interviewers and interviewer training was revised to clarify the kind of information that was most useful to the coders.

The new procedures were implemented in the field and the responses improved dramatically, which in turn enabled the coders to code much more accurately. Many other changes were also made to the coding system as a result of the teamwork. The weekly meetings were devoted to the Pareto analysis mentioned but also to personal error listings, where coders received a listing of up to five cases that they had not coded correctly. The listings displayed the entire text of the response as the coder viewed it originally. The listing also showed the code number assigned by the other coder and the one assigned by the adjudicator, and any comments made by the latter. During the meetings coders were able to look at these examples and discuss how they had arrived at the incorrect code number. Supervisors could then provide explanations and retraining to reduce misunderstandings about the application of these code numbers and to increase the likelihood that these code numbers would be used correctly in the future. The adjudicators could also provide their rationale for assigning a particular code number to a case.

The weekly meetings were not restricted to Pareto analysis and personal listings, however. Coders and adjudicators were encouraged to bring up problematic issues related to their work environment, the quality of the information they worked with, and other demands on their time, which impinged on their ability to work efficiently. The quality expert would, when necessary, work as a liaison to upper management at RTI and staff in other RTI divisions whose decisions and work procedures affected the coding operation. In this way, the coding operation could also be improved by raising the individual coding skills and by improving the infrastructure in which the coding operation takes place.

All these changes brought about rather dramatic quality improvements. Figure 7.7 shows the result for industry and occupation coding in terms of error rates. What is plotted on the Y-axis is the coding error rate (CER), the number of cases with a difference between the production code number and the one assigned by the adjudicator, divided by the total number of cases verified. For the first two quarters of the year, one can see that CER for industry varied

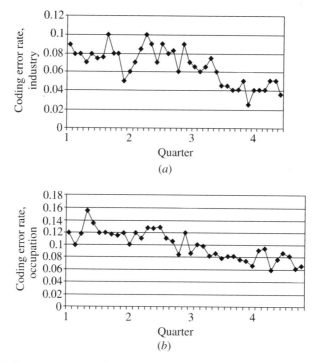

Figure 7.7 Coding error rates for (*a*) industry and (*b*) occupation during a one-year trial with continuous quality improvement. [From Biemer and Caspar (1974).]

around 8%. At the beginning of the third quarter, the CQI was implemented and ultimately the CER was reduced to 4%, a 50% reduction.

The results for occupation are similar. The first two quarters show no significant improvement pattern. The CQI was implemented prior to the third quarter, and we see a highly significant improvement during the last two quarters. The reduction in CER continued and without any cost increases. Over time, the cost might even be reduced, due to the fact that when CER decreases, fewer cases go to the adjudicator. The reduction of adjudication paid for the team meetings. In addition, the changes made to the coding system increased productivity to levels as high as at any time prior to the implementation of CQI.

Recently, the U.S. Bureau of the Census (2002) decided to develop coding verification procedures and checklists according to the methods discussed in Biemer and Caspar (1994). Of course, the CQI methods discussed can be used in other survey steps as well (see Chapter 10).

Despite what has been said about relying too heavily on acceptance sampling, there are situations where that kind of inspection is the only reasonable way to assure the outgoing quality (Mudryk et al., 2001b, 2002). If the coding process is unstable, perhaps because of a high coding staff turnover, there might simply be not enough time to develop new work procedures based on CQI. Even if CQI is used, we must be able to control the output of new coders or poor coders from going into the production stream at the time of coding. In this situation we can use acceptance sampling to control the output of new and poor coders while using the error information to provide the necessary feedback and training to eliminate the special causes of error. Then we could use the aggregated quality control results for two-way group feedback that would address common causes of error as well.

An approach like this might be seen as CQI within a quality control framework. Deming might have had that scenario in mind when he proposed Shewhart's control chart as the statistical tool to distinguish between special and common causes of variability. Juran might have had similar thoughts when he developed his Trilogy Quality Management Model, with quality control being at the center of this model as a way of generating quality improvements (Mudryk et al., 2002).

7.8 INTEGRATION ACTIVITIES

A number of authors have discussed the need for integrating the operations conducted within the data processing step as well as across all stages of the survey process; see, for example, Bethlehem (1997), Shanks (1989), Keller (1994, 1995), Pierzchala (1990), Baker (1994), and Weeks (1992). Traditionally, data processing is carried out in a centralized facility with separate processes for each operation. The processing is done sequentially, similar to an assembly line. Each operation may have a set of manuals with specifications describing how the operators and the operations are supposed to function. Numerous groups of people involved transfer substantial amounts of data down the line from process to process.

The advent of new technology makes it easier to carry out the data processing in a decentralized fashion, and as discussed above, technology can reduce the need for monotonous manual work. However, new technology has enormous potential to change the way that organizations process survey data in the future. We believe interactive processing will supplant batch processing as the operating standard. Rather than batches, work will flow through the process in streams and with much smaller processing cycles. Also, we should see much greater integration of processing steps, such as data entry, editing, tabulation, and estimation. Statistics Netherlands has a control center which is a user-friendly shell for taking survey data through all the required processing steps. In addition, the control center can produce both data and metadata files (see Chapter 10) in virtually any format, thus facilitating process analysis. Similar work on developing generalized survey processing software is being conducted at other agencies as well, for instance Statistics Canada (Turner, 1994) and Statistics Sweden (Blom and Lyberg, 1998).

CHAPTER 8

Overview of Survey Error Evaluation Methods

In this chapter we focus on methods and techniques for evaluating the non-sampling errors in survey work. Evaluation of nonsampling errors is a process whereby the contributions to total error of one or more survey error sources is assessed. The topic of survey error evaluation should not be unfamiliar to the reader of earlier chapters since throughout the book we have presented the results of various types of evaluation studies in discussing the methods for collecting survey data. For example, in Chapter 3 we considered the results of studies that evaluated the coverage bias and nonresponse bias in a survey estimator. In Chapter 4 we used the results of evaluations of the response process to illustrate various problems that can arise during the survey interview. Chapter 5 considered the results of interviewer variance studies and interviewer performance investigations. Chapter 6 considered the results of mode comparison studies, and finally, in Chapter 7 we presented some examples from survey evaluation studies to illustrate the types of errors arising from the post–data collection survey operations. The examples in earlier chapters provide ample evidence that survey error evaluation is an integral part of survey methodology since it provides a means of comparing one method with another to arrive at the best methods for conducting surveys.

8.1 PURPOSES OF SURVEY ERROR EVALUATION

As we have seen, the collection of survey data involves numerous compromises between survey costs and survey errors. In the end, all survey data are subject to errors from numerous sources. The goal of survey error evaluation is to assess, control, and/or compensate for the errors arising from these sources.

The purposes of a survey error evaluation are many and varied, depending on the stage of the survey process at which the evaluation is conducted. During the design stage of a survey, an evaluation may be conducted to compare competing design alternatives in order to choose the combination of design parameters that provide the highest data quality for the survey budget available. For example, an evaluation study may compare response rates, data completeness rates, response reliability, and costs for two modes of data collection with the goal of deciding on the preferred mode for a major study that is being planned. Such studies are often referred to as *pilot studies* and are a critical component in the planning of many major surveys.

Many surveys employ a number of evaluation techniques prior to fielding the study in order to *pretest* the survey procedures. The purposes of evaluations at the pretest stage are to identify potential problems in the survey design and to determine the best way to alleviate the problems. For example, the survey questionnaire might be pretested by interviewing a small number of persons in the target population. In this chapter we discuss a number of techniques for identifying flawed questions and determining how to improve them. In addition, the pretest may assess respondent attitudes toward the subject matter of the interview, interviewers' experiences as they conducted the interviews, difficulties encountered in completing the survey tasks, time required to complete the interview, and so on (see also Chapter 10).

Once the survey is under way, the evaluation efforts take the form of quality control and *quality assurance* techniques. Here the purpose is to control the survey errors as the data are being collected and to ensure that the production data meet certain standards for data quality. We may monitor the interviewers and verify their work to ensure that the interview guidelines are being followed. The questionnaire data may be inspected and responses checked for consistency with other responses on the same questionnaire. Various checks on the data may also be performed to ensure that the computerized survey instrument is branching correctly and otherwise guiding the interviewer correctly through the interview.

Finally, after the survey is conducted, a number of *postsurvey evaluations* may be conducted to fulfill a number of different purposes. First, we may want to collect data that can be used to optimize future survey designs. For this purpose, a survey designer would like to know how much of the total survey error is contributed by each of the major sources of error in a survey, such as nonresponse, frame coverage, the questionnaire, the interviewer, the mode of data collection, the respondent, data editing, and so on. However, determining the best allocation of survey resources to reduce nonsampling error not only requires information on the error components but also on how these are jointly affected by the many budget allocation alternatives that the designer may consider. Quite often, evaluation studies that have this objective will compare the data quality and costs from several competing design alternatives in order to arrive at the combination of design features that produces the most accurate data for the least cost.

A second purpose of postsurvey evaluations is to provide data users with information on data quality. Measures of nonsampling error indicating excellent or very good data quality create high user confidence in the quality of the data, while measures that imply only fair-to-poor data quality tend to have the opposite effect. Public reports on data quality that accompany the release of survey data are sometimes called *quality declarations* (see Chapter 10).

A third, but infrequently utilized purpose of postsurvey evaluations is to obtain data for use in adjusting the survey estimates for nonsampling bias. Postsurvey adjustments for nonresponse and coverage error are quite common in survey work (see Chapter 3). However, postsurvey adjustments to compensate for measurement bias are much less common. As we discussed in Chapter 2, evaluating the measurement bias in a survey estimate requires knowledge of the truth (i.e., where the bull's-eye is on the survey process target), and thus the survey data by themselves are not sufficient. Rather, estimation of measurement bias requires that additional data be collected, sometimes referred to as *gold standard data*, which for evaluation purposes are considered to be the truth. Some methods and techniques for collecting gold standard data are considered in this chapter.

Unfortunately, quite often, the total MSE of some of the adjusted estimates may actually be larger than the MSE of the corresponding unadjusted estimates. This occurs when the adjustment reduces the bias less than it increases the sampling variance. In addition, postsurvey adjustments require much more time in the schedule to conduct the evaluation, analyze the evaluation data, and then decide how best to incorporate adjustments into the estimates. For these reasons, there are few examples in the literature where these adjustments have been carried out successfully. Table 8.1 summarizes some of the purposes of survey evaluation at each stage of the survey process. A number of methods are available to the survey methodologist for addressing these objectives of survey error evaluation. Table 8.1 lists the most frequently used methods and the methods we study in this chapter. The methods in the table are organized in the order in which they might be used in the survey process (i.e., the survey design stage, the pretesting stage, the survey data collection stage, and the postsurvey stage).

Regardless of the method or the stage of the process where survey evaluation is implemented, a useful principle to apply for improving surveys is the well-known plan–do–check–act (PDCA) cycle found in the quality control literature (see, e.g., Scholtes et al., 1994). According to the PDCA principle, plans are first developed for the evaluation that specifies the evaluation objectives, the methods to be used in the evaluation and the analysis of the evaluation results, and the schedule and budget for the evaluation. Then the evaluation is conducted, the results are analyzed, and changes are made to the survey process on the basis of the evaluation results.

The next step in the PDCA cycle, assessment of the changes, is extremely important, although it is sometimes forgotten in practice. Prior to finalizing the enhancements to the survey process, further study is needed to ensure that

Table 8.1 Some Methods and Techniques for Survey Quality Evaluation

Stage of the Survey Process	Evaluation Method	Purpose
Design	Expert review of questionnaire • Unstructured • Structured	Identify problems with questionnaire layout, format, question wording, question order, and instructions
Design/pretest	Cognitive methods • Behavior coding • Cognitive interviewing • Other cognitive lab methods	Evaluate one or more stages of the response process
Pretest/survey/ postsurvey	Debriefings • Interviewer group discussions • Respondent focus groups	Evaluate questionnaire and data collection procedures
Pretest/survey	Observation • Supervisor observation • Telephone monitoring • Tape recording/CARI	Evaluate interviewer performance Identify questionnaire problems
Postsurvey	Postsurvey analysis • Experimentation • Nonrandom observation • Internal consistency • External validation	Compare alternative methods of data collection Estimate MSE components; validate survey estimates
	Postsurvey data collection • Gold standard methods (e.g., reinterview surveys and record check studies) • Nonresponse follow-up studies	Estimate one or more components of the MSE

the modifications that have been implemented achieve the intended results. Quite often, changes implemented to address one problem can cause other problems which may go undetected until the next time an evaluation of the process is conducted. For example, modifications to the questionnaire to enhance concept clarity could create additional burdens on the respondent which also increase nonresponse. Modifications to the CAPI or CATI instruments should be checked thoroughly to ensure that branching, data capture, and data editing features function properly.

Finally, on the basis of evaluation of the changes or modification, the changes are implemented survey-wide and the cycle begins again to plan for further improvement.

8.2   EVALUATION METHODS FOR DESIGNING AND PRETESTING SURVEYS

8.2.1   Expert Reviews

It is quite common practice for researchers developing survey questionnaires to seek the opinions of their colleagues on issues of questionnaire design. These *desktop pretests* typically involve unstructured reviews of the questionnaire design, wording of the questions, and the various tasks that respondents and interviewers are asked to complete during the interview. Desktop pretests may be carried out with individuals or in group settings using survey "experts" or even less expert fellow researchers who are asked to comment on the questionnaire from the perspective of a respondent. Even an informal review can be effective in identifying questions that may be misunderstood, confusing layout, misleading or complicated instructions, typographical errors, and other questionnaire problems. However, even when such reviews are conducted by experts in questionnaire design, there is a high risk that some important problems will still not be identified.

A more comprehensive review can be obtained using a *structured questionnaire review* approach like the one developed by Lessler et al. (1992). In the structured review approach, each question on the questionnaire is examined relative to a set of criteria based on the cognitive response process we studied in Chapter 4. Each criterion is assigned a code that is used to indicate the type of problem exhibited by each question. The coding system is used as a guide in the systematic appraisal of survey questions and helps the reviewer identify potential problems in the wording or structure of questions that may lead to difficulties in question administration, interpretation, or cognitive processing. The reviewer examines each question by considering specific categories of question characteristics in a stepwise fashion, and at each step, decides whether the question exhibits features that are likely to cause problems, such as problems with reading, instructions, item clarity, assumptions, knowledge or memory, sensitivity, response categories, and so on. At each step, the question coding system provides a series of codes (or shorthand descriptions) that identify the types of problems that are frequently found. In completing the appraisal, the reviewer indicates whether the problem is present by marking the corresponding code for the problem on the coding form, noting the reason the code was assigned. The reviewer may also record other comments on the item that do not fall under existing coding categories. A more complete list of the coding categories for an interviewer-administered questionnaire is provided in Figure 8.1.

Structured expert review is a relatively inexpensive method to implement during the design or pretest stage of the survey and can potentially improve data quality substantially. One disadvantage of the method is that it requires experts in questionnaire design who know how and when to apply a particular code. Further, sometimes the less experienced reviewers will identify issues

1. PROBLEMS WITH READING: Determine if it is difficult for the interviewers to read the question uniformly to all respondents.
    1a - WHAT TO READ: Interviewers may have difficulty determining what parts of the question are to be read.
    1b - MISSING INFORMATION: Information the interviewer needs to administer the question is not contained in the question.
    1c - HOW TO READ: Question is not fully scripted and therefore difficult to read.
2. PROBLEMS WITH INSTRUCTIONS: Look for problems with any introductions, instructions, or explanations from the respondent's point of view.
    2a - CONFLICTING OR INACCURATE INSTRUCTIONS, introductions, or explanations.
    2b - COMPLICATED INSTRUCTIONS, introductions, or explanations.
3. PROBLEMS WITH ITEM CLARITY: Identify problems related to communicating the intent or meaning of the question to the respondent.
    3a - WORDING: The question is lengthy, awkward, ungrammatical, or contains complicated syntax.
    3b - TECHNICAL TERMS are undefined, unclear, or complex.
    3c - VAGUE: The question is vague because there are multiple ways in which to interpret it or to determine what is to be included and excluded.
    3d -REFERENCE PERIODS are missing, not well specified, or are in conflict.
4. PROBLEMS WITH ASSUMPTIONS: Determine if there are problems with assumptions made or the underlying logic.
    4a - INAPPROPRIATE ASSUMPTIONS are made about the respondent or his/her living situation.
    4b -ASSUMES CONSTANT BEHAVIOR: The question inappropriately assumes a constant pattern of behavior or experience for situations that in fact vary.
    4c -DOUBLE-BARRELED question that contains multiple implicit questions.
5. PROBLEMS WITH KNOWLEDGE/MEMORY: Check whether respondents are likely to not know or have trouble remembering information.
    5a - KNOWLEDGE: The respondent is unlikely to know the answer.
    5b - An ATTITUDE that is asked about may not exist.
    5c - RECALL failure.
    5d - COMPUTATION or calculation problem.
6. PROBLEMS WITH SENSITIVITY/BIAS: Assess questions for sensitive nature or wording, and for bias.
    6a - SENSITIVE CONTENT: The question is on a topic that people will generally be uncomfortable talking about.
    6b - A SOCIALLY ACCEPTABLE response is implied.
7. PROBLEMS WITH RESPONSE CATEGORIES: Assess the adequacy of the range of responses to be recorded.
    7a - OPEN-ENDED QUESTION that is inappropriate or difficult.
    7b - MISMATCH between question and answer categories.
    7c - TECHNICAL TERMS are undefined, unclear, or complex.
    7d - VAGUE response categories.
    7e - OVERLAPPING response categories.
    7f - MISSING response categories.
    7g - ILLOGICAL ORDER of response categories.

Figure 8.1 Coding categories for a structured expert review system.

in the questionnaire that may not have very serious consequences for data quality, but even expert reviewers will vary in how well they avoid this problem. An experienced questionnaire designer should be able to decide what types of problems are important and how to solve problems that have been identified.

An expert review of the questionnaire is usually conducted as the initial phase of the questionnaire pretesting process. Once the questionnaire has been revised to address the problems identified in the appraisal, it may undergo another structured or unstructured review, and this review–revise–review process may continue for several iterations until the questionnaire designer is satisfied that all the major problems have been identified that can be by an expert review. At this stage, the questionnaire may undergo other forms of testing, such as cognitive testing using one or more of the methods to be described in the next section.

Expert reviews of questionnaires may be unstructured or structured. In an *unstructured review*, the reviewer reports the findings in any convenient format. A *structured review* is a more systematic consideration of all possible ways that the response process may be affected adversely by the questionnaire design.

8.2.2 Cognitive Methods

In the last two decades, cognitive methods have become an important component of survey pretesting activities (see Snijkers, 2002; Tucker, 1997). These methods involve getting input from respondents about the processes they use to arrive at survey responses. Like expert evaluations, these methods also make use of the model of the response process we studied in Chapter 4. By observing respondents as they are being interviewed or otherwise completing the questionnaires and getting them to talk about how they interpret questions, recall information, decide what information is relevant, and formulate answers, survey methodologists can learn much about the problems with survey questions.

The methods we study in this section are behavior coding, cognitive interviewing, and several special-purpose techniques: paraphrasing, vignette classification, and response latency. Although some of the methods are more appropriate for interviewer-assisted survey modes, most of the methods we discuss can be used to evaluate various aspects of the response process, regardless of the mode of data collection that will be used in the main survey.

Behavior Coding

Behavior coding is a systematic method for recording the frequency of certain types of behaviors that interviewers and/or respondents exhibit during an interview. Of particular interest are behaviors believed to be linked to survey error. Behavior coding is a technique that is particularly informative for systematic study of the interaction between the respondent and the interviewer during a telephone or face-to-face survey. As an example, the researcher may be interested in counting the number of times that an interviewer changes the wording of survey questions or fails to probe when a response to a question requires clarification or elaboration. The researcher may also wish to count the number of times that respondents ask for clarification of terms used in the questionnaire or express confusion about the meaning of a question. By analyzing these behaviors over many interviews, methodologists can learn much about problems with specific survey questions or the need to revise the interview guidelines and the various tasks associated with the interview.

Behavior coding is an important evaluation tool that can be used either at the pretesting stage or during production data collection. The focus in this section, however, is on behavior coding as a tool for questionnaire design and pretesting. Regardless of the stage of the survey process, the techniques for behavior coding are essentially the same. Survey questions can be behavior coded from either live or tape-recorded interviews; however, the latter method is usually more efficient and accurate. Tape-recorded interviews allow the analyst to code the interview at his or her convenience rather than having to rely on the interviewer's schedule, as is necessary for live coding. Since the interchange between the interviewer and the respondent can be replayed repeatedly by the analyst as necessary, coding reliability is increased. In addition, much more information about the interview can be documented when coding from tape than in real-time coding since with the latter, a much simpler coding scheme must be used to allow the coders to keep pace with the interviewers.

Behavior coders should have completed the basic interviewer training in order to understand the principles of interviewing and to know what is acceptable interviewer behavior and what is not. With adequate training, coders can code tape-recorded interviews with a high degree of consistency. Coding from live interviews, however, is usually much less reliable since the analyst is unable to "replay" the interchange between the interviewer and respondent in order to determine how best to code a particular behavior.

An example of a typical question-by-question behavior coding scheme appears in Table 8.2. To apply this scheme, the analyst listens to an interviewer reading a question and the respondent's response, then assigns one or more of the codes in the table. For example, if the interviewer reads the question with a major wording change, which was followed by a request for clarification from the respondent, the coder would code W2 and C for the interchange. The process continues in this manner for each question.

Table 8.2 Examples of Behavior Codes

Code	Description
Interviewer behavior codes	
E	Read exactly as worded
W1	Minor wording change
W2	Major wording change
P	Appropriate probe
F	Failure to probe/inadequate probe
Respondent behavior codes	
C	Respondent requests clarification
I	Respondent interrupts initial question reading with an answer

Table 8.3   Comparison of the Frequency of Behavior Codes for Questions A and B (Percent)

Behavior Code	Question A	Question B
E: exactly as worded	93	40
W1: minor wording change	2	23
W2: major wording change	3	13
I: reading interruption	0	17
C: clarification request	2	7
Total	100	100

Codes can also be recorded on coding sheets that include the code categories along the top and the question numbers down the left margin. Coding is then accomplished simply by checking a box under the appropriate code. An even faster method is to code directly into a computer database. This allows the results to be tabulated at any point during the coding operation. Table 8.3 presents the results of a behavior coding study consisting of 164 interviews. About 60 questions were coded using a simplified version of the coding scheme in Table 8.2. Table 8.3 contrasts the coding patterns for two of these questions. The first question, question A, was:

What was the purpose of your visit [to a health care provider]?

The question was quite simple and as can be seen from the table, interviewers read the question exactly as worded 93% of the time, and only 2% of the respondents asked for clarification.

The second question, question B, was much more complex. It asked:

How much did you pay or will you have to pay out-of-pocket for your most recent visit? Do not include the amount that insurance has paid for or will pay for. If you don't know the exact amount, please give me your best estimate.

There were considerably more problems with this question as reflected in the high rates of question rewording, respondent interruption, and requests for clarification. This suggests that the question may be too complex for respondents to comprehend in one reading. One way to remedy the problem is to divide the question into two or three questions as follows:

How much did you pay or will you have to pay out-of-pocket for your most recent visit? If you don't know the exact amount, please give me your best estimate.

Will you be reimbursed for any part of this payment by your insurance? (IF YES:) How much of this amount will your insurance cover?

Although this approach requires more questions, it may actually be less work for the respondent. Rather than requiring that the respondent understand all of what question B is asking and calculate mentally any out-of-pocket expenses not covered by insurance, only the information needed for this computation, not the computation itself, is required. Findings such as these are quite typical in behavior coding studies, especially for questions that have not been pretested before. However, behavior coding can also find problems with questions that have been pretested using other pretesting techniques, such as expert review and cognitive interviewing. For this reason, it is usually a good idea to use two or more methods for pretesting questions rather than relying on a single method to find all the problems.

The results from behavior coding investigations can often suggest the type of problem respondents may be having with the question and, therefore, how the question should be revised. For example, questions that are frequently repeated or reworded by the interviewer may be phrased awkwardly or may include words that are difficult to pronounce. Questions that are often interrupted may contain unexpected explanations or qualifiers at the end of the question that can be too easily ignored or forgotten by the interviewer. Questions that lead to requests for clarification may also be worded awkwardly, may contain vague or poorly defined terms, or may require a response that does not fit with the respondent's experience or frame of reference. However, since the method does not allow respondents to be questioned about the problems they encounter in the response process, identifying the appropriate actions to address the problems is still quite subjective. Thus, further testing and evaluation is usually required to ensure that the remedial actions have had the desired effect. In that sense, behavior coding should be viewed as an initial stage in the questionnaire evaluation and revision process.

Finally, as we saw in Chapter 5, behavior coding can also be used to evaluate interviewer performance on such interview tasks as question delivery, probing, feedback, following the instructions on the questionnaire, answering the respondent's questions about the survey questions, courtesy and politeness, and so on. The types of information that can be captured during a behavior coding review of interviewer performance is limited only by the coding scheme and the ability of the coders to assign the codes accurately.

Behavior coding involves the use of a coding scheme to classify various types of interviewer–respondent interactions during the interview into a small number of behavior categories. The purpose is essentially to identify types and frequencies of behaviors that are believed to have some effect on response quality.

Cognitive Interviewing

Cognitive interviewing refers to a set of methods for interviewing respondents so that the errors arising from specific stages of the response process (i.e., encoding, comprehension, information retrieval, response formatting, and communication) can be identified by the survey designer. Cognitive interviews are usually conducted in somewhat controlled settings such as the survey methodologist's office or a cognitive laboratory which is a room specially equipped for tape recording or video taping the interview. However, cognitive interviews can also be conducted in the respondent's home either in-person or by telephone. The interviews are also conducted during one-on-one sessions with the respondents usually for interviewer-administered questionnaires, although self-administered questionnaires can also be used. Usually, a recording is made of the interviews so that they can be studied in greater detail after the sessions.

As an example, Beatty et al. (1996) describe cognitive interviews that were conducted for the U.S. Behavior Risk Factors Surveillance System (BRFSS). In these interviews, a technique called *concurrent scripted probes* was used to elicit the respondents' understanding of the questions and the methods they employed in constructing a response. With this technique, the interviewer probes for information about the respondent's thought processes immediately following the response to a particular question.

The interviewer in a cognitive interview asked: "Now thinking about your physical health, which includes physical illness and injury, for how many days during the past 30 days was your physical health not good?" After some thought, the respondent replied: "About two days." Then the interviewer asked a series of questions about the process the respondent used in arriving at this answer, such as:

- How did you decide on that number of days?
- Describe for me the illnesses and injuries that you included in your answer.
- Did you have any difficulty deciding whether days were "good" or "not good"?

Interviewers were instructed to rely on their own judgments regarding which of the scripted probes to use. Unscripted probes were also allowed at the interviewers' discretion.

Another form of cognitive interview is the *think-aloud interview*, in which respondents are asked to think out loud regarding the process they use to interpret a question, retrieve the information needed to respond, formulate a response, and select a response alternative. For the concurrent think-aloud technique, respondents are instructed to report their thoughts at the same time that they answer the survey questions. For example, for the foregoing question from the BFRSS survey, the interviewer would ask the question in the same manner followed immediately by an instruction such as the following: "Now, tell me exactly what you are thinking as you try to answer that question. What is the first thing you are trying to think of?" . . . and so on.

With the concurrent think-aloud approach, interviewers obtain information about the response process as the respondent proceeds through the process.

In this manner, concurrent methods of cognitive interviewing can be an effective method for obtaining detailed information on the respondent's thought processes as he or she attempts to construct answers to survey questions. However, concurrent methods have the disadvantage of disrupting the normal flow of the interview. Therefore, generalizing the results from concurrent methods to typical survey settings is risky. In addition, there may be problems in execution of the method, since thinking out loud as one tries to respond to a question is difficult for some respondents.

Retrospective interviewing can be used to study the response process under conditions which are more similar to those of an actual survey. Corresponding to concurrent probing and think-alouds are the two retrospective interviewing variations: retrospective probing and retrospective think-aloud interviewing. With *retrospective probing and think-alouds*, respondents first complete an interview under conditions that are similar to the actual survey. Then the survey responses are reviewed and respondents are asked how they arrived at their answers. This may be done by using such probes as:

- What were you thinking when you responded to that question?
- How did you arrive at four days as your answer?
- What did the term *not good* mean to you?
- Did you have difficulty recalling whether you were ill in the last 30 days? How did you go about remembering your illnesses?

Think-aloud interviews and probing methods are widely used devices for studying the measurement errors arising from the response process, particularly:

- Comprehension errors due to difficult terms or ambiguous questions;
- Problems in adhering to the question format, for example, by probing verbal descriptions of health problems when a count of the number of health problems is required;
- Problems in identifying an answer category that fits the response; and
- Problems in recalling the requested information.

Problems discovered using this technique can then be addressed by questionnaire revisions, modification of data collection methods, interviewer training, and so on. Some examples will illustrate these uses of cognitive interviewing.

Cognitive interviewing is a collection of methods for studying the cognitive processes of respondents during the interview that use expanded or intensive interviewing approaches.

Example 8.2.1: Comprehension Error.   In an evaluation of the questions in the tobacco use supplement to the U.S. Current Population Survey (DeMaio and Rothgeb, 1996), the following question was tested using a retrospective probes protocol: "How many times during the past 12 months have you stopped smoking for one day or longer?" The intent of this question was to measure attempts to quit smoking. However, quite often, respondents thought of instances when they stopped smoking because they were ill, drank excessively the preceding day or did not have money to buy cigarettes, and other reasons besides an intent to quit smoking. The recommendation from the cognitive interviewing study was to revise the question simply by adding the phrase "because you were trying to quit smoking" to the end of the original question.

Example 8.2.2: Problem in Adhering to the Question Format.   In the study by Beatty et al. (1996) described earlier, the question "Now thinking about your physical health, which includes physical illness and injury, for how many days during the past 30 days was your physical health not good?" was evaluated cognitively. Although some respondents responded with a numerical answer (e.g., "two or three days") or answered in a way that implied "0," most gave answers such as "My health is good. I haven't seen a doctor in the last 30 days," or "I have a lot of aches and pains." The methodologists studying these results were quite surprised by how many respondents failed to respond with a numerical answer. The results suggested that the interviewers may be required to probe to a large extent to obtain an appropriate answer from respondents, thus introducing a potentially high level of interviewer subjectivity into the responses. These findings led to other investigations to identify the cause and remedy for the apparent failure to respond to the question as required by the question format.

Example 8.2.3: Problems with the Response Categories.   For a survey of medical practices, the self-administered questionnaire was evaluated cognitively using a retrospective think-aloud protocol. One of the survey questions asked "Is the repair and maintenance of your desktop computers centralized in one department or decentralized such that individual departments are responsible for their own repair and maintenance?" with two response categories: (1) centralized or (2) decentralized. Many respondents said that neither response category accurately described the organization of their computer services system and needed clarification as to what specific aspects of "repair and maintenance" were of interest. Particularly in large medical practices, some aspects of the services are centralized whereas others are decentralized. To remedy this problem, the response categories and the question were changed by replacing "centralized" and "decentralized" with "mostly centralized" and "mostly decentralized."

Example 8.2.4: Problems with Information Retrieval.   Both retrospective and concurrent think-alouds and probing techniques have been used to determine the mechanisms that respondents use to retrieve information. For example, retrospective probing conducted for the 1977 U.S. Census of Manufactures determined that many respondents were providing estimates of the number of employees working in the company based on memory rather than obtaining the figures from their companies' records. Sudman et al. (1996) provide additional examples where cognitive interviewing can help determine whether respondents are estimating the frequency of events, counting them, or some combination of both. They also provide examples where cognitive interviewing has been used to determine how respondents organize their thought processes to remember events and provide dates for them.

Although cognitive interviewing methods are powerful techniques for pre-venting important measurement errors in the data, some of their limitations should be mentioned. One obvious limitation of cognitive interviewing is the setting within which the interviews are conducted. Quite often, cognitive interviews are conducted in an office or "laboratory" setting. However, even when in a production setting, the evaluative nature of the interview is very dif-ferent from a real production interview. The cognitive interviewer is usually a researcher rather than an interviewer, and the probes, queries, and style of interviewing used during the interview may elicit cognitive processes that are not part of regular interviewing. Further, errors that occur during the response process that are due to satisficing are not likely to be replicated in a cognitive interview, which typically uses paid volunteers who are highly focused on the response task. For these reasons, combining cognitive interviewing with one or more other design and pretesting methods is usually best.

Other Cognitive Laboratory Methods
There are a number of other methods for learning about the difficulties that respondents encounter when attempting to respond to questions in a survey. This collection of methods, which includes cognitive interviewing, is referred to as *cognitive laboratory methods*. The methods are used to study one or more stages of the response process, usually focusing on a particular aspect of the process, such as comprehension, recall, or communication. Forsyth and Lessler (1991) provide an overview of many of these methods; however, we discuss only three of the more commonly used methods in this section: response latency, vignettes, and sorting.

Response Latency
Response latency refers to a set of techniques for measuring and analyzing the time it takes a respondent to answer a particular question. Correctly inter-preted, these data have been shown to be useful for providing insights into the response process. Computerized methods for conducting interviews, such as CATI and CAPI, have greatly facilitated our ability to measure response latency precisely. The computer can record the exact time the question is first presented on the computer screen as well as the exact time a response to the

question is entered. The difference between these two times can be used as a measure of response latency. As a result, response latency methods have become more generally available to survey researchers.

Bassili (1996) provides a number of examples of how response latency methods have been used to identify problems in the interpretation of questions, memory retrieval, and response selection. For example, by using the time that respondents take to respond to straightforward questions as a baseline, Bassili was able to identify questions that produced long delays in response and were therefore suspected of comprehension difficulties. One such question was: "Do you think that large companies should have quotas to ensure that a fixed percentage of women are hired, or should women get no special treatment?" For this question the response took 2.2 seconds on average compared with a lowest response latency observed of 1.4 seconds.

The interpretation of long and short response latencies depends on the type of question being studied. For example, longer response latencies to questions requiring memory retrieval are usually taken as indicators of more extensive cognitive processing and therefore more accurate responses. However, for questions that do not require recall, longer latencies may be indicative of excessive time required for comprehension and response formatting and, therefore, greater question difficulty. As we shall see, the effectiveness of the response latency method depends, to a large extent, on the techniques that are used to analyze the response-time data.

As an example, Bassili studied the relationship between latency times and the response alternatives that respondents selected. One of the things he discovered is the inordinate amount of time that respondents took to respond "don't know" to a question. A longer response latency followed by a "don't know" may be indicative of a true "don't know" rather than a "don't know" that is given in lieu of refusing to provide the requested information (i.e., a hidden refusal). Bassili used this same type of reasoning and analytical technique to assess the strength of attitudes or opinions in responses to opinion surveys.

The results of response latency studies can be quite useful for identifying potential problems in the questionnaire; however, extensive follow-up of the results may be required to verify the problems suggested by the latency results and to identify an appropriate remedy. Thus, we see this method working best when combined with one or more other methods listed in Table 8.1.

> *Response latency* refers to a set of techniques for measuring and analyzing the time it takes a respondent to answer a particular question.

Vignettes

For the hypothetical vignette technique, respondents are asked to read short descriptions of hypothetical situations that are presented to them as vignettes or stories. Then they are asked questions about how the hypothetical situation described in the vignette relates to a particular survey question or concept. For example, a hypothetical vignette may describe a behavior or activity of a fictitious person, and the respondent is then asked: "How should the person described in this scenario respond to the following question?"

A study at RTI conducted for the U.S. National Household Survey on Drug Abuse (NHSDA) used this technique to compare alternative wordings of questions that ask respondents about their possible use of prescription drugs for nonmedical purposes. In this experiment, subjects were presented with a question intended to identify the nonmedical use of various drugs available only by prescription. Then they were asked to read a series of vignettes and respond to the question for the character described in each vignette. For example, one version of the question was:

> Have you ever, even once, used a drug that is available only by a doctor's prescription and that was (1) either not prescribed for you OR (2) taken for the experience or feeling that it causes?

Three of the 11 vignettes used in the experiment follow.

Vignette 1. Jim Gillman has had some teeth pulled and the dentist gives him a prescription for Tylenol with codeine. He is supposed to take two pills every four hours, but because his mouth hurts so badly, he takes three pills every four hours. For Jim, what would be the correct answer to this question?

Vignette 2. Greg Wagner has become addicted to cough syrup. He goes to different doctors all over town with a fake cough, and gets a number of prescriptions for cough syrup. What is the correct answer to the question for Greg?

Vignette 3. While on vacation in Mexico, Elizabeth Clark gets an attack of diarrhea and buys some codeine, which does not require a prescription in Mexico. The next day she returns to the United States, where codeine does require a prescription. Elizabeth continues to take the medicine until the symptoms have gone away. How do you think Elizabeth Clark should answer this question?

In each case, the vignette was followed by an instruction such as the following:

> Mark the YES box if [name of person in the vignette] has ever used [name of drug in the vignette] without a prescription from a doctor, even once. Mark the NO box if he [or she] has never, even once, used [name of drug in the vignette] without his [or her] own prescription from a doctor.

YES ☐   NO ☐

By determining the question wording that resulted in the most correct classifications by laboratory subjects of the 11 vignettes according to the operational definition of nonmedical use of a prescription drug, researchers were able to determine the question wording that best conveyed the intended meaning of the question (Caspar et al., 1993).

The *vignette technique* asks respondents to read short descriptions of hypothetical situations that are presented to them as vignettes or stories and then to answer questions from a survey for the hypothetical situation described in the vignette.

One disadvantage of the vignette approach is the artificiality of the situation imposed on the respondent. Although respondents are asked to imagine themselves in various hypothetical situations, they are reporting about other persons rather than themselves. Respondents who are actually in the situations posed in the vignettes may respond quite differently. For example, a respondent who does not use illicit drugs may have no fear of disclosure or generate any social desirability bias when responding to questions about hypothetical drug use, whereas an actual drug user very well may.

Sorting

Sorting techniques can be applied to assist in a variety of questionnaire design decisions. The method for sorting is relatively simple. Respondents are given a set of objects that are arranged in no particular order and are asked to sort them according to some criteria. The data presented in Table 4.5 provide a potential application for sorting. Rather than asking respondents how uneasy they would feel about discussing certain topics in a survey, they might be asked to sort the topics from "most desirable" to "least desirable" topics to discuss. This would provide very similar information regarding topics that are considered "sensitive" by respondents. This method of sorting is sometimes referred to as *dimensional* since it asks respondents to sort a set of objects according to some criterion or dimension.

Free sorting allows the respondents to sort the set of objects by whatever criterion or method makes sense to them. For example, respondents might be given a set of questions to be placed on a survey questionnaire and asked to arrange them into groups of "similar" questions. For example, respondents may choose to group the questions by topic—health, income, activities, family structure, and so on—or by type—behavioral, attitudinal, demographic, and so on. The type of grouping preferred by a majority of respondents may suggest a natural method for organizing the questions on the questionnaire.

More sophisticated uses of the free sorting technique are reported in Brewer and Lui (1996). In one study, interest focused on the list of chronic conditions used in the U.S. National Health Interview Survey (NHIS). In that survey, respondents are given one of six possible lists of 30 or so chronic conditions and asked to indicate which conditions they or members of their families currently suffer. The question addressed by the research was whether the method of dividing the 100 items into six lists seemed logical and natural to respondents. Thus, a representative set of 68 chronic illnesses were selected from the NHIS checklist, and a label for each condition was typed on index cards. Then 70 subjects were asked to sort the conditions using the free-sort technique. Analysis was performed on the resulting groupings to determine how well they agreed with the current grouping of the items into the six lists. The authors concluded that the respondents' own natural groupings agreed well with the current NHIS grouping.

For the *sorting technique*, subjects are given a set of objects that are arranged in no particular order and are asked to sort them according to some criterion.

8.2.3 Debriefings and Pretest Observations

In this section we describe several additional techniques for designing and pretesting survey methods and questionnaires: debriefing methods, including focus groups, and pretest observation methods. *Debriefing* methods can obtain useful information for survey evaluation from both the respondents and the interviewers. Debriefings of interviewers are conducted routinely as part of the survey pretesting activities. A typical pretest might involve six to 12 interviewers who may conduct a dozen or so interviews each. At completion of the fieldwork, interviewers are brought together to describe and discuss their experiences with survey operations, the questionnaire, respondents' reactions to survey requests, and other aspects of the survey design. Their experiences in collecting the survey data (e.g., problems with the questionnaire, questions respondents asked about the survey, difficulties in accessing certain types of sample members) are usually very helpful in deciding how to change the data collection procedures to avoid the most common problems in future implementations of the survey.

A more formal and structured version of the interviewer group debriefing is the *focus group*. In a typical focus group, a moderator leads the discussion according to a predetermined discussion guide and raises topics for discussion with the group. The moderator may allow the participants to speak freely and step in to guide the discussion only if the group strays too far afield of the specific aspect of the survey being discussed at the time. Interviewers are invited to identify any problems or successes they encountered for each aspect of the survey and are encouraged to offer suggestions for addressing the problems.

Although interviewer debriefings and focus groups can provide valuable information regarding what problems may be encountered in the main survey with the present design, they can also be misleading. For example, one or two outspoken interviewers who are quite negative about some aspect of the survey can lead the group to suggest that some relatively minor problem is a major problem for the survey.

For example, in one such focus group, a rather vocal interviewer complained that in her experience, respondents were not receptive to the topic of the survey and that this is a signal that the survey will not achieve its response-rate goals. The other focus group interviewers were reluctant to disagree with the outspoken interviewer's strong opinion on the subject. Yet the main survey did not encounter this problem, counter to the prediction based on the focus group. This suggests that a single interviewer debriefing session should not constitute the sole method for identifying potential problems in the actual survey implementation.

Another difficulty is deciding how important some interviewer comments and criticisms are. Interviewers often raise issues about the reactions of respondents or the performance of the questionnaire that occur relatively infrequently. Therefore, it is important to determine how often interviewers encounter a particular problem to determine if it is widespread or just a fluke. In many cases, the problems raised can be deemed to be very unlikely to occur with any frequency. Despite the possibility of focus group contamination, interviewer debriefings are an excellent source of information regarding the strengths and weaknesses of a survey design.

Respondent group debriefings and focus groups can also be effective means of discovering areas of the questionnaire design that need improvement. As in interviewer debriefings, respondent focus groups usually involve a moderator to guide the discussion and six to 12 participants. Respondent focus groups and debriefing meetings may identify areas and issues in the questionnaire design that may otherwise have been ignored by the survey designers. These issues can then be explored further using think-aloud interviews or observations. However, as for interviewer debriefings, one or two vocal participants in a respondent focus group can dominate the discussion and lead the group to false conclusions regarding problems with the questionnaire, despite the efforts of the moderator to allow each respondent to participate equally in the discussions. Therefore, two or more focus groups should be conducted to corroborate the findings of the groups as means of guarding against focus group bias. In addition, this technique should be used in conjunction with other methods rather than as a stand-alone pretest evaluation technique. Ideally, input from interviewers and respondents should be complemented by more quantitative data.

Observing interviewers as they conduct their interviews is another frequently used pretest evaluation technique. For face-to-face interviews, this may involve field supervisor or senior members of the survey design team accompanying the interviewers as they conduct interviews. The observers may keep notes on any problems they observe with the interviewer's performance, the survey procedures, the questionnaire, the respondents' reactions to the questions, and any other activities required of the interviewers or respondents during the interview.

One difficulty with this method is the effect of the presence of the observer during the interview. This could cause the interviewer or respondent to react very differently to the survey procedures than they would in a normal survey situation. Therefore, some problems that may occur frequently in a more typical interview situation could be missed, while other atypical problems may arise due solely to the presence of the observer. Still, the method can be quite useful for identifying problems and major issues with the survey procedures.

A less obtrusive method of observation is to tape record the interviews for later analysis, possibly using behavior coding techniques to summarize the indicators of potential problems with the questionnaire or interview guidelines. This method has been made even less obtrusive using digital recording technology which works in conjunction with CAPI, a method referred to as *computer audio-recorded interviewing* (CARI) (Biemer et al., 2001). With CARI, an interviewer conducts a CAPI interview using the usual laptop computer, which also serves as a digital tape recorder. CARI's digital recording system uses the microphone built into the laptop and can be programmed to record the entire interview or any portion of it that may be of interest to survey designers.

To date, CARI has been used primarily for interview verification purposes (i.e., to detect and deter interviewer fabrication of interview data). However, Biemer et al. (2001) report that CARI has also been used successfully for questionnaire evaluation as well as for evaluation of interviewer performance in the U.S. National Household Survey of Adolescent Well-being. Studies suggest that the quality of CARI audio recordings is at least as good as recordings from tape recorders, without many of the problems inherent in using tape recorders in the field.

For pretesting surveys conducted by centralized telephone interviewing, telephone call monitoring is a frequently used method of observation. As in the case of face-to-face interview observation, the focus of telephone monitoring observations during the pretest stage is on the questionnaire, interview guidelines, respondent reactions to the survey requests, and overall feasibility and functioning of the survey *protocol*. Behavior coding methods may be used to summarize the findings, where the coders may code either live interviews or tape and digital recordings of the interviews.

The interviewer observation techniques as well as debriefing methods for pretesting survey questionnaires can also be employed during the survey to monitor data quality in the field or in a telephone facility. The only difference in how the techniques are applied may be the size of the observation sample and the focus of the observers. Uses of these techniques for survey quality evaluation are described in more detail in the next section.

8.3 METHODS FOR MONITORING AND CONTROLLING DATA QUALITY

8.3.1 Quality Control Methods

Quality control methods refer to a collection of techniques that are applied during the various phases of a survey to improve data quality, to ensure that it meets acceptable standards, and to collect information about survey data quality. Some of the quality control techniques have been discussed in other chapters of this book: for example, in Chapters 5 and 7.

Given the risk of nonsampling error at each stage of the survey process, the need for quality control methods is obvious. To be controlled, survey errors must be measured continuously during the course of the survey so that actions can be taken to prevent frequently occurring or critical errors from undermining the quality of the survey operations. When information on data quality is available on a real-time basis, survey managers can determine the degree to which errors are being controlled and can intervene when necessary to reduce errors.

Although it would be ideal to monitor the effect of nonsampling errors on the mean squared error as the survey progresses, this is rarely done. Estimating components of the mean squared error, such as reliability, measurement bias, interviewer variance, and so on, is possible only when special designs are imposed on the survey and the survey data are supplemented with other data, such as reinterview or administrative records data. Therefore, it is seldom possible to monitor every component of the mean squared error associated with a particular survey process (e.g., interviewer error) directly and continuously while the process is under way.

Instead, various indicators of data quality—response rate, data consistency rate, interviewer compliance rate, edit failure rate, item nonresponse rate, and so on—are typically measured and monitored. These quality indicators, called *key process variables* elsewhere in the book, can serve as proxy variables for the mean squared error components of interest. Since they are correlated to some extent with mean squared error components, maintaining the key process variables at desirable levels will usually ensure that the mean squared error components they reflect are also at desirable levels. Thus, the best indicators of quality are process variables that can be observed conveniently and continuously during the survey process and that are highly correlated with the components of error that need to be controlled.

As an example, as the survey data are being received from the field, the survey managers may monitor the item nonresponse for key questionnaire items to ensure that the items are being answered appropriately. A large item nonresponse rate could be an indication of a branching error in the instrument, a flaw in interviewing instructions, unanticipated respondent sensitivity to the item, or some other system defect that, hopefully, can be corrected before more data are lost to nonresponse. Actions taken during the survey to reduce the item nonresponse will pay dividends later in the analysis stage by reducing the bias and variance of the survey estimates derived from the item. This is especially important for critical survey items that will be used as dependent variables or independent variables for addressing a number of research questions.

Quality control methods implemented during the data processing phase— for example, during data capture, data editing, coding, and record linkage— produce a number of statistics that can also serve as quality indicators. For example, for data keying and coding, the verification procedures that are used in quality control (see Chapter 7) produce statistics on the agreement rates between the original keyer and the rekeyer or the original coder and the verification coder, as a by-product of the verification process. These data processing statistics are useful indicators of the quality of the data as they exit from these operations. Excessively high keyer or coder disagreement rates may indicate a problem with the data collection procedures or may suggest that the data processing personnel were not well trained. However, very low keyer and coder disagreement rates provide some assurance that these operations produced high-quality outputs.

In the next section we describe several methods for monitoring the quality of the interviewing in telephone and face-to-face surveys. These methods are useful not only to control the errors made by interviewers but also to assess the performance of the survey questionnaire.

8.3.2 Supervisory Observations During Data Collection

Rarely can errors that arise during the interview be detected simply by inspecting the completed questionnaires. The data may look quite plausible and satisfy all visible criteria for good data. Yet there may be hidden errors in the data caused by failures by the interviewers to read the questions as worded, probe appropriately, provide appropriate feedback to the respondents, respond accurately to questions from the respondents, and record responses correctly. Although interviewers may be well trained to conduct the interviews and to comply with the interview guidelines, they may fall into bad habits over time or may decide to take shortcuts with the procedures when pressed for time.

Observing the interviewers directly as they conduct their interviews can be quite effective at identifying certain types of interviewer errors so that corrective action can be taken to retrain the interviewers as necessary. Direct observation of interviewing can also detect other errors that are not caused by the interviewer. For example, the observations may reveal that respondents are encountering difficulties with interpreting the questions, recalling the requested information, or frequently may ask that certain questions be repeated. Respondent reactions to questions that were not anticipated in the questionnaire design is another common finding from direct observation of the interviews.

Monitoring and objectively coding the interactions between the interviewer and the respondent during the interview provide a systematic method for detecting many of the "invisible" errors that arise during the interview. When the behaviors of interviewers and respondents are summarized in this way, methods for addressing the errors in the process can be developed and the quality of the interviewing can be improved. The best methods for observing interviewers and respondents in a natural interviewing environment is through the use of unobtrusive interview monitoring.

Telephone Call Monitoring

For telephone surveys conducted in centralized facilities, call monitoring is a widely used technique for monitoring interviewer performance and controlling interviewer error. Usually, this type of monitoring is conducted by either a supervisor or a supervisory assistant in an area set apart from the interviewing area. Call monitors listen to the interviewers as they conduct their interviews and take notes on their observations. Usually, interviewers inform respondents at the start of the interview that a supervisor may listen in "to make sure I am doing my job properly." However, exactly when the call is being monitored is normally unknown to both the respondent and the interviewer. This is referred to as *unobtrusive call monitoring*.

In some cases it may be preferable to notify the interviewers when they are being monitored. For example, interviewers may recently have been trained on how to handle callbacks to reluctant respondents. An extension of the classroom training may involve intensive monitoring of the interviewers to observe their performances after training. In that situation it is more efficient to ask the interviewers to concentrate on these types of cases for a period of time so that monitors are able to observe a reasonable number of such cases; otherwise, the monitor may have to observe for longer periods of time to hear the same number of initial refusal cases unobtrusively.

Some monitoring systems employ a behavior coding scheme such as the one described in Section 8.2.2 to code systematically question delivery, probing, feedback behavior, and so on, on a question-by-question basis. Other coding schemes simply record various attributes of the interviewer's behavior and the interview, such as interviewer courtesy, knowledge, professionalism, manner as well as pace of the interview, tone of voice, and so on. An important aspect of call monitoring is timely feedback of monitoring outcomes to the interviewers so that any noncompliant behaviors observed during monitoring can be corrected promptly. As we saw in Chapter 5, the practice of monitoring telephone interviewers and providing feedback to them on their performances is believed to reduce the effects of interviewer variance on the survey estimates.

One advantage of the use of question-by-question behavior coding for call monitoring is the information obtained on the performance of specific items on the questionnaire. This information can be extremely useful for identifying needed improvements to the questionnaire for future implementations of the survey. Although questionnaire problems may be reported by the interviewers, systematically coding such problems provides an objective measure of the frequency with which such problems occur. In some cases, these problems can be identified during the current data collection, and changes can be implemented to rectify the questionnaire for all future interviews. Thus, telephone call monitoring can have immediate effects on the data quality by correcting the interviewers' behavior as well as important problems encountered in using the questionnaire.

Telephone survey organizations engaged in social science research routinely use unobtrusive call monitoring with timely feedback to the interviewers to control errors in the interviewing operation. A typical ratio of monitoring to interviewing is about one hour of monitoring for every 10 hours of interviewing. This ratio may vary during a survey so that interviewers are monitored more intensively (say, one hour of monitoring for each five to seven hours of interviewing) immediately following training, until the results from monitoring are stable and favorable, at which time it may drop to a lower level (say, a 1:10 or 1:15 ratio of monitoring to interviewing).

Field Interviewer Monitoring and Observation

For face-to-face interviewing, observations of the interviews have traditionally been conducted by essentially two methods: tape recording and supervisory observation (Lepkowski et al., 1998). The first method involves having the interviewers tape record some or all of their interviews so that these tape recordings can later be reviewed by their supervisors. More often, tape recording of interviews is reserved for special studies of interviewing procedures or the questionnaire. It is seldom used as a routine method of monitoring and controlling interviewer errors. The second method is supervisory observation where the supervisor (or supervisory representative) is physically present with the interviewer as the interviewer completes all or a portion of his or her assigned cases. The supervisor records (or codes) and evaluates the interviewer's performance during each interview, noting in particular any issues concerning the interviewer's compliance with the interview guidelines.

Both of these traditional methods suffer from two limitations. First, they are quite intrusive since both the interviewer and respondent are aware that they are being observed or tape recorded during the entire interview. Therefore, the methods change the interview setting in such a way that the behaviors observed may not be representative of typical interviewer and respondent interactions. In addition, in most countries field observations of interviewers can be costly and time consuming for large national surveys since supervisors must sometimes travel considerable distances to the interviewers' locations to observe them. For tape recording, the problems are related to the logistics of carrying, maintaining, and operating tape recorders in the field. The risks of mechanical failures and lost recordings are fairly high with this method.

Recent advances in computer-assisted interviewing technologies have now made it feasible to digitally record interviews unobtrusively directly on the hard disk of a CAPI interviewer's laptop computer using only the microphone that is built in to the laptop. The system is completely under software control such that at any predetermined section in the instrument, the recording can be switched on or off for all interviews or for interviews selected randomly. This function provides a capability that has previously been unavailable in a field environment: the capacity to provide an audio record of the interviewer–respondent verbal interaction without disrupting the normal interview process. This capability is very similar to the call monitoring used in telephone facilities for purposes of telephone interview call monitoring except that the feedback to interviewers regarding results of the monitoring requires more time, since recordings must be sent from the field to a centralized site where the monitoring is conducted. With these systems, unobtrusive monitoring of interviews is possible both for face-to-face interviewing in respondents' homes as well as decentralized telephone interviewing conducted from the interviewers' homes.

As mentioned in Section 8.2.3, the system for digitally recording the interview–respondent interactions during CAPI interviews is called computer audio-recorded interviewing (CARI). CARI potentially meets several critical needs specific to field interviewing. In general, it provides a means for monitoring the quality of the field interview, including the behavior of the interviewer during the interview and the reactions of the respondent to survey questions. CARI can potentially be used for a range of applications, including:

- Detecting interview fabrication and interview errors
- Evaluating interviewer performance and providing feedback to interviewers
- Collecting audio-based information for use in identifying questionnaire problems and for coding the interviewer–respondent interaction
- Recording information in response to open-ended questions

Research has indicated that the audio quality from computer recordings is at least on a par with recordings produced by tape recorders, but without many of the logistical problems associated with using high-maintenance equipment external to the standard CAPI laptop computer. Another advantage is the ability to program the computer to start and stop recording at specific points during the interview; for example, it may be important to record particular questions or sections of the questionnaire or just to have the computer begin recording at random points during the interview. In addition, the computer can control which interviewers and types of cases are recorded. This would allow, for example, more frequent recording of less experienced interviewers or respondents who satisfy certain prespecified criteria.

A disadvantage of recording is the potential effects on response rates. Before recording can begin, it is standard procedure to obtain permission from the respondent to record parts of the interview "for quality control purposes" or "to let my supervisor hear how I am performing my job." The evidence available from interviewer and respondent debriefings suggests that requesting permission to record using CARI has no appreciable effect on response rates (Biemer et al., 2001). However, an experiment to estimate the effect of recording on response rates has not yet been conducted and may be an issue for some surveys.

However, notwithstanding the potential effect on response rates and response quality, we believe that as the use of digitally recording interviewing becomes more widespread, many of the benefits realized from centralized telephone call monitoring will become accessible to dispersed field interviewing on a routine basis.

8.4 POSTSURVEY EVALUATIONS

While small-scale studies and cognitive methods can provide valuable insights into the response process and the sources of error in a survey, it is often desirable to estimate the magnitude of mean-squared-error components such as the various nonsampling error biases, interviewer variance, and other nonsampling error variance components. These objectives call for larger-scale investigations involving experimental designs or measurement methods that allow the estimation of the components of the mean squared error under investigation. Such investigations are discussed in the next several sections.

8.4.1 Experiments

Whether it be an investigation of alternative modes of interview, a test of alternative question wordings, or the estimation of interviewer effects, experimentation has played a pivotal role in the evaluation of survey data quality and the improvement of survey methods. An experiment is a data collection activity carried out under controlled conditions to discover an unknown effect or principle, to test or establish a hypothesis, or to illustrate a known principle. Typically, survey experiments involve randomly assigning the sample members or units to the various treatments (modes of interview, method of data collection, etc.) to be evaluated, tested, or compared.

One of the earliest uses of experimental design in survey work occurred in India in the mid-1940s when Mahalanobis (1946) randomly assigned interviewers in an agricultural survey in order to obtain evidence of interviewer biases. He referred to this randomization method as interpenetration, which we discussed in Chapter 5.

The use of randomized experimental designs in full-scale surveys is usually referred to as *embedded experiments* (van den Brakel, 2001). For example, in previous chapters, we described a number of experiments that varied some factor of a survey design, such as the interview mode, use of an advance letter, different methods of interviewer training, and so on. Such experiments are the mainstay of survey methods research. Experimental designs have also been used in small-scale field tests or pilot studies. For example, prior to conducting the main survey, we may wish to experiment with various levels of incentives to determine how best to use incentives in the main survey.

In the laboratory, where the experimental manipulation of survey conditions is easier, designs with very complex treatment structures are possible. As an example, O'Reilly et al. (1994) describe a laboratory study to investigate three modes of self-administration: paper-and-pencil interviewing (PAPI), computer-assisted self-interviewing (CASI), and audio computer-assisted self-interviewing (ACASI). In their design, respondents were interviewed twice with one of three combinations: PAPI–CASI, PAPI–ACASI, and CASI–ACASI. In addition, the order of administration was controlled so that, in effect, six treatment combinations were tested.

In the field, treatment structures are typically simpler, manipulating only one or two experimental conditions simultaneously. For example, the classic split-ballot experiment tests only two treatment conditions: the current data collection method (i.e., questionnaire, mode of data collection, etc.) and an experimental version of the method. However, the randomization of sample units to treatments can be quite complex. As an example, in a test of two types of interviewer training and two modes of data collection, interviewer training types were randomly assigned to groups of interviewers, while the mode of data collection treatment was randomly assigned to the households within the interviewer assignments. Such multilevel randomizations are usually necessary out of data collection concerns rather than for statistical efficiency.

Groves and Couper (1998) note the importance of going beyond single-factor experiments, such as the split-ballot design, for survey evaluations. Throughout the book we have seen many examples of multiple survey factors interacting in ways that affect various aspects of data quality. For example, interviewer characteristics can interact with respondent characteristics in ways that can be altered by a change in the mode of data collection. Such complex three-way interactions cannot be studied when only one factor is varied at a time. Further, when the factors interact, it is misleading to interpret the main effects (i.e., the effect of just the mode or some interviewer characteristic) without considering the levels of other interacting variables (e.g., the way the mode effect may change according to interviewer experience.) For these reasons, experimenters should carefully consider whether multiple factors should be manipulated in survey evaluation experiments. [Box et al. (1978) provide a good description of the interpretation of main effects and interactions in experiments.]

8.4.2 Observational Studies

Randomized experimentation is not always possible or desirable in survey work. Suppose, for example, we wish to determine the effect of establishment size on response rates. In particular, we wish to compare the response rates of establishments with fewer than 50 employees with those with 50 employees or more. To conduct a purely randomized experiment, the sample units in the experiment should be randomly assigned to the treatment groups to be compared in the experiment so that each treatment group constitutes a random sample of the original sample. However, this type of random assignment is not possible for comparing the smaller establishments with the larger establishments since size is a predetermined characteristic of the establishment. Nevertheless, we would still like to say something about the differences in response rates for larger and smaller establishments. This is possible using the methods that have been developed for observational studies.

In an *observational study*, the treatment assignments are predetermined prior to sampling and are not within the experimenter's control. That is, our interest is in making comparisons among groups of units where the groups are formed by the natural characteristics of the units, without any experimental manipulation. The treatments of interest in an observational study may be characteristics of the unit, such as age, race, or gender or characteristics of the data collection operation such as time of interview, mode of data collection, the length of the interview, and so on. Some of the characteristics can be manipulated by the experiment (e.g., we can randomly assign units to an interview mode) while other characteristics (e.g., race) are preassigned and not randomized. In the establishment survey experiment, we could still compare the response rates between smaller and larger establishments even though size was not randomly assigned. However, we can never be assured that the differences we observed are due strictly to size or to any number of other factors that may be associated with the size of an establishment. In an observational study, there is no statistical justification for attributing to the factors in the experiment the causes of the differences observed.

The analysis methods for observational studies are often very similar to those for randomized experiments; as examples, regression analysis, ANOVA, categorical data analysis, and correlation analysis can be used. However, the effects of various "disturbance" variables that can be eliminated in a designed experiment through randomization are present in an observational study. Consequently, it is critical for the analyst to control explicitly for these "confounding" effects in the modeling and estimation process.

Example 8.4.1 Many studies of the differences between proxy response and self-response are based on nonrandomized observation. Rather than the researcher determining which sample members should receive the proxy response treatment and which should receive the self-response treatment, the survey proceeds normally and the type of response (proxy or self) is recorded. The characteristics of persons providing responses by proxy in a survey may be quite different from those who provide self-responses. However, it is quite difficult to say why they are different. For example, the average household size for proxy respondents is usually larger than for self-respondents. In addition, compared with self-respondents, proxy respondents are disproportionately married, older, and female. Thus, the demographic differences between proxy and self-respondents could explain as much of the differences in their responses as the choice of respondent rule explains.

Variables such as marital status, age, and gender are sometimes referred to as disturbance variables, and the estimates of respondent rule effects are said to be "confounded" with these disturbance variables. In this case, the estimates of the effect of the respondent rule will be biased, sometimes referred to as *selection bias*. If the study design were completely randomized (i.e., if respondents were assigned randomly to each type of respondent rule by the researcher), the demographics for the two groups would be roughly the same except for variations due to an imperfect random assignment.

In an observational study, researchers may attempt to equalize the characteristics for the two respondent rule samples by weighting the observations so that the proportion of males, married persons, older persons, and so on, is the same for the groups of weighted observations. The hope is that these adjustments in the disturbance variables that are observed in the study will equalize all the variables that may disturb the comparisons, both observed and unobserved, so that comparisons are no longer confounded and the pure effect of proxy versus self-response can be estimated. Since the weighting adjustments will not remove all the effects of the disturbance variables, some selection bias may remain. Therefore, it is necessary to interpret the results of observational studies carefully as biases due to nonrandomization may still be present. Moore (1988) provides a critical review of a number of observational studies that attempt to assess differences between proxy and self-respondents.

8.4.3 Internal Consistency Studies

Another type of analysis that is often used to evaluate data quality is internal consistency analysis. This analysis involves comparing two or more variables collected in the same survey that are known to be strongly intercorrelated to determine whether they are indeed correlated to the extent expected. The variables may be redundant in that they attempt to measure the same characteristic. As an example, a survey on income may first ask a global question about the total household income from all sources and then ask more detailed questions about the income from specific sources. This dual strategy of collecting income provides an opportunity to compare the income report from the global question to the sum of the individual income reports. A large discrepancy between the two amounts would indicate a problem with the quality of the income data. Of course, in the absence of other information, there is no way of knowing which is more accurate.

Another type of consistency analysis involves gathering data on variables collected in a survey that measure very different characteristics but for which the relationship between the variables is known. For example, in an estab-

Table 8.4   Sex Ratios by Age for Black Persons 18 Years and Older

Gender of Respondent	Respondent's Age			Total
	18–29	30–44	45+	
Females	426	438	440	1304
Males	338	321	339	998
Sex ratio	1.26	1.36	1.30	1.31

Source: Maklan and Waksberg (1988), Table 1.

lishment survey, it may be known that the ratio of benefits costs paid to employees to total salaries should be between 25 and 45% for almost all firms. Any firm having a ratio of total benefits to total salaries outside this range would exhibit an inconsistency warranting further investigation.

Internal consistency studies may be ad hoc or planned. If they are ad hoc, the variables that are available for analysis may be few since redundancy is usually avoided in questionnaire design to reduce respondent burden. Ideally, internal consistency checks should be planned and built into the survey instrument, particularly if measuring data quality is a high priority for the survey. However, even when the measurement of data quality is not anticipated in the design stage, some level of consistency analysis is still possible for most surveys.

At a minimum, one can always compare the sex ratios for various demographic groups (assuming that the gender of the respondent is obtained in the survey) to determine if they are as expected. Tables of sex ratios by demographic group are usually available from the most recent census data. If the sex ratios for the survey differ significantly from the official numbers, this may be evidence of a bias in the achieved sample. Table 8.4 illustrates this type of analysis for a study of oral and pharyngeal cancer conducted by an RDD survey in 1985 in a restricted set of geographic areas in the United States. This analysis suggests a tendency for the survey to underrepresent black males.

However, internal consistency studies are usually more informative when they are planned and replicate measures of some of the key items in the survey are embedded in the questionnaire. For example, prior to 1999, the U.S. National Household Survey on Drug Abuse (NHSDA) used embedded replicate measures to check the consistency of reports of drug use. For marijuana use, one question asked: "How long has it been since you last used marijuana or hashish?" Then later in the questionnaire, the following question was asked: "Now think about the past 12 months from your 12-month reference date through today. On how many days in the past 12 months did you use marijuana or hashish?" If a respondent answered one day or more to the latter questions, the response to the former question should be no more than 12 months since the last use of marijuana. However, studies of the NHSDA have shown that about 1.5% of respondents respond inconsistently to this question. Since the proportion of the U.S. population that has used marijuana in the past year is roughly 8%, it is possible that 1.5/8.0, or roughly 19%, of past-year marijuana users answer these two questions inconsistently.

The results from internal consistency studies can be used by survey designers to identify problems with the interpretation or wording of questions as problems in the data collection process. In the case of the NHSDA, statisticians use the responses to both drug questions to estimate the prevalence of past-year drug use. For example, if the respondent indicates use of the drug in response to either question, he or she would be classified as a user for purposes of prevalence estimation.

However, studies of internal consistency can also be used to edit the data. For example, the Subcommittee on Measurement of Quality in Establishment Surveys (U.S. Federal Committee on Statistical Methodology, 1988) found that most federal surveys used some type of internal consistency analysis to edit the data and to control response error. About three-fourths of these surveys compute edit failure rates, and about half compute interviewer error rates as an indirect technique to measure response error. Information on the potential for measurement errors to affect the survey results can be gleaned from analysis of internal consistency rates as well as from debriefings with respondents and interviewers. These data can be very helpful in targeting areas that should be further evaluated using the methods described in this chapter.

Internal consistency analysis is quite often applied to data from panel surveys to check that data obtained at a previous time are consistent with the data from the current time point. For example, for monthly labor force surveys it is quite standard to compare the reported industry and occupation for the current month to the prior month's report. Too many changes from month to month, particularly for a person's industry classification, could suggest that there are response errors in the industry and occupation data. In fact, the U.S. Census Bureau observed this type of inconsistency in the industry and occupation data collection in the Current Population Survey. This prompted a change in the methodology for collecting these data to a dependent interviewing approach.

With the *dependent interviewing* approach, the preceding month's report on industry is preloaded into the CAPI instrument for every respondent interviewed in the previous month (Brown et al., 1998). Thus, when the respondent provides a report for the current month, it is checked against the prior month's report and discrepancies between the two reports are verified by the interviewer immediately to ensure that the industry actually changed. For example, the interviewer asks: "Have you changed jobs since the last interview?" A similar approach is used by the U.S. Department of Agriculture, where farmer-reported planted acreage from the spring agricultural survey is checked against responses for the same question and the same farmer from the previous fall survey.

Table 8.5 Comparison of Household Size for the RDD Survey and the March 1986 CPS

Household Size	RDD (%)	CPS (%)
1 person	20.9	22.7
2 persons	34.1	31.7
3 persons	18.6	18.1
4+ persons	26.4	27.5

Source: Maklan and Waksberg (1988), Table 5.

8.4.4 External Validation Studies

External validation is a process for evaluating survey estimates in which estimates from the survey are compared to external estimates, which are considered to be more accurate. *External estimates* may be estimates from the most recent census or from a survey that is considered to be a gold standard for characteristics to be evaluated. They may also be estimates derived from administrative records that are considered to be highly accurate: for example, income estimates from the tax authorities for evaluating survey estimates of income. Thus, to the extent that the external estimates are accurate, the bias in the survey for the estimates in the comparison can be assessed.

As an example, in the United States, the Decennial Census and the Current Population Survey are considered to be highly accurate surveys for assessing the bias in a wide range of survey estimates. For example, for national surveys, the survey estimates of gender ratios, distribution of age and race groups, and many other demographic characteristics can be compared to the corresponding estimates from the Census or the CPS to evaluate any distortions in the survey sample. This is shown in Table 8.5, for an RDD study on adult smoking behavior. The table compares the household size, including persons of all ages, for the RDD survey and the CPS for the comparable time frame. The data show a tendency for the RDD survey to underrepresent single-person and 4+ households while overrepresenting households around the median size of 2.7 persons. We might speculate that the former result is due to unit nonresponse, particularly noncontacts for single-person households in the telephone survey, and the latter result is due to a potential problem in the telephone household rostering procedures, particularly for larger households. However, further investigation would be needed to understand the root causes of the results in the table and to address the problems indicated by these data.

8.4.5 Administrative Record Check Studies

In this section we discuss a method for evaluating individual responses from a survey by comparing the survey response to a value for the individual unit obtained from an administrative record of some type. In many countries throughout the world, data from records and population registries are accessible to researchers, not only for evaluation purposes, but also for population estimation. In that case, some of the estimates in a survey program may be based solely on administrative records data. It may not even be necessary to conduct a survey if the characteristics of interest in a study can be obtained from administrative records.

In this case, use of administrative records can be quite economical since no new data need be collected. Unfortunately, administrative records for many characteristics of interest in a survey are not readily available and accessible or may not exist. When they are available, the expense of obtaining administrative records data for evaluation purposes may be considerable due to the difficulties of finding records for specific individuals in the sample and creating variables from the records that are comparable to the survey variables to be evaluated. Consequently, record check evaluation studies occur relatively infrequently in the survey methodology literature. Nevertheless, they are a very important method for assessing survey error.

For this technique, the survey responses for units in the survey for some characteristic (e.g., age) are compared with the corresponding values for this characteristic obtained from administrative records (e.g., birth certificates). The usual assumption is that the administrative record value is the most accurate or *gold standard value*, and thus any difference between the survey response and the administrative record value is attributable to error in the survey response. In this way, the bias in the survey estimates of the characteristic can be estimated. Administrative records that have been used in survey evaluations include federal/state income or sales tax reports, birth certificates, licensing information, population and government welfare registers, police records, and other special-purpose records.

In one type of record check study, a sample of survey units is drawn from the sampling frame and then the corresponding administrative records for the units are located. For example, the researcher may begin with a sample of households, collect information on the characteristic of interest in the survey, and then attempt to find information for the respondent in the appropriate administrative record system in order to compare the survey and record values. This type of study is referred to as a *forward record check study*. Alternatively, the researcher may start with a sample of records and then interview the corresponding households to obtain the information that is contained on the records. Referred to as a *reverse record check study*, the latter study design is usually more efficient because the precision of the estimates for rare items can be better controlled. For example, if the study design calls for 500 persons who have incomes in the lowest 5% of the population, a very large household sample would be needed to ensure that 500 are selected. However, with a reverse record check sample, exactly this number could be chosen from administrative records containing income data.

As mentioned previously, a key assumption in the use of administrative records for evaluation purposes is that the records contain essentially error-free information (i.e., true values) for the survey characteristics of interest. Under this assumption, estimates of the measurement bias can be obtained by forming an estimate from the survey data and subtracting the corresponding estimate from the record data. However, four problems typically plague the method and limit the usefulness of the record data:

- The time periods for the record data and the survey data may not coincide.
- The characteristic(s) being reported in the record system may not be exactly the same as the characteristic(s) being measured in the surveys.
- To save evaluation study costs, the record study may be confined to a very restricted geographic area and inferences beyond this restricted population may be invalid.
- Administrative records can be prone to error, sometimes considerably so.

Privacy limitations limit access by the research community to administrative records. Thus, the method may not be feasible even when suitable records for evaluation exist. Administrative records are subject to nonsampling errors as well; however, they may still be useful for evaluation purposes as long as the errors are small relative to the bias in the data being evaluated. That is, the records provide a "silver" rather than a "gold" standard. There are examples, however, of official statistics whose accuracy is inadequate for evaluation purposes. A study conducted by the U.S. Office of Management and Budget (U.S. Federal Committee on Statistical Methodology, 1980) provides some guidance on methodological requirements to conduct record checks.

An example of a use of administrative records for comparing biases associated with collecting income by telephone and face-to-face interviewing is provided by Körmendi (1988). In a study conducted in Denmark, Körmendi obtained information on gross and net income from tax authorities for respondents from a telephone and a face-to-face survey to examine the extent to which the two interviewing methods resulted in systematically different answers. Table 8.6 displays the comparisons. Treating the amounts from the tax authorities as the gold standard, the difference between the survey estimate and the tax authority amount is an estimate of the bias in the survey estimates. The data in the table suggest that there is a statistically significant bias in the reports of income for women in the telephone survey (−4000 kr) and for men in the face-to-face survey (7000 kr).

8.4.6 Reinterview Studies

One of the most widely used methods for estimating the components of measurement error in a survey is the reinterview survey. In a *reinterview survey*, respondents from the original survey sample are revisited and reasked some

Table 8.6   Estimates of Gross Income in a Telephone and a Face-to-Face Survey Compared with Income Information from Tax Authorities

Mode of Interview	Survey Estimate (1000 kr)	Tax Authority Amount (1000 kr)	Difference (1000 kr)
Telephone	106	107	−1
Men	139	137	2
Women	70	74	−4**
Face-to-face	108	107	1
Men	138	131	7*
Women	78	81	−3

Source: Körmendi (1988).
* Significant at the 5% level.
** Significant at the 1% level.

of the same questions that were asked in the original survey. The objective of reinterview is to collect information for the same characteristics as the original survey, so reinterview questions are designed to reference the same time points and time intervals that were referenced in the first interview. Under various assumptions about the error in the reinterview, components of the error associated with the first response can be estimated.

As shown in Table 8.7, there are two general types of reinterviews that we discuss: gold standard and replicated reinterviews. The *gold standard reinterview*, as the name implies, attempts to collect information that may be considered highly accurate for the purposes of estimating measurement bias or for evaluating survey methods; that is, it seeks to measure the truth. The *replicated reinterview* may be of two types: a test–retest reinterview or simply a repeated measurement reinterview. *The test–retest type of reinterview* seeks to replicate the original survey process; that is, it is intended to obtain a second set of measurements that is subject to the same nonsampling errors as the original survey measurements. The objective of the test–retest reinterview is to estimate simple response variance and response reliability, often referred to as *test–retest reliability* (see Chapter 2). For *repeated measurement reinterview* designs, the only requirement is that the nonsampling errors in the second measurement be uncorrelated with the nonsampling errors in the original survey measurement. This type of reinterview is the most general of all reinterview designs. For example, the first measurement may come from a mail survey and the second measurement from a telephone reinterview survey of the mail survey respondents. Due to the different modes of data collection, the second measurement is neither a replication of the first nor a gold standard measurement. These reinterview designs are now discussed in more detail.

Gold Standard Reinterview Studies
When accessing administrative records is not feasible because of their unavailability, high cost, privacy concerns, or other reasons, a reinterview approach

Table 8.7   Comparison of Gold Standard and Replicated Reinterview Survey Designs

Type of Reinterview	Goal of the Design	Objective of the Analysis
Gold standard		
Reconciled reinterview, no determination of causes	To obtain highly accurate responses	Estimation of measurement bias
Reconciled reinterview, determination of causes	To determine the source of the error	Interviewer performance evaluation; evaluation of survey procedures
Probing or expanded reinterview methods	To obtain detailed information that can be used to infer a true value	Estimation of measurement bias; evaluation of survey procedures
Replicated reinterview		
Test–retest reinterview	To replicate the original interview process	Estimation of simple response variance and reliability
Repeated measurements	To obtain additional measurements of the original interview characteristics, possibly using different methods	Estimation of measurement variance and/or measurement bias using model-based methods such as latent class analysis

that aims to obtain the true values of the survey variable may be a feasible alternative. Usually, this type of reinterview is confined to a small number of "key" survey characteristics, so that more time can be taken during the interview and greater attention can be given to the reporting task in an effort to obtain highly accurate data.

> The *gold standard reinterview* is a reinterview aimed at obtaining highly accurate (i.e., essentially error-free) measures of a subset of survey items for the purpose of estimating measurement error bias.

One form of the gold standard reinterview is the reconciled reinterview. In a typical *reconciled reinterview survey*, a subsample of respondents who were originally interviewed days or weeks before is recontacted by a reinterviewer who is typically a supervisor or other senior interviewer. The reinterviewer asks the respondent a subset of the original set of questions and then compares these responses with the responses from the original interview. If the responses all agree, the reinterview is completed. If one or more responses do

Table 8.8   Basic Interview–Reinterview Table for a Dichotomous Variable

	Interview Response		
Reinterview Response	1	0	
1	a	b	$a + b$
0	c	d	$c + d$
	$a + c$	$b + d$	$n = a + b + c + d$

not agree, the reinterviewer is instructed to inform the respondent of the discrepancy and to attempt to determine the most correct response to the question, a process called *reconciliation*. The reconciled response may be the original response, the reinterview response, or an entirely new response that the respondent (or reinterviewer) considers to be most accurate. Following the reconciliation process, the reinterviewer may be instructed to determine the main reason for the discrepancy and to record these reasons for each discrepancy encountered. This information can be quite useful for understanding the causes of measurement error.

Table 8.8 shows a typical interview–reinterview table for a dichotomous variable (i.e., a variable that takes on the values 1 or 0). For example, "1" may denote "in the labor force" and "0" may denote "not in the labor force." The table classifies each person who responds to both the interview and reinterview by his or her interview response (columns) and reconciled reinterview response (rows). Thus, a in the table denotes the number of persons whose interview and reinterview classifications are both "1." Similarly, d in the table denotes the number of persons whose interview and reinterview classifications are both "0." The number of persons whose interview responses are different is $b + c$. Therefore, the proportion of persons in the reinterview analysis whose interview and reinterview classifications disagree is given by

$$g = \frac{b + c}{n}$$

which is referred to as the *gross difference rate* or *disagreement rate*. This statistic is not particularly relevant for gold standard reinterviews but is a key measure for replicated reinterviews, as described below.

For gold standard reinterviews, the key statistic is the *net difference rate* (ndr), which is the difference between the estimate of the population proportion based on the original interview and the estimate of the same proportion based on the reconciled reinterview. Note from Table 8.8 that the proportion of the sample classified in category 1 in the interview is $p_1 = (a + b)/n$. For the reinterview, the proportion is $p_2 = (a + c)/n$. If we take these two proportions as estimates of the population proportions, p_1 is an estimate that is subject to the measurement biases of the original interview, whereas p_2

Table 8.9   Interview–Reinterview Table for 1981–1990 CPS Reconciled Reinterview Data

Reinterview Response	Interview Response		
	In Labor Force	Not in Labor Force	
In Labor Force	230,559	3,604	234,163
Not in Labor Force	1,224	138,077	139,301
	231,783	141,681	

Source: Sinclair and Gastwirth (1998).

is, by the gold standard assumption, not biased by measurement error. Thus, the difference, $p_1 - p_2$, is an estimate of the bias in the original survey estimate. It is easy to verify that

$$\text{ndr} = p_1 - p_2 = \frac{b-c}{n}$$

As an example, consider the data in Table 8.9 from the 1981–1990 CPS reinterview program. In this table, $b = 3604$, $c = 1224$, $n = 373,464$. Using the formula above for the ndr, we see that the bias in the original classification is ndr $= -0.00637$ or -0.637%. The bias in the labor force response can also be expressed as a relative bias (see Chapter 2) by dividing the bias by the estimate of the population proportion from the original interview; that is,

$$\text{relative bias} = \frac{\text{ndr}}{p_1}$$

which for Table 8.9 is -0.0102 or -1.02%.

In some designs, the reconciled reinterview is used to evaluate the performance of the original survey interviewer. In that case, the reinterviewer may be asked to determine whether the discrepancy between two responses is the result of an error that was made by the original respondent, the current respondent, the original interviewer, the reinterviewer, or some other factor. Errors that are determined to be the fault of the original interviewer are then discussed with the interviewer to understand how the error occurred and what actions the interviewer could take to avoid such errors in future interviews. Errors that are caused by the questionnaire or one of the respondents may be ignored.

The U.S. Census Bureau has used the reconciled reinterview to evaluate interviewer performance and questionnaire performance and to estimate the response bias. However, several published studies have provided evidence that the reconciled reinterview approach may not be accurate enough to provide good estimates. Biemer and Forsman (1992) provide evidence that data from the reconciled reinterview can be just as erroneous as data from the original survey. The system of reinterview they considered was the one in use by the U.S. Census Bureau in the 1980s, which combined the objectives of bias estimation and interviewer performance evaluation into one reinterview. Apparently, the field supervisors who conducted the reinterview were more concerned in providing an objective assessment of the field interviewer's performance than in obtaining a "true" value of respondent characteristics for the purpose of bias estimation. Consequently, some discrepancies that were determined not to be the fault of the original interviewer were not reported. However, with appropriate attention to design, the reconciled reinterview approach can provide good measures of measurement bias for some characteristics. Forsman and Schreiner (1991) discuss additional strengths and weaknesses of the reconciled reinterview approach.

Since the 1950 census, the U.S. Census Bureau has used a type of reconciled reinterview as a means of obtaining a more complete household roster for the census postenumeration survey (PES). The PES provides a second count of the persons who reside in a sample of areas in the census for the purposes of adjusting the original census counts for persons who are missed or included mistakenly. The most commonly used procedures for obtaining the PES roster is first to obtain a new roster of all persons who resided at the address on census day (April 1) and then to reconcile this list of persons with the persons reported on the census roster.

Fecso and Pafford (1988) describe a reinterview study used to measure the bias of an agricultural survey. The U.S. National Agricultural Statistics Service (NASS) had begun collecting data on livestock and crop inventories using CATI instead of face-to-face interviewing as done previously. Obtaining accurate responses by telephone was considered a problem because of the detailed nature of these data and because the centralized state telephoning crews lack familiarity with farm terms.

To evaluate the bias in the new CATI data collection process, NASS conducted reconciled reinterviews with the CATI respondents using face-to-face interviewing methods. For the face-to-face reinterview, experienced supervisory field enumerators were used who were presumably knowledgeable about the survey procedures and content and would best be able to probe to obtain the best response during the reconciliation process. Approximately 1000 farm operations were reinterviewed in the study, and these results are reported in the following tables, specifically for the grain stocks items (corn and soybean stocks).

Table 8.10 shows the reconciled reinterview estimates of the relative bias expressed as a percentage of the true value. The biases in these data were statistically significant for all but one item (soybean stocks in Indiana). The direction of the bias indicates that the CATI data collection mode tends to underestimate stocks of corn and soybeans. In the process of reconciliation, the reasons for differences were collected. Table 8.11 indicates that an overwhelming percent of differences, 41.1%, could be related to definitional

Table 8.10 Estimates of Percent Relative Bias in CATI Survey Estimates Using a Face-to-Face Reinterview as the Gold Standard

State	Corn Stocks	Soybean Stocks
Minnesota	10.4*	14.9*
Indiana	17.9*	5.9
Ohio	12.0*	13.7*

Source: Fecso and Pafford (1988).

* Indicates that CATI and final reconciled responses were significantly different at $\alpha = 0.05$.

Table 8.11 Distribution of Reasons for Differences in CATI and Reinterview Responses for Corn Stocks in Minnesota

Reason	Number	Total (%)
Estimated/rounding	28	31.1
"Definitional"	37	41.1
Other	25	27.8
Total	90	100.0

Source: Fecso and Pafford (1988).

problems (bias-related discrepancies), and not those of simple response variance (random fluctuation). Examples of these definitional problems are rented bins not included, confusion with reporting government reserve grains, failed to include grain belonging to someone else, and bins on son's farm included mistakenly.

Another form of gold standard reinterview uses an in-depth, probing approach rather than reconciliation to obtain much more accurate measurements than the original survey measurements. For example, in the 1980 U.S. Census, a gold standard reinterview survey was conducted that used this intensive interviewing approach to evaluate the bias in a selection of census questionnaire items. For example, to determine the respondent's Hispanic or Spanish origin or descent more accurately, detailed questions about the respondent's ancestry were asked going back two generations for both parents. Then, during the data processing stage, this information was used to classify the respondent as to his or her Hispanic/Spanish origin.

Replicated Reinterview

As mentioned previously, replicated reinterview encompasses two types of reinterview designs: test–retest reinterview and repeated measures. The difference between these two designs is that a test–retest reinterview design attempts to replicate the response process in the reinterview, and thus the design must adhere to strict rules as to how the reinterview is conducted. As the name implies, a repeated measures reinterview design simply repeats the survey for a sample of respondents without trying to strictly replicate the

original survey process. Repeated measures designs may still be useful for estimating measurement variance (i.e., reliability) as well as measurement bias.
However, estimates of nonsampling error components derived from this type
of reinterview depend on statistical models and latent variable assumptions.
Consideration of these complex modeling approaches is beyond the scope
of this book, however. Therefore, we confine our discussion of replicated
reinterviews to test–retest reinterviews.

The *test–retest reinterview* aims to replicate the original survey process. The
goal is to obtain new responses from the original survey respondents under
the same survey conditions as those of the original survey.

When the goal of reinterview is to estimate measurement reliability rather
than measurement bias, a test–retest reinterview design is required. For the
estimator of reliability to be valid, the response process that the respondent
uses for the reinterview to arrive at a survey response to some question should
be essentially identical to the response process that he or she used in the
original interview for that question. This ensures that the variable errors that
might have been generated during the interview process have the same chance
of occurring in the reinterview. In this way, the measurement error variance
associated with the original survey response can be estimated by the variation
between the original and reinterview responses.

Thus, the objective of the test–retest reinterview design is to try to
recreate the original survey conditions so that the response processes used in
both interviews can be assumed to be the same. For this reason, the reinterview is usually designed according to the following five design principles:

1. The reinterviewers should be selected from the same interviewer labor
   pool as the original interviewers.
2. The reinterviewing procedures should, as nearly as possible, be identical
   to the original interviewing procedures.
3. The same survey questions should be used when possible. Some alter-
   ations may be necessary to adjust the reinterview reference periods to
   coincide with the original interview reference periods. However, changes
   in the survey questions should be minimized.
4. The reinterview respondent should be the same as the original interview
   respondent.
5. The time between the interview and reinterview should be short enough
   to minimize recall error and real changes in the characteristics whose
   reliability ratios are to be estimated and long enough to avoid the

situation where respondents simply recall their original survey response and repeat it in the reinterview without replicating the response process that generated the original survey response. This time period is usually expected to be between 5 and 10 days.

Despite strict adherence to these design principles, it is unlikely that the original survey conditions can ever be replicated precisely in the reinterview. For example, the original interview may have had some effect on the respondent that would influence responses to the questions when asked at a later time. These are referred to in the literature as *conditioning effects*. For example, suppose that the respondent is asked about the number of hours that he or she worked in the preceding week. This number may have been reported somewhat imprecisely in the interview. However, after the interview, the respondent checked the number out of curiosity and reports the revised number in the reinterview. Obviously, the response process was altered by the original interview.

In the same example, suppose that the reinterview occurs about three weeks following the original interview. Since the reference week must remain the same in the reinterview, respondents must now recall how many hours they worked four weeks ago rather than last week as they did in the original interview. Thus, the reinterview response may be subject to more recall error than the original interview. On the other hand, if the reinterviewer conducts the reinterview the day after the original interview was conducted, there is a good chance that the respondent may simply remember the answer from the day before and repeat it rather than trying to think about the previous week and arrive at an answer independent of the first interview response. When this occurs, responses appear more reliable on the basis of reinterview than they really are.

Many other factors may intervene to violate the assumption that the reinterview replicates precisely the original interview response process. For example, to save cost and respondent burden, only a subset of the original questionnaire may be used, thus making the reinterview much shorter than the original interview. If the original survey was conducted by face-to-face interviewing, the reinterview may be conducted by telephone interviewing, which could change the response process. These and other reinterview design issues are discussed in more detail in Forsman and Schreiner (1991). Despite its limitations, the test–retest reinterview can closely approximate the original survey process, and the estimates of measurement reliability will still be quite useful for evaluation purposes.

There are several methods for estimating the reliability ratio, R, described in Chapter 2 using data from test–retest reinterviews. For continuous data, one simple method for estimating R is simply to compute the correlation between the original interview and reinterview responses. For example, if the correlation between the original and reinterview survey responses for some charac-

Table 8.12   Interview–Reinterview Table for CPS Unreconciled Reinterview Data

| Reinterview Response | Interview Response | | |
	In Labor Force	Not in Labor Force	
In Labor Force	4664	157	4821
Not in Labor Force	213	2613	2826
	4877	2770	7647

teristic, say income, is 0.80, we say that the reliability of the income variable is 80%.

For categorical data (i.e., data from questions with discrete response categories), the methods are more complicated. The method used by the U.S. Census Bureau is to compute the *index of inconsistency*, which is an estimate of 1 minus the reliability ratio, or $1 - R$. Whereas the reliability ratio is the ratio of the variance of the true values of the characteristics in the population to the total variance of the observed values, the index of inconsistency is the ratio of the total error variance in the observations to the total variance. Thus, the index of inconsistency is a measure of the proportion of the total variance that is measurement error variance.

The formula for the index of inconsistency is

$$I = \frac{g}{p_1(1 - p_2) + p_2(1 - p_1)}$$

As an example, consider the data in Table 8.12 from the CPS. This time the data are from the test–retest reinterview part of the 1996 CPS reinterview program. The index is computed as follows:

$$g = \frac{157 + 213}{7647} = 0.0484$$

$$p_1 = \frac{4877}{7647} = 0.638$$

$$p_2 = \frac{4821}{7647} = 0.630$$

$$p_1(1 - p_2) = 0.231$$

$$p_2(1 - p_1) = 0.233$$

$$I = \frac{0.0484}{0.231 + 0.233} = 0.104$$

The value of I from this computation is 0.104 or 10.4%. This is equivalent to a reliability ratio, R, of $(1 - 0.104) = 0.896$ or 89.6%. The U.S. Census Bureau has developed a rule of thumb for judging the acceptability of the index of

> **Good:** $I \leq 0.20$ or $R \geq 0.80$
>
> **Fair:** I between 0.20 and 0.50, or
> R between 0.80 and 0.50
>
> **Poor:** $I \geq 0.50$
> $R \leq 0.50$

Figure 8.2   Rule of thumb for judging the acceptability of response inconsistency and reliability.

inconsistency or reliability ratio. As shown in Figure 8.2, an index of inconsistency of approximately 10% (reliability of 90%) is considered good.

Finally, another measure of data quality that is often reported in test–retest reinterview survey analysis is the *agreement rate* or proportion of persons whose reinterview classification agrees with their original interview classification. In the notation of Table 8.7, the agreement rate is $A = (a + d)/n$ or, equivalently, 1 minus the gross difference rate, g. For the data in Table 8.11, the agreement rate is $A = (4664 + 2613)/7647 = 0.952$ or 95.2%. Thus, about 95% of reinterview respondents agreed with their interview classification, which is good agreement for these data.

8.5   SUMMARY OF EVALUATION METHODS

In this chapter we described a number of commonly used methods for evaluating the error in surveys and described how they might be applied at various stages of the survey process. Methods applied during the design and pretesting stages can guide the designer in optimizing the survey design to reduce the total survey error subject to cost constraints. Methods that are applied concurrently with data collection and data processing can be used to monitor the quality of the data and to sound a warning when important errors enter the survey process. Finally, methods applied after the survey is completed are useful for describing the error for users and providing information to survey designers for improving future surveys. Thus, we see that quality evaluation can be a continuous process that is carried from the design stage to the post-survey analysis stage.

Quality evaluation is a critical branch of the field of survey research since without it, the process of survey quality improvement is pure guesswork. In fact, earlier chapters of the book could not have been written without survey evaluation since so much of what we know about survey methodology is based on the results of survey quality evaluations.

For most surveys, data quality evaluation is limited to pretesting of the questionnaire and data collection methods and some quality control measures for the interviewing and data processing operations. Embedded experiments and postsurvey evaluations to estimate specific components of the mean squared error are seldom conducted. However, as response rates to surveys continue to fall, particularly for telephone surveys, evaluations of the nonresponse bias in the survey estimates are becoming more common. These evaluations may involve following up on nonrespondents to the survey for the purpose of persuading to respond. More typical are descriptive studies of the nonrespondents. For these studies, analysts use whatever variables are available for both nonrespondents and respondents, either from the survey frame or the fieldwork, to describe how the nonrespondents differ from respondents for these variables.

For surveys that are conducted on a continuing basis or are repeated periodically, survey evaluation plays a critical role. It is important not only for continuous quality improvement, but also as a vehicle for communicating the usefulness and limitations of the data series. For this reason, a number of U.S. federal surveys have developed quality profiles as a way of summarizing all that is known, as well as pointing out the gaps in knowledge about the sampling and nonsampling errors in the surveys.

A quality profile is a report that provides a comprehensive picture of the quality of a survey, addressing each potential source of error: specification, nonresponse, frame, measurement, and data processing. The quality profile is characterized by a review and synthesis of all the information that exists for a survey that has accumulated over the years that the survey has been conducted. The goal of the survey quality profile is:

- To describe in some detail the survey design, estimation, and data collection procedures for the survey
- To provide a comprehensive summary of what is known for the survey for all sources of error, sampling as well as nonsampling error
- To identify areas of the survey process where knowledge about nonsampling errors is deficient
- To recommend areas of the survey process for improvements to reduce survey error
- To suggest areas where further evaluation and methodological research are needed in order to extend and enhance knowledge of the total mean squared error of key survey estimates and data series.

The quality profile is supplemental to the regular survey documentation and should be based on information that is available in many different forms, such as survey methodology reports, user manuals on how to use microdata files, and technical reports providing details about specifics. A continuing survey allows accumulation of this type of information over time, and hence quality profiles are almost always restricted to continuing surveys.

In the United States, quality profiles have been developed for the Current Population Survey (Brooks and Bailar, 1978), the Survey of Income and Program Participation (Jabine et al., 1990), U.S. Schools and Staffing Survey (SASS; Kalton et al., 2000), American Housing Survey (AHS; Chakrabarty and Torres, 1996), and the U.S Residential Energy Consumption Survey (RECS; U.S. Energy Information Administration, 1996). Kasprzyk and Kalton (2001) review the use of quality profiles in U.S. statistical agencies and discuss their strengths and weaknesses for survey improvement and quality declaration purposes. Following are two examples of topics considered in quality profiles: one from the SASS and one from the AHS.

EXCERPT OF A QUALITY PROFILE OUTLINE FOR THE SASS DATA COLLECTION SECTION

- Procedures
  - Mode
  - Schedule
  - Telephone follow-up of nonrespondents
  - Analysis of characteristics of schools in need of follow-up
  - Mode effects
  - Supervision and quality assurance
  - Length of time to complete questionnaire
- Cognitive research and pretests
- Reinterviews and response variance
  - Reinterview sample
  - Reinterview procedure
  - Measures of response variance
  - Questions evaluated per round
  - Effect of mode change in reinterview
- Nonresponse
  - School nonresponse rate
  - Weighted school nonresponse rate
  - Item nonresponse rate

EXCERPT OF A QUALITY PROFILE OUTLINE FOR THE AHS DATA PROCESSING SECTION

- Editing
- Quality control operations in data processing
  — Clerical edit
  — Data keying
  — Preedit
  — Computer edit
- Quality assurance results for keying
  — Methodology
  — Error rates and rejection rates
  — Results of keying verification
  — Rectification
- Results of research on regional office preedit
  — Status of rejects
  — Type of error
  — Computer edit action

CHAPTER 9

Sampling Error

In Chapter 2 we examined the total error in a survey estimate and showed how the total error is made up of two types of error: sampling error and non-sampling error. Chapters 3 through 8 were devoted to topics involving non-sampling errors, their causes, and the various methods for evaluating and reducing their effects. This chapter is devoted to the topic of sampling error in a survey estimate with particular emphasis on the following questions:

- Why do we sample, and what is the importance of randomization in sampling?
- What is sampling error, and how does it arise?
- How is sampling error measured in a survey?
- What design factors affect the magnitude of the sampling error?
- What are the meanings of some important concepts found in the survey sampling literature, such as standard error, design effect, and stratified sampling?

This chapter is not intended to provide comprehensive coverage of the topic of sampling error, nor is it a guide on how to design or select a sample, estimate a standard error, or choose an estimator. There are many excellent books devoted to sampling theory and practice that cover these topics in great detail, and some are included in the reference list. The emphasis here is on the *concepts* involved in a study of sampling methods. Our goal is to provide a brief overview of the major topics in sampling, primarily to familiarize the reader with the concepts, terminology, and rationale for sampling. We also strive to discuss each topic using only the most basic mathematical concepts and terminology, so that persons with a limited background in mathematics and statistics will still find the material generally accessible and understandable.

We have attempted to minimize the use of mathematical notation, although the use of some equations and mathematical symbols was unavoidable. We justify this by noting that much of the literature on survey methodology is rooted in statistical theory, so some familiarity with the language of statistics and mathematics (i.e., equations and mathematical symbols) is necessary to access this literature fully. Consequently, this chapter is the most technical chapter in the book. As an aid to the least mathematically inclined among us, we have tried to include verbal descriptions of the concepts to accompany and clarify the mathematical expressions used in the technical descriptions, where appropriate.

In Section 9.1 we provide a brief discussion of the history of sampling and describe the progression of sampling from convenience samples, to the creative construction of purposive samples, and finally, to the methods of random sampling used almost exclusively in modern survey work. This section complements the historical discussion provided in Chapter 1 but focuses on sampling developments rather than survey methodology as a whole. In Section 9.2 we discuss the basic motivation for randomization and why it is preferred over other nonrandom methods. In Section 9.3 we develop the basic ideas of sampling using a very basic method referred to as simple random sampling and illustrates how this method provides the essential building blocks for more complex forms of sampling in use today. In Section 9.4 we discuss statistical inference from sample surveys. Finally, in Section 9.5 we discuss some of the more advanced concepts and terminology encountered in the sampling and survey methods literature.

9.1 BRIEF HISTORY OF SAMPLING

Surveys of human populations can trace their origins at least to biblical times, when censuses were conducted to determine the size of a country's population for military and taxation purposes. However, students of survey methodology are often surprised to learn that random sampling is a relatively new concept, with origins in the early twentieth century. The genesis of modern sampling theory can be traced to three key statisticians who are credited with developing the essential ideas of sampling: Kiaer, Bowley, and Neyman. However, many other statisticians who followed these early innovators have also contributed enormously to the field and helped to shape the current methodology, which is survey sampling. A few of these will also be mentioned.

Around 1900, a Norwegian statistician named Kiaer first put forth the idea that a small sample of population units could be used to estimate accurately the parameters of a large population. Until then, most statisticians thought that only a complete census of the population could provide a truly accurate estimate of a population parameter. Kiaer called his idea the *representative method* since the aim was to represent the total population with a relatively small, carefully selected sample from it. Kiaer's method did not involve randomization of any type and, in fact, the concept of randomization was not well understood by the statisticians of the time. Rather, his method relied solely on the judgment and skill of the sampler to create a sample that could be accepted as representing the population. The idea was to use expert judgment and knowledge of the population to create a miniature population that would be easier to study, due to its much reduced size, than the entire population. Even though the size of this miniaturized copy of the population is a small fraction of the size of the entire population, if designed properly, it could have many of the same characteristics as the original population.

Kiaer relied on data from the most recent census to aid his purposive selection of population members for the sample, matching both the population and sample on those characteristics he thought were most important for the purposes of the survey at hand. For example, he might construct his sample to reflect the mix of urban and rural areas in the country accurately if he believed that urbanicity was a particularly important explanatory variable in his survey research.

By building on and extending Kiaer's ideas, an Englishman named Bowley is credited with formally introducing the idea of randomization to the field of survey sampling. Bowley reasoned that rather than relying on one's knowledge of the population and purposive selection to ensure that a sample will be representative, as Kiaer did, random selection of units could be used to more effectively achieve representativeness. Bowley argued that random sampling does not rely on complex theories and descriptions of the population. Thus representative samples could be selected by anyone, not just experts and persons having extensive and specialized knowledge of the populations to be sampled. Bowley's randomization method assigned an equal probability of being selected to every unit in the population. He showed that a representative sample of any size could be selected essentially simply by pulling numbers out of a hat. Later, this method became known as *simple random sampling*.

Then, in 1934, the survey world was forever changed when a statistician named Jerzy Neyman published a famous paper that provided a theory for random sampling, which allowed either equal or unequal probabilities of selection. Prior to Neyman's paper, there was an ongoing debate as to whether purposive sampling or random sampling was preferred for survey work. Each has its advantages and disadvantages and some argued that purposive sampling could be much more efficient than random sampling. Neyman's paper put that debate to rest by providing a unifying theory for statistical inference based on the principles of randomization. Neyman showed that through randomization it is possible to make inferences about the population with known probabilities of being correct. He showed further that there can be no corresponding statistical theory of inference for purposive sampling. For example, it is possible to construct statistical confidence intervals using random sampling, and it is impossible to do so with purposive sampling.

Neyman's landmark paper generated considerable interest within the statistical community, and in the 1940s a number of papers were published that extended his ideas. Statisticians such as Hansen, Hurwitz, Madow, Cochran,

Kish, and Deming formulated the framework that is the basis for modern sampling theory. Other statisticians, such as Yates, Sukhatme, Murthy, and Des Raj, authored textbooks on sampling that eventually advanced the field further. Statisticians such as Dalenius, Godambe, Horvitz, Thompson, Jessen, McCarthy, Waksberg, Pritzker, Stephan, Tepping, and Mahalanobis made substantial contributions to the development of sampling theory and practice during the period 1940–1965.

History provides a clear motivation for sampling. The original idea of collecting survey data on a small subset of the population rather than for the entire population (i.e., a complete census) arose from the need to save costs while still obtaining estimates that are accurate enough for the purposes intended. However, early statisticians also realized another very important benefit that sample surveys often have over censuses: that of improved data quality. They reasoned that because the data are collected on only a fraction of the population, the scale of the data collection effort can be more manageable, which often means that data quality can be better controlled. However, a major disadvantage of basing inference on only a sample of the population is, of course, sampling error. This can be a particularly important disadvantage if estimates are needed for many small areas, such as counties, townships, and civil districts. If high accuracy for small areas is a priority for data collection, a sample survey may not be the best option.

Today, sample surveys are accepted as yielding valid results, and it is difficult to believe that only about 50 years ago sampling was perceived as a radical idea by many statistical offices. Yet widespread acceptance of sampling was quite slow until Neyman provided a rigorous statistical theory proving that inferences based on sample surveys are valid and that the error in sampling can easily be quantified using the sample data. Neyman showed that through randomization (i.e., selecting the units for the sample with known probabilities of selection) it is possible to construct confidence intervals around the estimates and to use these confidence intervals to infer the value of the population parameter.

However, there is a serious shortcoming in what is now referred to as *classical sampling theory* that was not fully appreciated in Neyman's day. Classical sampling theory assumes that all units in the sample are observed and that the observations are without error (i.e., there are no nonsampling errors in the data). Only if this assumption holds is classical sampling theory applicable for a survey. But as we have demonstrated time and again throughout this book, nonsampling errors almost always occur, and consequently, estimates that are supposed to be unbiased according to sampling theory are not unbiased. Indeed, inferences about the population based on Neyman's theory are substantially more erroneous than the theory suggests. Over the years, statisticians have developed methods to adjust and adapt the theory to compensate for the inevitable nonsampling errors in the data. Some illustrations of these methods are provided later in the chapter after we have described some of the basic concepts of classical sampling theory.

> *Modern sampling theory* evolved from purposive sampling, to simple random sampling, and ultimately to unequal probability sampling using complex designs.

9.2 NONRANDOM SAMPLING METHODS

Despite the widespread use of random sampling in survey research, nonrandom sampling methods are still in use. There are special situations when nonrandom methods may even be preferred over random sampling or may offer efficient and effective alternatives to random sampling methods. However, it is not uncommon to find nonrandom methods being used inappropriately. For example, a nonrandom sample is selected but the analyst "pretends" it was selected at random for the purposes of statistical inference. We discuss some abuses of the methods later in this section.

One frequently used method of nonrandom sampling is *convenience sampling* or *haphazard sampling*. This type of sampling may involve any of the following:

- Students who volunteer to participate in a laboratory experiment in response to flyers posted on school bulletin boards
- Hotel guests who respond to a customer satisfaction questionnaire left in their hotel rooms
- Friends and relatives whom you ask to complete a questionnaire
- TV viewers who call in response to TV "polls" to vote their opinions on a particular topic

Although there is no basis for statistical inference for such samples, convenience samples can still be quite useful for special purposes, such as evaluating survey questions or testing survey procedures, as described in Chapters 8 and 10. Unfortunately, convenience samples are often used inappropriately as the basis for inference to some larger population. As an example, a hotel chain might infer that its guests are not very satisfied with the service at its hotels on the basis of customer satisfaction questionnaires that are placed in every hotel room. But often, such questionnaires are completed by guests who have had some problem with the service and, as a result, are motivated to complete the survey to record their complaint. Such samples of respondents cannot be considered as representative of all guests of the hotel and thus inferences derived from them can be quite erroneous. Certainly, there is no statistical theory for convenience samples which suggests that such inferences can be statistically valid.

Another type of nonrandom sampling is known as *purposive sampling*, also called *judgment sampling*. It is essentially the approach proposed by Kiaer many years ago that we described in Section 9.1. Purposive selection involves constructing samples using expert judgment to select units that the sampler believes will represent the target population. As an example, suppose that the president of a large company wants to find out what her employees think about some company policy. She decides to select a purposive sample of 10 employees of the company. Based on her knowledge of the company's workforce, she selects five managers and five nonmanagers at various salary grade levels so that all grade levels are represented. She also decides to select only persons who she knows are willing to speak honestly and openly about the company's policies.

It is quite possible for this sampling strategy to provide valid and useful information regarding the policy in question, and the company president will obtain a fairly accurate reading on the prevailing opinion in the company. However, it is also quite possible that persons who are asked to participate do not reflect the general view of the company's workforce. Much depends on the president's knowledge of the personnel in the company, her ability to select people who represent the company's view, and the interpretation of the opinions she hears. Particularly with a very small sample of only 10 persons, purposive sampling may provide better and more accurate information than a random sample of the same size. Certainly, a random sample of size 10 from a large company would be subject to considerable sampling error, even though the opportunity for sampling bias is much reduced over purposive sampling. For example, if only 10 persons are selected completely at random from the company, the risk is quite high that employees will be included who are not particularly knowledgeable or who hold rather unique views.

However, as we shall see later in the chapter, samples of somewhat larger sizes can be quite accurate since sampling error decreases as the sample size increases. For example, if the sample size were increased from 10 to, say, 50 employees, there is a much greater chance that inferences from random samples will be more accurate than those from purposive samples of the same size. Purposive samples of 50 workers are likely to have larger total error as a result of the selection biases of the survey designers. Further, unlike random samples, purposive samples contain no information on how close the sample estimate is to the true value of the population parameter.

The U.S. Census Bureau has used purposive selection for choosing locations to conduct pretests of the census procedures. Such pretests usually involve only two or three sites (cities or rural areas), due to the high cost of conducting a census of each site. Since the objective of a census pretest is primarily to test and compare alternative data collection and estimation procedures rather than for making national inferences, purposive sampling is preferred since it allows more control over the types of sites that are selected. The Census Bureau might choose a city that represents a very difficult-to-enumerate area of the country, another that represents rural areas, which pose very different problems for enumeration, and still another that represents areas having high concentrations of a minority group, such as Native Americans. Leaving these selections to random chance would probably not provide the bureau with the areas they need to test thoroughly many critical aspects of the census procedures.

A third type of nonrandom sampling is *quota sampling*. Quota sampling is a term used to describe a number of sampling techniques where sampling continues until a predetermined number of persons satisfying prespecified sampling criteria are interviewed. For example, an automobile manufacturer interested in assessing customer satisfaction with various types of vehicles may decide to conduct a mall intercept study. For this approach, interviewers are instructed to continue to interview persons in a shopping mall until they have conducted interviews with 10 persons who own vans, 20 persons who own cars, and 15 persons who own trucks.

Quota sampling is often used by market researchers because it is easy to administer, is usually less expensive than probabilistic sampling methods, and can be completed more quickly than random methods. However, quota sampling suffers from essentially the same limitations as convenience, judgment, and purposive sampling (i.e., it has no probabilistic basis for statistical inference). The quota sample survey interviewers are free to choose persons who are readily available and who satisfy the sampling criteria. Often, these respondents are not representative of the target population, but rather, may reflect the attitudes and beliefs of the interviewers who are conducting the data collection. In addition, they may simply be respondents who were the most willing to participate or most available to express their views to interviewers.

Some variations of quota sampling contain elements of random sampling but are still not statistically valid methods. For example, a quota sample may be initiated by selecting a large random sample of the population, which is then assigned to the interviewers for data collection. Interviewing continues until certain quota requirements are met. For example, the requirements may state that there should be x persons in the age group 12 to 15, y persons 16 to 25 years of age, and z persons older than 25 years of age. If a sample member refuses, is unavailable, or cannot be contacted on the first attempt, the interviewers are instructed simply to move on to the next sample member, and so on, until all quotas are filled. Such samples will tend to underrepresent persons who are difficult to contact or who are reluctant to participate, since, in effect, the method substitutes cooperative and available sample members for uncooperative and unavailable ones. In traditional random sampling, interviews are attempted for all members of the sample without regard to quotas. Reluctant, temporarily unavailable, and difficult-to-contact persons are revisited and attempts are made to complete interviews for every sample member. This type of data collection strategy ensures that even persons who are difficult to interview are still represented appropriately in the survey results.

Nonrandom sampling methods include convenience sampling, haphazard sampling, purposive sampling, and quota sampling.

An example of "random" quota sampling is a survey conducted for a random sample of recent college graduates. For this study, the client specified that interviews should be conducted for 1000 graduates. The survey designer determined from previous studies that the response rate to the survey would be about 50%, so a sample of 2000 graduates was selected to achieve the target number of 1000 interviews. However, the response rate to the survey was much higher than expected, and the 1000 interviews were completed before all the 2000 sample members had been contacted. Using a quota sampling strategy, sampling would be discontinued when 1000 interviews had been completed. However, this strategy results in a sample that overrepresents persons who are cooperative and easily reached since such people are usually among the first to be interviewed. Random sampling rules dictate that attempts to contact and interview all of 2000 sample members be made. Further, the response rate for the study should be based on the original 2000 members sampled. That is, if 1000 of the original 2000 sample members are interviewed, the response rate for the study would be 50% even if 200 or so of the sample members were never contacted by the interviewers after the 1000th completed interview was conducted.

To summarize this discussion, nonrandom methods have their place in survey research but should be used only when random sampling may produce unreliable results or when greater control over who should be surveyed is more important than the statistical validity of the survey inferences. However, in most cases, random sampling has important advantages over nonrandom sampling methods. The following is a list of some of these.

ADVANTAGES OF RANDOM SAMPLING OVER NONRANDOM SAMPLING METHODS

- Produces samples that are representative of the entire population (i.e., samples that are free from sample selection biases)
- Provides a theoretical basis for statistical inference
- Is relatively easy to apply; requires no expert judgment
- Uses methods that are reproducible (i.e., sampling approaches are well documented and can be carried out by others)
- Provides measures of precision of the estimates that can be predetermined fairly accurately by the researcher through the specification of the sample size for the survey

---

In the next section we describe one of the simplest methods for drawing a random sample—simple random sampling. This method is important—not because of its applicability in actual practice but because it forms the basis of other, more complex sampling schemes.

9.3 SIMPLE RANDOM SAMPLING

9.3.1 Terminology Used in Sampling

Our discussion of the methods of sampling are facilitated by first defining some terms that will be used throughout this chapter. Although some of these have been encountered in earlier chapters, we include them again here for the sake of completeness.

As discussed in Chapters 1 and 2, one of the first steps in survey design is to define the population to be studied. However, several types of populations may be defined for a single survey. For the purposes of defining the goals and objectives of the survey, it is useful to define the target and frame populations. The target population is the population to be studied in the survey and for which inferences from the survey will be made. However, a population that is more useful for sampling purposes is the *frame* or *survey population*. This is the population represented by the sampling frame used to select the sample. Due to the practical constraints of the sampling frame, sampling process, and data collection process, the target population may be quite different than the frame population, and in Chapter 2, we discussed examples that illustrate how these two populations may differ. Recall from Chapter 3 that differences between the frame population and the target population give rise to frame coverage errors and bias.

Once the population has been defined and the survey objectives are known, the question of how to design the sample to achieve the survey objectives can be addressed. For the purposes of sampling, the population can be thought of as containing elements and units. A *population element* is the basic entity for which the survey data will be collected. For example, in a household survey, this is usually a person living within a household if person-level data are to be collected or the housing unit in case of a housing survey. A *population unit* may be defined as a cluster of population elements where membership in the unit is by some natural criterion. For example, a population unit may be a housing unit composed of persons who live in the unit that comprise the population elements. In a business survey, the unit may be the company composed of elements which are the establishment or branch offices within the company. In an agricultural survey, the unit may be the entire farm and all the land contained within it; however, the population elements are parcels of land that comprise the entire farm.

An essential requirement of any form of probability sampling is the existence of a *sampling frame* from which the population elements can be selected. In the simplest case, the frame is a list of all the population elements; for example, a list of all members of a professional society to be surveyed. However, quite often the frame is a list of population units, such as addresses of households in a survey of area residents or a list of companies in an establishment survey. In some cases, rather than a physical list of population units, the frame may be defined implicitly. A good example of this is the list of all residential telephone numbers in a random-digit-dial (RDD) survey.

In many practical situations, the frame does not cover the entire target population, and certain subgroups of the population are knowingly excluded. For example, in an RDD survey, persons living in nontelephone households are excluded due to the limitations of this mode of data collection. If inference is to be made to a target population that is defined as all persons living in the area to be surveyed, including persons in nontelephone households, the target population will not be the same as the frame population and the potential for coverage bias must be considered. However, if the target population is defined as only persons in telephone households, the target and frame populations are the same and thus there is no risk of a coverage bias due to the exclusion of nontelephone households.

Similarly, an area frame survey may exclude persons living in very small communities due to the costs of traveling to these areas to conduct only one or two interviews. In this situation, the frame intentionally excludes the smallest areas of the target population. Nevertheless, we may still define the target population as all persons in the area, including those living in areas deleted from the frame. If the population living in the excluded areas account for only 1 or 2% of the target population, the risk of coverage bias is very small (see Chapter 3). There are many other situations where it is good survey practice to accept a small coverage bias in exchange for a substantial increase in sampling and/or data collection efficiency.

The summary characteristics that describe the population are called population *parameters*. The goal of a survey is to estimate one or more parameters of the population. The sample statistics that are used for this purpose are called *estimators*. For example, the population parameter may be the average income of the population members and the estimator may be the sample average. Or the population parameter could be the correlation between education and income for a population, whereas the estimator is the sample correlation for these characteristics. Realizations of estimators from the survey data are called *estimates*. In other words, estimators are essentially formulas for computing the sample estimates.

All sample estimates are subject to *sampling error*. Sampling error is the error in a sample estimate (i.e., the difference between the estimate and the value of the population parameter) that is due to the selection of only a subset of the total population rather than the entire population. The usual measure of sampling error is the *sampling variance*, essentially the squared deviations between the value of the estimator and the value of the population parameter for a particular sample averaged over all possible samples that can be drawn from the population. The variance of an estimator contains information regarding how close the estimator is to the population parameter. The squared root of the sampling variance of an estimator is the *standard error* of the estimator. The estimates of the population parameters and the estimates of the standard error of the parameter estimates are the basic quantities that are needed for inference (i.e., confidence intervals and hypothesis testing). As we noted previously, through the random selection of the sample, estimates of the variances and the standard errors of the survey estimates can be computed directly from the sample data.

Two properties of estimators that will be useful in our discussion are unbiasedness and efficiency. An estimator is *unbiased* if the average value of the estimator computed for all possible samples from the population is equal to the value of the population parameter. That is, unbiased estimators, on average, yield the population parameter even though the estimates may differ from the parameter across the various samples that can be drawn from the population. An estimator is *efficient* if it has a small standard error (i.e., good *precision*). In some cases, there may be several estimators that can be used to estimate a particular population parameter. If all the estimators are unbiased, the best estimator is the one that is most *efficient*.

SUMMARY OF SOME TECHNICAL TERMS USED IN SAMPLING

Population element. Basic entity on which measurement is taken.

Population unit. Naturally occurring entity composed of one or more population elements and including all population elements that are to be sampled.

Target population. Collection of all units we want to study.

Frame. List of population units that will be used for sampling.

Frame or survey population. Collection of units represented on the frame.

Population parameter. Summary value characterizing the population to be estimated.

Estimator. A function of the sample observations that estimate a population parameter.

Estimate. A realized value of the estimator.

Sampling distribution. A frequency histogram that shows the value of an estimator for each sample that can possibly be drawn from the population.

Unbiased estimator. The mean of the sampling distribution of the estimator is equal to the population parameter.

Efficient estimator. Estimator with the smallest variance among all estimators of a parameter that can be formed from the sample.

epsem design. Equal probability selection method design that gives an equal chance of selection to all units in the population.

In the remainder of Section 9.3, we describe some methods for selecting population elements from a sampling frame using probabilistic methods. We begin with simple random sampling, which has limited practical utility but is nevertheless covered in some detail because of its theoretical importance. We then discuss several other methods for selecting samples so that all population units have the same probability of selection. These sampling schemes, called *equal probability of selection methods* (epsem), are regarded as very efficient in terms of sample size (i.e., cost) and precision.

9.3.2 Definition of Simple Random Sampling

A simple random sample of size n is a method where every possible sample of n distinct units that can be drawn from the sampling frame has the same probability of selection. As the name implies, simple random sampling (SRS) is one of the simplest sampling schemes available for drawing random samples from populations. However, it is not a widely used method for large-scale business and social surveys, since one can usually find more efficient sampling schemes (i.e., sampling schemes that yield estimators with better precision). For small-scale studies or for informal randomization purposes, SRS is still very useful.

As an example, to pick the winners in a raffle, the names of the raffle ticket holders are often put into a large basket and mixed thoroughly. Then the name of a winner is drawn by reaching into the basket and grabbing a ticket "at random." To draw additional winners, the process is repeated without replacing the tickets drawn previously. In this manner, every raffle ticket has the same probability of being selected. Variations of this method have been to select lottery winners or the starting lineup or order of play for games and sports competitions.

SRS is a useful method of sampling for some types of surveys, although it is usually possible to construct more complex sample designs that provide greater precision in the estimates. Nevertheless, SRS is still quite useful in the study of sampling methods since it provides a statistical foundation for these more complex forms of sampling. Quite often, the efficiency of a more complex sampling method is stated in terms that compare it to SRS. For example, the *design effect* for an estimator, which we discuss in Section 9.5.6, is expressed in terms that are relative to SRS. Moreover, the estimation formulas and variance estimates for SRS are the simplest and easiest to understand, so these formulas often serve as the building blocks for more complex sampling schemes. Even very complex sampling schemes may use SRS at some point in the sampling process and then the SRS sampling formulas can be applied.

In the next section we describe some methods for drawing SRS samples and provide the basic formulas for making inferences to the target population using the method. We also discuss a number of issues in the design of survey samples, including the determination of sample size, selection of the "best" estimator for inference, and effect of nonsampling errors on statistical inference.

> A *simple random sample* (SRS) of size n is a method where every possible sample of n distinct units that can be drawn from the sampling frame has the same probability of selection.

9.3.3 Basic Approach to Simple Random Sampling

To illustrate the basic approach of SRS, we consider how we would draw a sample from a very small artificial population: say, a population consisting of only 10 units. Of course, a population this small would probably not be sampled in practice; however, it is easier to illustrate the concepts of SRS using such a small population, as we will see. Suppose that we wish to draw a sample from a population to estimate the average income of individuals in the population. The values of all 10 members of the population for this illustration are listed in Table 9.1. The first column of the table lists the label of the population unit and the second column lists the income for the unit, which is labeled as Y_i, where i identifies the ith unit.

Consider how one would draw a sample of size $n = 2$ from this population. Several methods are available for drawing a simple random sample. Each method is quite general and will produce an SRS from any population regardless of the population size, N, or the sample size, n. However, some are easier to implement than others. All the methods we discuss are referred to as *SRS without replacement* methods. This means that once a population member is drawn from the population it is essentially removed from the frame and cannot be drawn again. *SRS with replacement*, which is not covered in detail in this book, is very similar to SRS without replacement except that a unit selected remains on the frame after it is drawn, possibly to be selected again, with a subsequent draw. Thus, using sampling with replacement, it is possible to draw the same population unit many times for a single sample. In very large populations, however, the chance of drawing the sample unit twice is quite small. In most practical survey situations, sampling is either without replacement or is assumed to be without replacement.

Lottery Method
This is the method we discussed previously for selecting the winners of a raffle. The lottery method draws the two sample members as one would draw names from a pool of names in a lottery. The names of the 10 population members are written on slips of paper which are then shuffled and mixed well in a bowl. Then two names are drawn from the bowl without looking. Although it may not be immediately obvious, this method satisfies the definition of SRS in that

every sample of size 2 that can be formed from the 10 population elements has the same probability of selection.

Random Number Method

The random number method is a more accurate method of selecting an SRS, especially for large, computerized sampling frames. To select a sample of size 2 from our population with this method, the population members are first numbered from 1 to 10, and then two numbers between 1 and 10 are drawn at random from a random number table. (Random number tables can be found in most elementary statistics textbooks.) The two persons in the population who are assigned numbers corresponding to the two random numbers are selected for the sample.

All Possible Samples Method

This method is based on the definition of simple random sampling. This method begins by forming all samples of size 2 that can be constructed from the population of size 10. There are a total of 45 samples of size 2 that can be selected without replacement from a population of size 10. Then one of the 45 samples is selected at random. This can be done by numbering the samples from 1 to 45 and then using a random number table to select a number between 1 and 45. It is fairly obvious that this method gives all 45 possible samples the same chance (1 out of 45) of being selected.

Since the latter method is fundamental to our understanding of SRS, we consider in some detail how the 45 samples of size $n = 2$ are formed. Let us begin by counting the number of different samples that contain population unit P1. There are a total of nine of these: {P1, P2}, {P1, P3}, and so on, to {P1, P10}. Next consider the samples that contain population unit P2, not counting the sample {P1, P2}, which has already been counted. There are eight of these samples: {P2, P3}, {P2, P4}, and so on, to {P2, P10}. Continuing to count in this manner, one can verify that there are seven samples containing P3 that have not been counted previously, six for P4, five for P5, and so on, until we reach the one sample that can been drawn for P9 that has not been counted previously. There are no samples containing P10 that have not already been counted. Thus, the total number of possible samples that can be drawn is just the sum of these: $9 + 8 + 7 + 6 + 5 + 4 + 3 + 2 + 1$, or 45, samples. In the first two columns of Table 9.2 we list these 45 samples of size 2.

Although it may not be obvious, all three methods just described are equivalent and will produce SRSs of size 2. Further, each method can easily be generalized to select an SRS sample of size 3 or 4, or any number (up to 9 for a population of size 10). Of course, a sample of size 10 is a complete census, so sampling is not needed for that sample size. As mentioned previously, the random number method is the most common method for drawing SRS samples for most survey work; however, we will use the all possible samples method in the following just to illustrate some of the concepts of random sampling.

9.3.4   Estimating the Population Mean from Simple Random Samples

Having selected a simple random sample of some size, say n, let us consider how to compute an estimate of the population mean, for example, the mean income of the population. To simplify the discussion, assume that the income of every person selected for the sample is obtained (i.e., there are no missing data due to nonresponse). Further assume that the true income is recorded for each person in the sample (i.e., there are no measurement errors). Later in this section we discuss some ways to compensate for nonresponse and some types of measurement errors in the estimation process.

When the sample is selected using SRS, an estimator of the population mean that has very good statistical properties is the sample mean. Later we discuss some of these statistical properties and why the sample mean is a good estimator. For now, let us consider how to write the computational formula for the sample mean and the population mean. This requires some mathematical notation. The symbols we define are used fairly consistently throughout the sampling literature, and therefore, it is essential to understand what they mean.

Let i denote the ith person in the sample and let j denote the jth person in the population. For example, $i = 1, \ldots, n$ denotes the n elements (or units) in the sample and $j = 1, \ldots, N$ denotes the N units in the population. The Greek letter sigma (i.e., Σ) denotes the summation of the quantities following the summation sign. We use lowercase letters to denote the sample data and uppercase letters to denote the values of the characteristic for the population units. Using this notation, the sample values of some characteristic, y, for the n sample members are denoted $y_1, y_2, \ldots, y_n$ and the values of the characteristic for the population members are denoted by $Y_1, Y_2, \ldots, Y_N$. For example, $\Sigma\, y_i$ denotes the sum of all n sample quantities and $\Sigma\, Y_j$ denotes the sum over the N population quantities. The sample mean is denoted by the symbol $\bar{y}$ (referred to as "y bar") and is defined by the formula

$$\bar{y} = \frac{y_1 + y_2 + \cdots + y_n}{n} = \frac{\sum y_i}{n} \tag{9.1}$$

This formula states that the mean is equal to the sum of the n units in the sample divided by n. Similarly, the population mean is the sum of the N population units divided by N, which can be written mathematically as

$$\bar{Y} = \frac{Y_1 + Y_2 + \cdots + Y_N}{N} = \frac{\sum Y_j}{N} \tag{9.2}$$

Thus, we say that $\bar{y}$ is an estimator of $\bar{Y}$.

As an example, for the population in Table 9.1, the population mean of the incomes of the 10 population members is

$$\overline{Y} = \frac{60,000 + 72,000 + \cdots + 160,000}{10} = \frac{1,100,000}{10} = 110,000$$

Of course, in a real survey, the population mean is unknown and must be estimated. Let us consider how to do this.

Suppose that we draw a simple random sample of size $n = 2$ from this population for this purpose and the two units selected are population units $j = 4$ and $j = 8$. Then the value of the first sample unit is $y_1 = 90,000$, corresponding to Y_4 in the population, and the value of the second sample unit is $y_2 = 135,000$, corresponding to Y_8 in the population. Thus, the estimate of $\overline{Y}$ is just the average of these two values: $\overline{y} = (\$90,000 + \$135,000)/2$, or $\$112,500$. This is our estimate of $\overline{Y} = \$110,000$ for this particular sample.

Note that the error in this estimate of $\overline{Y}$ is $\overline{y} - \overline{Y}$, or $\$2500$. Since we have no nonsampling error in this example, the error must be due solely to sampling error. However, the sampling error we obtain depends on the sample we draw. For example, suppose that instead of population units $j = 4$ and $j = 8$, we selected $j = 1$ and $j = 5$. Now our estimate of $\overline{Y}$ is $\overline{y} = (\$60,000 + \$102,000)/2 = \$81,000$. The sampling error is now much larger and has the opposite sign since $\overline{y} - \overline{Y} = \$81,000 - \$110,000$, or $-\$29,000$. This illustrates how different samples will deviate from the population mean by varying amounts, resulting in different amounts of error.

The amount of error associated with a particular sample is not known in practice, since to know the error we need to know the value of the parameter that is to be estimated. However, using the statistical methods to be described in the next section, it is possible to obtain a measure of the sampling error that is related to the sampling *process* rather than a particular sample we might draw. This measure is called the *sampling variance* of the estimator and is the average squared deviation of the sample mean from the population mean over all possible samples that could be drawn from the population. The squared root of this measure is called the *standard error* of the estimate of the mean.

9.3.5 Statistical Inference with Simple Random Sampling

In this section we discuss the nature of sampling error and how it relates to the standard error of an estimate. We also discuss how the standard error can be estimated from a single sample, how to interpret such estimates, and how these estimates can be used to derive a confidence interval for the parameter. Also discussed is an important property of an estimator of a population parameter, that of unbiasedness.

As mentioned previously, a key advantage of random sampling over non-random sampling methods is that an assessment of the amount of error in the sample estimates is possible with random sampling. As illustrated in Section 9.3.4, an estimate of the population parameter from a single sample can differ considerably from the population parameter. Although in practice we do not know how close an estimate is to the population parameter, it is possible to compute a *confidence interval* for the population parameter from the sample which gives us some idea of how close our estimate is to the parameter value. A confidence interval is simply a range of values that is likely to include the population parameter value that is computed from the values of the units we observe in the sample. A confidence interval is associated with a particular sample we draw. For example, when we compute a 95% confidence interval for the population mean from a particular sample, we are computing an interval that will contain the population mean for 95% of all possible samples we can draw from the population. That is, we are engaging in an inferential process that yields an interval that contains the population mean 95% of the time.

Statistical theory tells us how to construct these intervals such that a certain specified percentage of the samples we can draw from the population will contain the population parameter. Confidence intervals provide means for inferring the value of the population parameter from a single sample we draw from the population.

A key concept in understanding how inferences about population parameters can be made from a single sample is to consider the *process view of sampling and estimation*, which considers the act of drawing a sample and computing an estimate from it as a process than can be repeated many, many times. Each time the process is repeated, a new sample of the same size is drawn by the same sampling scheme, and the estimator is used to compute an estimate of the parameter of interest. This view of sampling is concerned more with the behavior of an estimator over hypothetical repetitions of the sampling process rather than the estimate produced by a single sample. Inferences about the population are based on statistical theory about the sampling and estimation process. The particular sample we draw for inference is designed to inform us about both the process and the population.

The process view of sampling and estimation at first appears to be at odds with the way we draw samples in practice. In practice, only one sample is drawn for a survey and the data are collected for this sample. Yet, to understand how it is possible to make inferences about the population based on observations of a relatively small part of it, we need to view the sample as only one of many that could have been drawn by the sampling process. These ideas will be treated in some detail in the following discussion.

Process View of Sampling

Let us consider how to apply the process view of sampling to the small population in Table 9.1. As described previously, there are a total of 45 possible samples of size 2 that can be drawn from this population, which are listed in Table 9.2. The second column of the table lists the population units for each sample, the third column gives the sample mean for this sample. For example, sample 1 from Table 9.2 consists of units P1 and P2. The sample mean for this sample is $(60,000 + 72,000)/2 = 66,000$ (from column 3).

Table 9.1   Income for a Small Population of 10 Persons

Person in the Population	Actual Income
P1	$Y_1 = 60,000$
P2	$Y_2 = 72,000$
P3	$Y_3 = 94,000$
P4	$Y_4 = 90,000$
P5	$Y_5 = 102,000$
P6	$Y_6 = 116,000$
P7	$Y_7 = 130,000$
P8	$Y_8 = 135,000$
P9	$Y_9 = 141,000$
P10	$Y_{10} = 160,000$
Mean	$\overline{Y} = 110,000$

Since 45 samples in Table 9.2 constitute all possible samples of size 2, any sample of size 2 that we might draw from the population must be one of these samples. As we can see, depending on which sample we draw, the sample mean can be much smaller than the population parameter value (e.g., $66,000 as in sample 1, or much larger as for sample 45, where the mean is $150,500. The average value of the sample mean is $110,000 (as shown in the last row of column 3), which is exactly equal to the population parameter value. Thus, the sampling process will produce means that equal the population mean, on average.

That the average value of the sample mean is exactly the population mean (as shown in the last column of Table 9.1, row 2) is no coincidence. This will always be the case with simple random sampling when there are no non-sampling errors. That is, for simple random sampling, the average value of the sample mean for all possible samples is equal to the population mean. The statistical term for this average mean value is the *expected value*. The expected value of the sample mean is equal to the population mean when simple random sampling is used to select the samples. The property of an estimator that its expected value is equal to the value of the target parameter is called *unbiasedness* in the statistical literature.

The property that the mean of a simple random sample is unbiased for the population mean will hold regardless of the sample size or the size of the population. However, the process can be quite variable since any given sample can be off by between $-44,000$ and $40,500$, as shown in column 4 of Table 9.2. For larger and larger sample sizes, however, the error in the sample mean becomes smaller and smaller. For example, if the sample size is 10 (a complete census), the error in the mean of the sample will be zero. Therefore, a survey designer can control the sampling error in an estimate by the choice of sample size.

Table 9.2   All Possible Samples of Size 2 from a Population of Size 10 (All entries have been divided by 1000)

Sample	Units	Sample Mean, $\overline{y}$	Deviation from $\overline{Y}$	Squared Deviation
1	P1,P2	66.00	−44.00	1936.00
2	P1,P3	77.00	−33.00	1089.00
3	P1,P4	75.00	−35.00	1225.00
4	P1,P5	81.00	−29.00	841.00
5	P1,P6	88.00	−22.00	484.00
6	P1,P7	95.00	−15.00	225.00
7	P1,P8	97.50	−12.50	156.25
8	P1,P9	100.50	−9.50	90.25
9	P1,P10	110.00	0.00	0.00
10	P2,P3	83.00	−27.00	729.00
11	P2,P4	81.00	−29.00	841.00
12	P2,P5	87.00	−23.00	529.00
13	P2,P6	94.00	−16.00	256.00
14	P2,P7	101.00	−9.00	81.00
15	P2,P8	103.50	−6.50	42.25
16	P2,P9	106.50	−3.50	12.25
17	P2,P10	116.00	6.00	36.00
18	P3,P4	92.00	−18.00	324.00
19	P3,P5	98.00	−12.00	144.00
20	P3,P6	105.00	−5.00	25.00
21	P3,P7	112.00	2.00	4.00
22	P3,P8	114.50	4.50	20.25
23	P3,P9	117.50	7.50	56.25
24	P3,P10	127.00	17.00	289.00
25	P4,P5	96.00	−14.00	196.00
26	P4,P6	103.00	−7.00	49.00
27	P4,P7	110.00	0.00	0.00
28	P4,P8	112.50	2.50	6.25
29	P4,P9	115.50	5.50	30.25
30	P4,P10	125.00	15.00	225.00
31	P5,P6	109.00	−1.00	1.00
32	P5,P7	116.00	6.00	36.00
33	P5,P8	118.50	8.50	72.25
34	P5,P9	121.50	11.50	132.25
35	P5,P10	131.00	21.00	441.00
36	P6,P7	123.00	13.00	169.00
37	P6,P8	125.50	15.50	240.25
38	P6,P9	128.50	18.50	342.25
39	P6,P10	138.00	28.00	784.00
40	P7,P8	132.50	22.50	506.25
41	P7,P9	135.50	25.50	650.25
42	P7,P10	145.00	35.00	1225.00
43	P8,P9	138.00	28.00	784.00
44	P8,P10	147.50	37.50	1406.25
45	P9,P10	150.50	40.50	1640.25
	Average	110.00	0.00	408.27

The *process view of sampling* views the sample selected for a survey as one of many that could have been selected by the sampling process. Statistical inference, then, uses this single sample to describe the characteristics of both the target population (e.g., the population mean) and the sampling process (e.g., the standard error of the estimate).

Sampling Distribution

Figure 9.1 shows a histogram of the 45 sample means (i.e., a plot of the frequencies of the values of $\bar{y}$). This histogram is known as the *sampling error distribution* of the sample mean. It is approximately symmetrical around the population mean of 110,000. This reinforces what we already know about the distribution of the sample mean, that on average it is equal to the population mean (i.e., $\bar{y}$ is unbiased under simple random sampling). In fact, the histogram is somewhat bell-shaped (i.e., it is humped in the middle with a gradual trailing off to each side). The symmetrical bell shape of the sampling distribution of the sample mean is another characteristic of simple random sampling.

The distribution of $\bar{y}$ will have this shape no matter what the sample size and population size are. If instead of samples of size 2, we considered samples of size 3 or 4 or more, we would see the shape of the sampling distribution change. It would remain centered around 110,000, but it would become narrower; that is, the tails of the distribution should begin to draw up toward the center of the histogram. Figure 9.2 shows this effect for increasing sample sizes (small, medium, and large) from a large population. The amount of spread of each curve varies with the change in sample size. Note that as the sample size increases, the spread or dispersion of the curve decreases. The implication of this effect for statistical inference is simply that as the sample

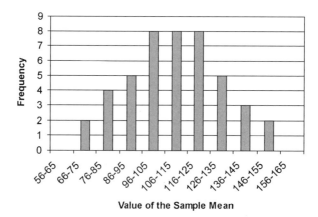

Figure 9.1 Sampling distribution of the sample mean for all possible samples of size 2 selected from the artificial population in Table 9.1.

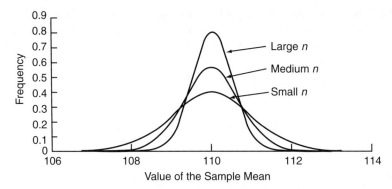

Figure 9.2 Effect of increasing the sample size on the shape of the sampling distribution of the sample mean.

size increases, the chance of drawing a sample whose mean differs substantially from the population mean decreases. In other words, bad samples (i.e., samples with means that differ considerably from the population mean) are drawn much less frequently when the sample size is larger than with smaller sample sizes.

Standard Error of the Estimate

The effect of larger and larger samples on the sample estimate is quantified by the standard error of the estimate. The standard error of the sample mean is a measure of the dispersion (or spread) of the sampling distribution of the sample mean. It is defined as the squared root of the variance of the sampling distribution of the sample mean. The variance of the sampling distribution of the sample mean is usually denoted by $\text{Var}(\bar{y})$ and referred to as the *variance of the sample mean*. It is defined as the average over all possible samples of the squared deviation between the sample mean and the population mean.

In Table 9.2, the fourth column displays the differences, $\bar{y} - \bar{Y}$, for the 45 samples of size 2. The fifth column displays the square of these deviations, and the last row of this column gives the average value of these squared differences, which is $\text{Var}(\bar{y})$ [i.e., $\text{Var}(\bar{y}) = 4{,}082{,}700$. The standard error of $\bar{y}$ is, therefore, $\sqrt{4{,}082{,}700}$ or 2021. Thus, the mathematical formula for $\text{Var}(\bar{y})$ is

$$\text{Var}(\bar{y}) = \frac{\sum (\bar{y} - \bar{Y})^2}{M} \tag{9.3}$$

where in this case the Σ denotes the sum of the squared deviations $(\bar{y} - \bar{Y})^2$ over all possible samples and M is the total number of possible samples. Thus, (9.3) states that the squared deviation between the sample mean and the population mean, averaged over all possible M samples that can be drawn from the population, is equal to the variance of the sample mean.

This formula is not very useful since it requires that we know the mean of every possible sample that can be drawn from the population. This could be an enormous number of samples. For example, for a population of 100, the number of samples of size 10 that can be drawn is over 10 trillion! However, using a little algebra, a much more usable formula can be derived.

It is possible to show that (9.3) is equivalent to

$$\text{Var}(\bar{y}) = \left(1 - \frac{n}{N}\right)\frac{1}{n}\frac{\sum(Y_j - \bar{Y})^2}{N-1} \tag{9.4}$$

$$= \left(1 - \frac{n}{N}\right)\frac{1}{n}S^2 \tag{9.5}$$

where the Σ term in the numerator denotes the sum of the squared deviations $(Y_j - \bar{Y})^2$, over all units in the population and where

$$S^2 = \frac{\sum(Y_j - \bar{Y})^2}{N-1} \tag{9.6}$$

is the variance of the Y-values in the population. The reader should verify that for the data in Table 9.1, (9.3) and (9.5) give the same answer (i.e., 408, 270).

Equation (9.4) states that the variance of the sample mean is equal to the sum of the squared deviations between a population unit value and the mean of the population "averaged" over all population units, divided by the sample size, and multiplied by a factor, $1 - n/N$. The word "averaged" is in quotes since the sum is divided by $N - 1$, not N as it would be for a true average.

The formula for $\text{Var}(\bar{y})$ in (9.5) is preferred over (9.3) since it is much simpler to compute the variance of the values in the population than it is to form all possible samples and compute the variance of these sample means. Further, it is very apparent from (9.5) that $\text{Var}(\bar{y})$ becomes smaller and smaller for larger and larger values of n.

The term $(1 - n/N)$ in (9.5) is called the *finite population correction factor* (*fpc*). For sampling very large populations such as populations of several million persons or more, and when the sample size is several thousand persons or less, the *fpc* is approximately 1 and can be ignored since it has very little effect on the computations. In fact, it is usually acceptable to ignore the *fpc* if the sampling fraction, $f = n/N$ is less than 0.1; that is, if less that 10% of the population is to be sampled. This makes the formulas a little simpler and easier to compute while not making an important difference in the value of the computed variance. Thus, if we can ignore the *fpc*, the variance of $\bar{y}$ is simply

$$\text{Var}(\bar{y}) = \frac{S^2}{n} \tag{9.7}$$

Since the standard error of an estimator is simply the squared root of the variances, the standard error (SE) of $\bar{y}$ is

$$SE(\bar{y}) = \sqrt{Var(\bar{y})} \qquad (9.8)$$

Although $Var(\bar{y})$ and $SE(\bar{y})$ are both measures of the dispersion or spread of the sampling distribution of $\bar{y}$, the standard error is the preferred measure in statistical reporting because its unit of measurement is the same as the unit of measurement of the characteristic of interest. For example, if the characteristic of interest is income in dollars, the $SE(\bar{y})$ is also in dollars, while the $Var(\bar{y})$ is in squared dollars. In addition, the $SE(\bar{y})$ is the quantity that is needed for constructing confidence intervals and for testing hypotheses. For example, a 95% two-sided confidence interval for the population mean, $\bar{Y}$, is

$$[\bar{y} - 2SE(\bar{y}), \bar{y} + 2SE(\bar{y})] \qquad (9.9)$$

The same formula applies to general $(1 - \alpha) \times 100\%$ confidence intervals except instead of multiplying $SE(\bar{y})$ by 2 (or more precisely, 1.96) in (9.9), we would use the $(1 - \alpha/2) \times 100$ percentile of a normal distribution. Tables of these values can be found in any standard textbook on statistical analysis.

The interpretation of such confidence intervals depends on the *central limit theorem*. According to this theorem, if we draw a sample from the population and compute the interval in (9.9), we have a 95% chance that the interval will contain the population mean, $\bar{Y}$. Another way of saying this is that 95% of all possible samples will produce intervals by the formula in (9.9) that contain $\bar{Y}$. Essentially, as long as the sample size is large enough, say $n = 30$ or more, it is usually valid to invoke this theorem for simple random sampling.

The terms *standard error* and *standard deviation* are very similar and are easily confused. A *standard deviation* is the squared root of a population variance. For example, the standard deviation of income values in a population is a measure of the dispersion of the values of income across the units in the population. A *standard error* can also be defined as the squared root of a variance (i.e., the variance of the sampling distribution of an estimator). In that regard, the standard error is a standard deviation of the sampling distribution of the estimator. The key difference is that the standard error always refers to an estimator, not a population.

9.3.6 Determining the Required Sample Size

We can use the formula for the 95% confidence interval in (9.9) to determine the sample size required to obtain a confidence interval of some desired length. A common practice is to specify a desired margin of error in the estimate and then determine the sample size required to achieve it. The *margin of error* of an estimate is simply the length of the 95% confidence interval divided by 2. The length of a 95% confidence interval is the upper confidence limit minus the lower confidence limit. If we subtract the expression for the upper confidence limit from the lower confidence limit in (9.9), we obtain the formula for the length of a 95% confidence interval [i.e., 4 SE($\bar{y}$)]. Therefore, the margin of error for the sample mean is 4 SE($\bar{y}$) divided by 2, or 2 SE($\bar{y}$).

To obtain a formula for the sample size that will yield a specified margin of error, we equate the formula for the margin of error to the numerical value of the margin of error desired and then solve for n to determine the desired sample size. Using the formula for SE($\bar{y}$) above and solving for n yields the following formula for the sample size:

$$n = \frac{4S^2}{d^2} \qquad (9.10)$$

where d is the desired margin of error for the estimate. If the *fpc* is nearly 1 (and can be ignored) (e.g., 0.9 or larger), (9.10) is sufficiently precise. However, if the *fpc* is less than 0.9, the result from (9.10) should be adjusted to account for the finite population.

To see how to do this, let n_0 denote the sample size obtained from applying (9.10). Then the sample size that takes into account the relative size of the sample to the population size is the following:

$$n = \frac{n_0}{1 + n_0/N} \qquad (9.11)$$

That is, we compute the sampling fraction using the initial sample size. Call this sampling fraction $f_0 = n_0/N$. Then the adjusted sample size is the initial sample size divided by $1 + f_0$.

Note that the formula in (9.10) assumes that we know the population variance, S^2. This is seldom the case and an approximate value of S^2 must be used instead. In some cases, a previous survey can provide information on this parameter. However, many times we only have a vague notion about the value of S^2. In that case we may have to guess at a value of S^2 or specify a range of values that will, of course, also produce a range of possible values of n for the survey. Then we can choose a value of n in this range that we feel is best, all things considered.

The simplest case of estimating the required sample size is when we are interested in estimating a single proportion in the population; for example, the proportion of persons who own their homes. Even if we have no idea what this proportion is, we still estimate a sample size requirement if we are willing to be somewhat conservative (i.e., if we are willing to accept a sample size that may be somewhat larger than we would get if we used the actual value of the proportion in the computations). To do this, we assume the proportion in the population is roughly 0.5, or 50%. If we use this value of the proportion, the sample size we compute will always be more than adequate to estimate a smaller or a larger proportion with the same precision. Thus, this is a conservative way of computing a sample size for a study when not much is known about the size of the proportion to be estimated or when there are many different proportions to be estimated, some large and some small.

To convert the formula in (9.10) to one that is suitable for estimating proportions, we denote by P the proportion to be estimated and by Q its complement, $1 - P$. A well-known result in survey theory is that

$$S^2 = PQ \qquad (9.12)$$

That is, the population variance is simply $P(1 - P)$. Now substituting this into (9.10), we get the following formula, which is specifically for proportions:

$$n = \frac{4PQ}{d^2} \qquad (9.13)$$

As an example, if $P = 0.5$, then $Q = 0.5$, then $PQ = 0.25$. Substituting this into (9.13), we obtain the following simple formula for a conservative estimate of the required sample size needed to estimate a proportion with desired margin of error equal to d:

$$n = \frac{1}{d^2} \qquad (9.14)$$

Example 9.3.1 Suppose that we wish to estimate some population proportion of unknown magnitude and we desire a margin of error of 5 percentage points or less. That is, we want a confidence interval for the estimated proportion, p, that is, ± 5 percentage points, no matter what the value of P is. Using the formula in (9.14), we see that this is $n_0 = 1/(0.05)^2 = 400$. Now suppose that the population size is only 2000 and thus the finite population correction factor is $1 - n_0/N = 1 - 400/2000$, or 0.80. Since this value is less than 0.9, we use (9.11) to adjust the initial estimate of n for the finite population size. Thus, the final sample size is $400/(1 + 0.2) = 333.33$. When the computed value of the sample size is not an integer, we round up to the nearest integer. Therefore, the sample size required to achieve a 5-percentage-point margin of error is 334. That is, if we complete 334 interviews and form a 95% confidence interval for P using (9.9), the length of this confidence interval will not exceed 5 percentage points.

9.3.7 Estimating the Standard Error of the Sample Mean

In Section 9.3.5 we provided a few basic formulas for statistical inference in sample surveys. However, recall that the formulas for the standard errors of the estimates required essentially that we know the values of the characteristic of interest for all units in the entire population. In this section we study methods for estimating the standard error from a single sample. An amazing result in sampling theory is that we can assess the usefulness of a particular sample for statistical inference by using only the information in the sample itself without knowledge of any other population values. This important feature of random sampling gives it a tremendous advantage over nonrandom methods.

The estimation formulas for the variance and standard error of $\bar{y}$ look very similar to the formulas for the population quantities they are intended to estimate [i.e., the formulas for the estimators of $\text{Var}(\bar{y})$ and $\text{SE}(\bar{y})$ resemble sample versions of the population formulas]. The primary difference in the formulas is the use of a sample in place of population values. We provide the estimation formula for $\text{Var}(\bar{y})$, then the formula for $\text{SE}(\bar{y})$, and finally, provide examples illustrating their use.

Recall the concept of unbiasedness of an estimator means that the average value of the estimates over all possible samples is equal to the value of the population parameter to be estimated. Thus, an unbiased estimator of $\text{Var}(\bar{y})$ is

$$var(\bar{y}) = \left(1 - \frac{n}{N}\right)\frac{s^2}{n} \tag{9.15}$$

where

$$s^2 = \frac{\sum(y_i - \bar{y})^2}{n-1} \tag{9.16}$$

Therefore, an estimator of $\text{SE}(\bar{y})$ is

$$se(\bar{y}) = \sqrt{var(\bar{y})} \tag{9.17}$$

Example 9.3.2   To illustrate how the formulas for the estimates of the variance and standard error are applied, consider the population in Table 9.1 once more. But this time, suppose that we draw a sample of size 4 (units P3, P5, P6, and P10). To apply (9.16) and obtain the estimate of the population variance, S^2, we first compute the mean of the sample, $\bar{y}$. The values of the four sample units are: 94,000, 102,000, 116,000, and 160,000, respectively. Thus, $\bar{y}$ is the average of these four values, or 118,000.

Next, we compute the differences $(y_i - \bar{y})$ for all four sample values. These differences are: −24,000, −16,000, −2000, and 42,000, respectively. As a check, these four differences should add to 0, which they do. Next, square the differences and sum them, to obtain 2.6×10^8. Then divide this number by $4 - 1$, or 3, as in (9.16) to obtain an estimate of s^2 of 0.8667×10^8. Now, computing

(9.15), we multiply this value by $(1 - 4/10) = 0.6$ and divide by 4 to obtain an estimate of $\mathrm{var}(\bar{y})$ of 0.13×10^8. Finally, to compute (9.17), we take the squared root of this value to obtain 0.36×10^4 or 3600. Thus, our estimate of $\mathrm{SE}(\bar{y})$ is $3600 and a 95% confidence interval for $\bar{Y}$ is $\bar{y} \pm 2 \times 3600$ or $[110,800, 125,200]$.

Keep in mind that Example 9.3.2 is a simple illustration of how the formulas work. In practice, we would not attempt to compute an estimate of either the population mean or the standard error with only a sample of size 4 since the estimates are likely to be very imprecise. Further, the validity of confidence intervals rely on the central limit theorem, as noted previously. The central limit theorem applies only for samples that are considerably larger than size 4, so even the validity of the confidence interval we computed is questionable with such a small sample.

9.3.8 Other Estimators of $\bar{Y}$ Under Simple Random Sampling

As we have discussed, sampling error can be controlled by the sample design and the sample size. However, another determinant of sampling error is the choice of estimator. For simple random sampling, the natural estimator of the population mean is $\bar{y}$, referred to as the *simple expansion estimator*. However, this estimator is not the only estimator of $\bar{Y}$ in simple random sampling. In fact, it may not even be the best estimator. Other estimators include the ratio estimator, the regression estimator, and the sample median. Usually, the choice of estimator is simply the estimator that has the smaller *mean squared error*. In this section we discuss one alternative to the simple expansion estimator for estimating means, proportions, and totals (i.e., the *ratio estimator*). Ratio estimators are quite important in practical sampling situations and, for complex sample designs, are one of the primary types of estimators used. We discuss this estimator for simple random sampling. However, extensions of the estimator for complex sampling are straightforward.

For the moment, suppose that we ignore the effects of nonsampling error on the mean squared error. Then, except for complete censuses of the population, the only error in an estimator is due to sampling. We discussed the sampling variance of an estimator and the relationship between sampling variance and sample size. A somewhat new concept is estimator bias, sometimes referred to as the *technical bias* in an estimator. This bias is not a result of nonsampling error, but rather, the result of choosing an estimator that is not unbiased. Ratio estimators have a small technical bias; however, in certain situations, they have a smaller variance than simple expansion estimators, so their mean squared errors are smaller. Further discussion of the bias in the ratio estimator is beyond the scope of this book. However, the reader is referred to any number of books on sampling (e.g., Lohr, 1999) for a discussion of this concept and how to choose the best estimator in various sampling situations. In this section we discuss briefly the ratio estimator because of its importance in sampling, particularly for complex sampling designs.

The ratio estimator is used in situations where some type of *auxiliary information* is available on both the sample units and the population units. Auxiliary information is essentially another characteristic of the sample units, usually one that is correlated with the characteristic of interest, y.

For example, suppose that we know the income of every member of a population as of five years ago and we wish to estimate this population's current average income. We draw an SRS of size n and obtain the current income of each sample member. Let $\bar{x}$ denote the mean income for this sample based on five-year-old data and let $\bar{y}$ denote the current mean income from the survey. Finally, let $\overline{X}$ denote the mean income for the entire population based on the old data. The data on a person's previous income is referred to as *auxiliary data* since they are data obtained from sources outside the survey. The ratio estimator is then defined as

$$\bar{y}_R = \frac{\bar{y}}{\bar{x}}\,\overline{X} \tag{9.18}$$

Since the previous income data, X, are expected to be highly correlated with the current income data, Y, the ratio estimator, $\bar{y}_R$, is likely to be a much more precise estimator of $\overline{Y}$ than the simple expansion estimator, $\bar{y}$. A proof of this and a comparison of simple expansion and ratio estimators can be found in any sampling theory textbook (see, e.g., Cochran, 1977).

As we will see in the next section, sampling error is affected not only by the choice of survey design but also by the choice of estimator. For example, suppose that we select a sample from the population by some method, such as simple random sampling. Once the data are collected, the statistician must decide what estimator to use to estimate the parameters of interest. As an example, if the statistician desires to estimate the mean income of the population, he or she may have several estimators to choose from, depending on what auxiliary data may be available. If no auxiliary data are available, the usual estimator of the population mean is the simple expansion mean, or the sample average. However, more precise estimators can sometimes be constructed if information on total or mean income is available from external sources, such as a previous census or survey.

9.4 STATISTICAL INFERENCE IN THE PRESENCE OF NONSAMPLING ERRORS

The statistical results presented in Section 9.3 are based on the assumption that the data are free of nonsampling errors. Classical sampling theory assumes that sampling is conducted with a perfect frame, that every unit selected for the sample responds fully and accurately, and that no errors are introduced in the data from data processing. Although these assumptions are unrealistic in practical applications, they are made to simplify the statistical

Table 9.3   Income for a Small Population of 10 Persons

Person in the Population	Actual Income (÷1000)	Reported Income (÷1000)
P1	$Y_1 = 60$	missing
P2	$Y_2 = 72$	missing
P3	$Y_3 = 94$	$X_3 = 80$
P4	$Y_4 = 90$	$X_4 = 76$
P5	$Y_5 = 102$	$X_5 = 90$
P6	$Y_6 = 116$	$X_6 = 100$
P7	$Y_7 = 130$	$X_7 = 132$
P8	$Y_8 = 135$	$X_8 = 142$
P9	$Y_9 = 141$	$X_9 = 157$
P10	$Y_{10} = 160$	$X_{10} = 135$
Mean	110	114

theory and, in many cases, the statistical results obtained under these "ideal-world" assumptions very often approximate the "real-world" results. However, it is still important to realize that sampling theory is based on idealistic assumptions that seldom hold in practice. Further, nonsampling errors, which are inevitable in any practical survey situation, can distort the sampling distributions in ways that may invalidate the results derived under classical sampling theory assumptions. In this section we provide some illustrations of the effects of nonsampling errors on statistical inference and demonstrate how nonsampling error can invalidate the results of classical survey inference.

To illustrate the basic ideas of survey inference in the presence of nonsampling errors, we return to the example considered in Table 9.1 of a population of size $N = 10$. To introduce nonsampling errors, suppose that only eight of the 10 units are listed on the sampling frame; for example, let us suppose that P1 and P2 are missing from the frame. Even though P1 and P2 are missing, they are still part of the target population and are eligible for the survey. Therefore, they are frame omissions in the sense of the discussions in Chapter 3. We further assume that there are measurement errors in the survey reports of income. That is, instead of reporting the actual income, Y_j, unit j in the population reports a slightly erroneous income, X_j, if unit j is selected for the survey.

Table 9.3 summarizes these assumptions about the frame and reporting accuracy for this population. The second column replicates the values in Table 9.1, and the third column lists the income values the population members will report if they are selected for the sample. Note that since P1 and P2 are missing, the frame only contains $N = 8$ units, corresponding to units P3 to P10. Using these data, let us examine the sampling distribution of the sample mean for samples of size $n = 2$ from this frame consisting of $N = 8$ units.

There are 28 possible, unique samples of size 2 that can be selected from a population of size 8. To see this, note that there are seven samples that contain P3 (i.e., {P3, P4}, {P3, P5}, . . . , {P3, P10}), six samples that contain P4 among

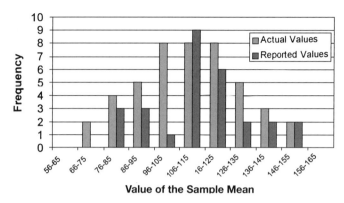

Figure 9.3 Comparison of sampling distributions for actual and reported values of the sample means for samples of size 2 selected from the artificial populations in Table 9.3. Note that the distribution of the reported values is shifted to the right slightly, suggesting a bias in the sample mean as an estimator of the population mean.

the samples that were previously uncounted, five that contain P5 not previously counted, and so on. Thus, the total number of unique samples is $7 + 6 + 5 + 4 + 3 + 2 + 1 = 28$. For each of these 28 samples, we can compute the sample means, $\bar{x} = (x_1 + x_2)/2$, using the X-values in Table 9.3 in the same manner as we did using the Y-values for Table 9.2. Then we can form a frequency distribution of the 28 values of $\bar{x}$ and compare it with the corresponding frequency distribution of the 45 values of $\bar{y}$. In this way we can compare the sampling distributions of the sample means for actual and reported values of income. This comparison is shown in Figure 9.3.

Recall that when we used the actual values of income, the average value of the sample means from the 45 possible samples (i.e., the expected value of $\bar{y}$) was 110,000, the same value as the population mean, $\bar{Y}$. This property of the sample mean for simple random sampling was referred to as *unbiasedness*. However, the unbiasedness property does not necessarily hold when there are nonsampling errors. If we use the X-values in Table 9.3 to compute the means of all possible samples, the average value of $\bar{x}$ for these 28 samples is 114,000, the same as the population mean of the X-values in Table 9.3, which we denote by $\bar{X}$. That is, the expected value of $\bar{x}$ is 114,000, or $\bar{X}$. The difference between the two expected values is the bias in the sample mean due to nonsampling errors. That is,

$$\text{Bias}(\bar{x}) = \bar{X} - \bar{Y} \tag{9.19}$$

or $114{,}000 - 110{,}000 = 4000$.

Although 4000 seems like a large bias, it is a very small percentage of the parameter value to be estimated (i.e., $4000/114{,}000 = 0.035$ or 3.5% of $\bar{X}$). The quantity 3.5%, called the *relative bias*, is the ratio of the bias to the population parameter. There is no general rule of thumb for determining whether a bias is large or small. Usually, a bias that is much larger than the standard error of the estimate is considered to be a large bias and, conversely, a bias that is much smaller than the standard error is considered to be small. Like the relative bias, the standard error can also be expressed as a percentage of the population parameter. This quantity is called the *relative standard error* or, more often, the *coefficient of variation* (CV). Thus, if the CV of $\bar{y}$ were 10%, a relative bias of 3.5% would be considered to be a small bias.

The bias in $\bar{y}$ can be seen in Figure 9.3 as well. Note that the sampling distribution of $\bar{x}$ is slightly shifted to the right of the sampling distribution of $\bar{y}$. That is because the distribution of $\bar{x}$ is centered around its mean of 114,000 and the distribution of $\bar{y}$ is centered around its mean of 110,000. This is more clearly illustrated in Figure 9.4, which is discussed in more detail subsequently. Both figures show the effect of nonsampling error on the sampling distribution of the sample mean. In the presence of nonsampling error, the distribution is flatter (more dispersed), which illustrates the increase in variance created by adding variable error to the data. It is also shifted to the right, illustrating the effect of systematic errors, which in this example create a positive bias in the estimate of the sample mean.

In practice, if we know the amount of this bias, we can correct $\bar{x}$ for the bias simply by subtracting it from $\bar{x}$. Note that the sampling distribution of $\bar{x} - 4000$ would shift the distribution to the left so that it would be centered over 110,000 or $\bar{Y}$. Unfortunately, Bias($\bar{x}$) is rarely known in practice, so correcting for the bias is not possible.

In general, nonsampling errors can bias the estimates of population parameters so that inferences about the parameters are no longer valid. Nonsampling errors can change the shape of the sampling distribution, which will affect the variances and variance estimates as well as the location of the sampling distribution that affects the bias. Usually, these effects are unknown to the survey analyst and therefore are not accounted for in the analysis.

In Figure 9.4, we see the potential effects of the nonsampling errors on the bias and variance of the sample mean. The bell-shaped curve on the right is the correct sampling distribution based on complete and accurate population data, while the curve on the right is the sampling distribution based on data that are subject to nonsampling error distortions. The figure shows the center of each distribution, which is also the expected value of the estimator. In the case of the true sampling distribution, the expected value of the estimator is the population parameter value. For the distorted distribution, the center of the distribution is offset from the actual distribution's center (i.e., the expected value of the estimator in the presence of nonsampling errors differs from the population parameter value). The difference between the two expected values is the bias in the estimator due to nonsampling errors.

Figure 9.4 also illustrates that the spread or dispersion of the sampling distribution can also be affected by nonsampling errors. In the illustration, the dispersion of the sampling distribution with nonsampling errors is larger than the standard error of the estimator without nonsampling errors. Since the

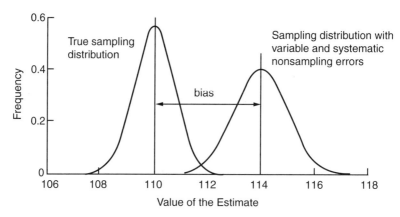

Figure 9.4 Bias and increased variance in the sampling distribution of an estimator in the presence of nonsampling errors.

standard error of the sample mean is related directly to the dispersion of the sampling distribution, the standard error of the sample mean with nonsampling errors is larger than the standard error without nonsampling errors in the illustration. Thus, statistical inferences can be affected by both distortions of the sampling variance estimates as well as biases in the estimator. As an example, 95% confidence intervals constructed according to (9.9) may in fact have coverage rates which are much less than 95%. That is, a 95% confidence interval computed in the standard way may actually be an 85% or even 70% confidence interval as a result of the distortions of the sampling distributions caused by nonsampling error. Unfortunately, this fact is seldom taken into account when making statistical inferences from survey data.

Very few methods are available to correct statistical inferences for the distortions of the sampling distributions caused by nonsampling errors. For example, as we have already discussed, if a good estimate of the bias is available, it can be used to adjust the estimate for the bias. However, care must be taken in making these adjustments since the bias estimates are themselves subject to nonsampling errors as well as sampling errors. Thus, the adjusted estimator could actually be no better or even worse, in terms of mean squared error, than the unadjusted estimate.

Some methods of standard error estimation quite successfully capture the sampling and nonsampling variance in the sampling distribution. For example, the estimator of the standard error in (9.17) will reflect nonsampling errors which are uncorrelated, such as measurement errors made by respondents. However, correlated errors such as interviewer variance are not reflected in the standard errors and, consequently, the standard error estimates will underestimate the true variance, which includes both sampling and nonsampling variance. Wolter (1985) discusses a number of variance estimation methods for reflecting the correlated as well as the uncorrelated nonsampling errors.

None of the methods for dealing with nonsampling errors are fully successful at compensating for these errors. Each adjustment method may address one source of error well but introduces other sources of error as a consequence of the adjustment. The best strategy for dealing with nonsampling error is error prevention rather than error adjustment. Of course, that is the motivation for writing this book! Still it is impossible to avoid some nonsampling errors, in which case some knowledge of their effects and the consequences of such errors on statistical inference is extremely valuable.

Example 9.4.1 In many RDD surveys, a random sample of households are first selected using stratified simple random sampling methods and then, for each of these households, a person is selected at random from all eligible persons in the household to complete the interview. The most precise method for selecting the random person is the *Kish roster method*. For this method, a list of all eligible persons is made and then a random person is selected from this list using some type of random number generator. Some researchers claim that this method contributes to nonresponse since many households are reluctant to provide to a stranger on the telephone information regarding all the persons who live in the household. Many of these researchers prefer the *last birthday method* of respondent selection.

To implement the last birthday method, the interviewer does not need to develop a roster of eligible persons within the household. Instead, he or she merely asks "May I speak to the person in the household who is 18 years of age or older and whose birthday was most recent?" Therefore, it is nonintrusive and quite brief, saving time and respondent burden. In some studies, the method compares very well to Kish's more scientific method. However, other studies suggest that the method skews the sample toward females and younger adults. One study (Lavrakas et al., 2000) showed that the method selects the wrong person in about 20% of households. This is particularly true in low-education households and large households. Further, respondents have mixed reactions about the method, but many do not find the method credible or scientific. For this reason, we recommend that the method not be used in rigorous scientific research.

Nonsampling errors distort the sampling distribution of an estimator so that classical methods of inference are no longer completely valid. For example, systematic errors create biases in the estimates, and consequently, a confidence interval that is nominally 95% may have a much lower effective confidence level. Variable errors can also create biases in estimates of regression coefficients, correlation coefficients, and other nonlinear estimators, resulting in similarly distorted sampling distributions and loss of confidence.

9.5   OTHER METHODS OF RANDOM SAMPLING

Simple random sampling is usually reserved for situations where a complete and accurate frame is readily available and where data collection does not require traveling to the sample units. For example, for a mail survey of a list of members of a professional society or an RDD telephone survey, simple random sampling may work very well since the cost of interviewing does not depend on distance between the sample units. However, the method is not very practical for surveys that involve traveling to the units. For example, in face-to-face surveys, the cost of sending interviewers out to the sampled units to collect the survey data is a substantial part of the total data collection costs. A simple random sample of all the households in a large country such as the United States or Canada would have enormous travel costs since interviewers could be traveling long distances, perhaps hundreds of miles, just to interview one household. In addition, to draw the sample, a frame containing all households in the country would have to be constructed for the survey to have complete population coverage. In some countries (notably, the Scandinavian countries), this can be done quite economically using national population registers. However, in the United States and many other countries, this effort would also be quite expensive, since getting a complete list of all the households in the country would require an effort on the scale of a national census.

In situations where no complete list of population elements is available, more efficient methods of drawing the sample that reduce the field costs by selecting clusters of units can be used. Moreover, even in situations where simple random sampling is appropriate, there may be important advantages in using an alternative method of sampling that is either simpler to use or that may result in more precise estimates.

In this section we provide a brief overview of some methods of sampling commonly used in survey work. Our objective is to explain briefly what these methods are and to provide some guidance regarding when their use might be considered. It is not our intention to provide a comprehensive coverage of each method. There are a number of excellent texts on sampling that can provide additional details (see, e.g., Cochran, 1977).

9.5.1   Systematic Sampling

When a physical list of the population is to be sampled, simple random sampling can be quite tedious. For example, to select a sample of 500 names from a list of 5000 names using the random number method would require generating 500 random numbers and then laboriously counting through the list to select the names at the 500 random positions on the list. A much simpler method is to take every tenth name on the list. Selection of the names is much easier to do by hand since the interval between selections on the list is constant. If the order of the items on the list is more or less random, this method is essentially equivalent to simple random sampling.

To select a sample of 500 items from 5000, one would first compute the "take every" number or skip interval, which is essentially 1 divided by the sampling fraction. Since the sampling fraction is 500/5000 or 0.1, taking every tenth name on the list will result in a systematic sample of 500 names. The next step is to choose the "start with" number or random start. This is a random number between 1 and the "take every" number, 10 in this example. Suppose that this random number is 5. Then the systematic sample will consist of the names at positions 5, 5 + 10 = 15, 5 + 20 = 25, and so on, until item 5 + 499 × 10 = 4995 is selected. Note that there are only 10 possible samples of size 500 from this list, corresponding to the 10 possible random starts, 1 through 10. In this manner, the selection of a random "start with" number determines the specific sample that will be drawn.

Strictly speaking, systematic sampling is not equivalent to simple random sampling except in the special case where the list is randomly ordered. Although this is seldom the case in practice, in many situations the list can be treated as though it were randomly ordered. For example, if a list of names is arranged alphabetically, the ordering may be completely uncorrelated to the characteristics of interest in the survey. Or the list may be ordered by some type of identification number that is assigned independent of the characteristics of interest in the survey. In such cases, treating the systematic sample as though it were a simple random sample would be justified.

However, one of the major benefits of a systematic sample is realized when ordering is imposed on the list. Selecting a systematic sample from a list by first sorting the list by variables that are related to characteristics of interest is a powerful technique for increasing the precision of the estimates. This method is related to the stratified random sampling method discussed in the next section. In fact, using systematic sampling after a well-designed sorting of the list can produce samples that are much more efficient in terms of statistical precision than simple random samples, even though the same estimators are used for both.

The goal of sorting is to try to order the file so that units that have very nearly the same value of the characteristic are adjacent in the list. This creates a type of grouping or stratification on the list so that as selection proceeds through the list, the units selected are as different from each other as possible. Another way of saying this is that we want an ordering so that one systematic sample will produce very nearly the same estimate as any other systematic sample of the same size from the list.

One of the risks of using systematic sampling in an unsorted list is the possibility that the list may be ordered in a nonrandom way which actually reduces the precision of the estimates. For example, suppose that the list is ordered by identification number, with odd numbers assigned to female population members and even numbers assigned to males. In that case, a systematic sample from the unsorted list could result in a sample that contains all males or all females. To protect against this, the list could be sorted by variables on the file other than identification number.

> *Systematic sampling* is one of the simplest sampling methods. It begins with a list of all the units in the frame population, which is sorted by criteria related to the characteristics of interest. Then, to select a sample of approximately size n from the N units on the list, choose unit r on the list and then every kth unit thereafter, where r is a random integer between 1 and k, and k is the closest integer to N/n.

9.5.2 Stratified Sampling

Quite often, the frame to be sampled contains characteristics of the units in the population that can be quite useful in sampling. As we saw in Section 9.5.1, the frame information can be used in a systematic sample by grouping together units that are similar with respect to the characteristics to be measured in the survey. However, stratified sampling is another way in which this information can be used to accomplish essentially the same thing. The goal in stratified sampling is to form groups or strata of units such that within a stratum, the units are very similar on the characteristics of interest. Then each stratum is sampled independently to obtain the sample for the survey.

Stratified sampling can produce estimates of the entire population which are much more precise than unstratified sampling. For example, suppose that the population to be sampled is composed of 25% blacks, 60% whites, and 15% other races. A simple random sample of this population could produce a sample that has a very different composition of these races than in the population. For example, only 5% of the sample may be in the other race category, or the sample may contain 75% white and only 15% black. This is possible because we are sampling from the population randomly without regard to race. As a consequence of this variation in the sample composition across different simple random samples, the variances of the sampling distributions can be quite large, resulting in reduced precision in the estimates.

However, if the sample takes into account the distribution of race in the population, the possibility of obtaining samples with these distortions is eliminated. To do this, we form three strata: one for blacks, one for whites, and one for other. Then we draw 25% of the sample from the black stratum, 60% from the white stratum, and 15% from the other stratum. Now every sample we draw will have exactly the distribution of race that we find in the population. This reduces sample-to-sample variability and thus reduces the sampling variance of the estimates. This type of sampling is referred to as *stratified random sampling with proportional allocation*. The sample size drawn (or allocated) to a particular stratum is proportional to the size of the stratum on the frame. To select a sample of size 1000 by stratified random sampling with proportional allocation, a simple random sample of 250 persons would be selected from the black stratum, 600 persons would be selected from the white stratum, and 150 persons would be selected from the other stratum.

Thus, the sampling variation inherent in the simple random sampling design can be eliminated using stratified sampling with proportional allocation. This can mean a considerable increase in precision for estimates of characteristics correlated with the stratification variable. Furthermore, no additional complexity is introduced for estimating means, totals, and proportions, since the same estimators used in simple random sampling are also appropriate for stratified random sampling with proportional allocation. However, the estimates of standard errors of the estimates is somewhat more complex than for SRS. Further, if the allocation of the sample to the strata is disproportionate, both the estimation of parameters and the estimation of standard errors are more complicated than for SRS.

It is possible to obtain very nearly the same gain in precision with systematic sampling as with stratified random sampling with proportional allocation using a method known as *systematic sampling with implicit stratification*. For this method, rather than sampling the strata independently, the list is sorted by stratum and then systematic sampling is used throughout the entire list using a single random start. The result will be approximately the same as proportional allocation. In the example above, the resulting sample will contain exactly 250 blacks, 600 whites, and 150 other race. For other examples, the numbers could differ by at most one or two from proportional allocation. As in the case of stratified random sampling with proportional allocation, the estimation formulas from simple random sampling are still appropriate for this sampling method as well.

Another important use of stratification is to ensure that the sample we select has a sufficient number of units that possess some fairly uncommon characteristic: for example, persons living in large households. If household size is known for all persons on the frame, persons who live in large households can be grouped together in one stratum and sampled separately from the rest of the frame population. In this manner, the exact number of persons of each type can be selected for the sample.

This is often referred to as *oversampling*, since we are sampling more units of some type than would come into the sample through a purely random process. The characteristic of interest may only be present in 5% of the units in the population. But to ensure we select a sufficient number of these types of units in the sample, we stratify the frame and select a higher percentage, say 30%, of the sample from the rare stratum.

Stratified sampling can also be used with other types of sampling, which we discuss later in this chapter. Once the strata are formed, any method of sampling can be used within the strata, and the methods can differ from stratum to stratum. For example, one type of stratum may contain population units that are not to be sampled at all (referred to as the *nonsampled stratum*). These are units, which may only comprise a very small percentage of the total population, that are assigned a zero probability of selection in order to achieve greater sampling efficiency and to reduce data collection costs. Another type of stratum may contain units that are to be selected with probability 1. That is, they are selected for the sample with certainty (referred to as the *certainty stratum*). These are units that are either very large or very important, so allowing their entry into the sample purely by chance is not an option. Finally, a third type of stratum (the *noncertainty stratum*) contains all units that are not in the other two strata. These units are sampled using whatever method is most efficient.

> *Stratification* is essentially a partitioning of the frame population into mutually exclusive and exhaustive groups which are internally homogeneous with respect to the characteristics to be measured in the survey. In *stratified random sampling*, an SRS sample of units is selected within each stratum.

9.5.3 Multistage and Cluster Sampling

So far we have considered three ways in which auxiliary information that may be available on the frame can be used to improve estimator precision: ratio estimation, systematic sampling, and stratified sampling. It is also possible to combine ratio estimation (or other methods of estimation that use auxiliary information) with the sampling methods we discuss here. Multistage sampling and cluster sampling are two other methods for making use of auxiliary information in the sampling process.

Multistage sampling and *cluster sampling* are methods related to stratified sampling in that all three methods begin by dividing the frame into groups defined by demographic characteristics, geographic location, or other characteristics of the frame units. However, there are several important differences between them:

- In stratified sampling, each group (or stratum) is sampled. In multistage sampling, only a random sample of the groups are sampled (see Figure 9.5).
- In stratified sampling, the groups are formed by the sampler with the objective of creating groups that are internally homogeneous with regard to the characteristics of interest. For example, if the characteristics to be measured in the survey are expected to vary considerably by race, the sampler tries to form strata that correspond to each race. If low-income respondents are expected to respond differently than high-income respondents, the sampler tries to create a low-income stratum and a high-income stratum.
- In cluster sampling, the groups are usually naturally occurring divisions of the population, like counties or neighborhoods for household surveys; land areas, and segments for agricultural surveys; school districts or even schools for school surveys; and so on.

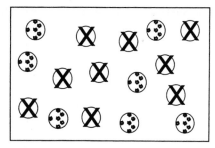 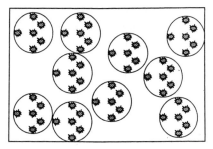

X = not sampled

(*a*) **Cluster Sampling**   (*b*) **Stratified Sampling**

Figure 9.5   Cluster sampling versus stratified sampling. For cluster sampling [part (*a*)], units in the area to be sampled (represented by the rectangle) are grouped into many small clusters of units (the circles) that are mutually exclusive (i.e., nonoverlapping) and exhaustive (i.e., include all members of the target population in the area). Then a sample of these clusters is selected and sampling continues to select the units within the clusters (stars within the circles). The clusters that are not selected are represented in the figure by the "×'s."

In stratified sampling [part (*b*)], a similar process is implemented. The area to be sampled is again divided into mutually exclusive and exhaustive groups. As shown in the figure, the groups, or strata, are defined somewhat differently than in cluster sampling. The key difference between cluster sampling and stratified sampling is that instead of sampling within only some of the groups, as in cluster sampling, sampling is conducted in *all* the strata. Thus, no sampling variation is introduced by the grouping process for stratified sampling as it is in cluster sampling.

- Typically, there are few strata relative to the number of clusters. For example, in the U.S. Current Population Survey, there are only a few hundred strata, whereas there are thousands of clusters formed by the U.S. counties and census blocks within counties.

> Stratified sampling and cluster sampling are often confused. The main difference is that with *stratified sampling*, all groups (strata) are sampled, whereas for *cluster sampling*, only a random sample of groups (clusters) are sampled.

Both single- and multistage cluster sampling are possible. If all the units in the randomly selected groups are included in the sample, the method is known as *one-stage cluster sampling*. For *multistage sampling*, sampling continues within the clusters and there may be a new process of cluster and sampling within higher-stage clusters. When the units sampled within the groups are the actual elements to be observed in the survey, the method is known as *two-stage sampling*. The first or *primary stage* is the sampling of clusters and the *secondary stage* is the selection of elements within the clusters. However, three- and four-stage designs are not uncommon.

As an example of a two-stage design, a survey of schools in a state might first select a sample of school districts and then a sample of schools within the districts selected. The sample of districts is the primary sampling stage and the sample of schools is the secondary sampling stage. If we wish to go on to select a sample of students from each of the schools sampled, the students become the tertiary sampling stage. Or we may select classes within the school as the tertiary stage and the students to be interviewed as the quarternary stage. However, if, instead, we take all the students in the classes selected, the design is referred to as a *three-stage cluster sample*, where the schools are now clusters. Often, multistage and cluster sampling is loosely referred to as *multistage cluster sampling* or just *cluster sampling*.

Although stratified sampling and cluster sampling both involve grouping the population elements prior to sampling, the two methods serve entirely different sampling purposes. *Stratified sampling* is a device that is used primarily to increase the precision of the estimates. This is because, by forming groups of units that are homogeneous with respect to the characteristics of interest in the survey, a relatively small sample from each of these homogeneous groups is needed to describe the group adequately. This means that the total sample size can be smaller when stratified sampling is used.

On the other hand, in cluster sampling, only some of the clusters will be selected. Therefore, we would like each cluster to contain as much information about the population as possible. In other words, we would prefer that clusters be internally *heterogeneous*, which is very different from the stratified sampling objective. This is not usually possible with cluster sampling since the clusters are often formed by natural divisions in the population: counties, neighborhoods, schools and school districts, land segments, and so on. Unfortunately for cluster sampling, this usually implies that the clusters will be internally homogeneous for the survey characteristics. Consequently, cluster sampling invariably will lead to a loss of precision compared with unclustered sampling methods such as simple random sampling and systematic sampling.

The primary justification for using cluster sampling is the economy it creates for sampling and data collection, particularly in face-to-face surveys. By clustering the units, travel costs between the sampling units can be reduced dramatically. For example, a simple random sample of 50,000 households drawn from the entire United States could result in interviewers traveling to all 3141 counties in the United States. Further, there may be many miles between households, and the travel costs would be prohibitively large. However, by clustering the households in, say, 400 to 500 counties, travel costs can be reduced dramatically.

However, the disadvantage of cluster sampling is that the variance of the estimates may be considerably larger than for a simple random sample of the same size. For example, a simple random sample of 50,000 units selected from the United States will yield a variance V_{SRS}. A cluster sample of the same size selected in only 400 counties will result in a variance, V_{CL}, which is $D \times V_{\text{SRS}}$

for some number D larger than 1 (D, the *design effect*, is discussed later in more detail). The cluster sampling variance will usually be larger than the simple random sampling variance as a result of *intracluster correlation*. That is, two units selected at random from some cluster are likely to be more similar than two units selected at random from the entire United States.

For example, in the 2000 U.S. presidential election, the United States was essentially evenly divided between presidential candidates Bush and Gore. Suppose that we wish to estimate the proportion of the voting population who will vote for each candidate. Which would be more precise: a sample of 1000 voters from 10 election precincts (100 voters per precinct) or an SRS sample of 1000 voters from the entire country?

In the 2000 elections, it was quite common for precincts to favor one candidate disproportionately; in many precincts, the proportion of voters favoring a particular candidate was 90% or higher. Thus, it is quite likely that a random sample of only 10 precincts out of thousands of precincts all happened to favor one of the two candidates by a large margin, thus predicting a similar result for the entire nation. However, it is quite unlikely that an SRS sample of 1000 voters tends to favor one candidate by the same wide margin.

Put in another way, 100 voters selected from the same precinct are more likely to favor one of the candidates by a wide margin than 100 voters selected from the entire United States when the margin at the U.S. level is less than 1%. Thus, the similarity of voters within precincts (i.e., the precinct clustering effects) equates to a loss of precision in the estimate of the margin of victory for either candidate.

Despite the loss in precision due to clustering effects, for a given survey budget, the best precision in the estimates may be obtained by taking a very large cluster sample rather than a much smaller unclustered sample. For example, suppose that the survey budget is sufficient to conduct a personal visit survey of 50,000 households when the sample is clustered by geographic area. However, because of enormous travel costs, only 10,000 households could be surveyed for the same budget if the sample were completely unclustered. Finally, suppose that D, the design effect, is 3.0 for the clustered sample. That is, $V_{CL} = 3 \times V_{SRS} = 3 \times \sigma^2/50,000$, where σ^2 is the population variance. This suggests that for the same budget, the cluster sample could be afforded, which would have a variance of $\sigma^2/16,667$ compared to an SRS with a variance of $\sigma^2/10,000$. Thus, in this case, cluster sampling is the better choice since it will produce an estimate with a smaller variance for the same budget.

The trade-offs between the various types of sampling have been well documented in many sampling texts. Also, a number of sampling books provide cost-error optimization formulas for deciding how to select a sample from a population to obtain the lower sampling variance for a given survey budget.

Two-stage cluster sampling begins with a sample of *primary sampling units* (PSUs), which may be selected with equal or unequal probability sampling. Within each PSU, a sample of *secondary stage units* (SSUs) is drawn. If all the units within every PSU is selected (i.e., no second stage sampling), we refer to the method as *single-stage cluster sampling*. In *multistage cluster sampling*, three or more stages of sampling are implemented with sub-sampling within each stage, down to the *ultimate stage unit* (USU) or population element.

9.5.4   Unequal Probability Sampling

SRS, systematic sampling, and stratified random sampling with proportionate allocation are known as *equal probability of selection methods* (epsem). That is, every element or unit in the population is given an equal chance of being selected with these designs. Similarly, a cluster sample where the clusters are chosen by SRS is also an epsem design. Multistage samples can be constructed so as to be epsem by assigning the probabilities of selection judiciously at each stage. For example, a two-stage design in which both the primary sampling units (PSUs) and secondary stage units (SSUs) are selected with SRS, and the sampling fraction within each PSU is the same for all PSUs, is also an epsem design. One nice property of the epsem design is that the sample is *self-weighting*; that is, the sample average is an unbiased estimator of the population mean just as it is in SRS.

However, many cluster samples, multistage samples, and all disproportionate stratified samples are not epsem. For example, suppose that we wish to draw a sample of 1000 persons containing a disproportionate number of persons, say 50% of the sample, who live in large households, whereas in the population only 5% of people live in such households. We construct two strata: one for persons who live in large households and the other for all other persons. Then 500 persons are selected by SRS from each stratum. As we discussed previously, this type of disproportionate stratified sampling is referred to as *oversampling* since the proportion of persons in large households is larger in the sample than in the population. The result of oversampling of persons in large households is that these persons have a higher probability of being selected for the survey than persons in the "other" stratum (i.e., the design is not epsem). As a consequence, the sample mean is no longer an unbiased estimator of the population mean.

In multistage sampling, the primary sampling units (PSU) quite often are selected by a method known as *probability proportional to size* (PPS) *sampling*. This sampling strategy assigns a higher probability of selection to the larger PSUs. For example, in a multistage sample of students where the PSUs are schools, a school having 5000 students would be assigned a probability of selection that is 10 times larger than the probability assigned to a school having

500 students. Even though the PSUs are selected with unequal probability, the overall probabilities of selection for the students may still be epsem. If the same number of students is selected for the sample in each school regardless of the probability of selection of the school, the sample will be epsem. However, if the within-school sample sizes vary by school, the overall selection probabilities for the students will not be epsem.

Finally, unequal probability sampling can occur for cluster sampling as well. If a cluster sample is selected with PPS, the design is not epsem since elements in larger clusters will have a higher probability of selection than will elements in smaller clusters.

9.5.5 Sample Weighting and Estimation

As noted previously, when the sample design is not epsem (i.e., it is not self-weighting), the sample average is not an unbiased estimator of the population mean. In this case, the values of the characteristics for sample units are weighted prior to averaging to compensate for the unequal selection probabilities. The weight to be applied to sample unit i, denoted w_i, is defined as the reciprocal of the probability of selecting the unit into the sample, π_i:

$$w_i = \frac{1}{\pi_i} \tag{9.20}$$

and the estimator of the population mean, $\overline{Y}$, is then

$$\overline{y}_w = \frac{\sum w_i y_i}{\sum w_i} \tag{9.21}$$

In addition to correcting the sample mean for unequal selection probabilities, weighting is also used for postsurvey adjustments to compensate for nonresponse and frame errors. This use of weighting was discussed to some extent in Chapters 2, 3, and 7. Thus, there are essentially two types of weights, sample selection weights which are also called *base weights* (denoted by w_{Bi}) and *postsurvey adjustment weights* (denoted by w_{Ai}). The final estimation weights, w_{Fi}, are simply the product of these two weights (i.e., $w_{Fi} = w_{Bi}w_{Ai}$). Then instead of using w_i in the estimator (9.21), we use w_{Fi} to obtain the estimator of $\overline{Y}$.

When adjustments for nonresponse and frame coverage are included in the weights, weighting can reduce the bias in the estimates resulting from these sources of nonsampling error. However, a drawback of weighting is that it can increase the variance of an estimate. Usually, the more variable the weights are across the units in the sample, the greater the potential for weighting to increase the variance. Sometimes, the largest weights are reduced, or trimmed, to somewhat attenuate the effects of weighting on the variance.

> *Weights* are required in estimation to compensate for unequal probabilities of selection as well as for nonresponse and frame coverage errors.

9.5.6 Standard Error Estimation and Design Effects

As illustrated in the preceding discussion, simple random sampling serves as a useful benchmark against which to compare other sample designs. A commonly used measure for this comparison is the design effect, the ratio of the variance of the estimator based on complex designs to the variance of the estimator based on an SRS of the same size. For example, let $\bar{y}_w$ denote the estimator of $\bar{Y}$ from some complex sample design, δ. Then the design effect for the estimator $\bar{y}_w$ is

$$\text{Deff}(\bar{y}_w) = \frac{\text{Var}_\delta(\bar{y}_w)}{\text{Var}_{\text{SRS}}(\bar{y})} \tag{9.22}$$

where $\text{Var}_{\text{SRS}}(\bar{y})$ is given by (9.5). The squared root of the Deff is the Deft:

$$\text{Deft}(\bar{y}_w) = \sqrt{\text{Deff}(\bar{y}_w)} \tag{9.23}$$

The Defts for complex survey designs can be applied directly to estimates of the standard errors for purposes of confidence interval estimation when the standard errors are computed using SRS formulas instead of the estimation formulas that are appropriate for the design.

Example 9.5.1   As an example, suppose that the estimate of the population unemployment rate from a labor force survey is 7%. The labor force survey sample design is a complex stratified, multistage design, so computing the variance of this estimate requires considerable computational effort. However, suppose we know from previous implementations of this survey that the Deff for the unemployment rate is approximately 1.83. What is a 95% confidence interval for this estimate if the estimate is based on 10,000 responses?

In (9.9) we gave the formula for an approximately 95% confidence interval. An estimate of the variance of a proportion for a simple random sample from a large population is simply pq/n, where p is the sample proportion, q is $1 - p$, and n is the sample size. Thus, the variance of the estimate from the labor force survey can be estimated by $\text{Deff} \times pq/n$. Substituting these estimates into (9.9), we obtain a 95% confidence interval for a proportion, p, under SRS as follows:

$$\left[p - 2 \times \text{Deft} \times \sqrt{\frac{pq}{n}},\ p + 2 \times \text{Deft} \times \sqrt{\frac{pq}{n}} \right] \tag{9.24}$$

We can use this formula to compute a 95% confidence interval for the labor force rate. The Deft is $\sqrt{1.83} = 1.35$ and $\sqrt{pq/n} = \sqrt{(0.07)(0.93)/10,000}$ = 0.0026. Thus, the lower confidence limit is $0.07 - 2 \times 1.35 \times 0.0026 = 0.07 - 0.0069 = 0.063$, and the upper confidence limit is $0.07 + 0.0069 = 0.077$. An approximate 95% confidence interval expressed as percentages is therefore [6.3, 7.7].

Deff, the design effect, is the ratio of two variances. The numerator variance is the variance of the estimate using formulas for the actual sample design used. The denominator variance is the variance of the estimate using formulas which assume that the sample was selected using SRS. The Deff provides a shortcut to variance estimation since, if we know the Deff, we can use SRS formulas to compute the variance and multiply the result by the Deff.

Finally, design effects can also be applied for estimates of subpopulations and population subgroups. For example, we may wish to estimate unemployment for women or for persons within the age group 18 to 25. In general, Deffs computed for the total population estimates will usually be larger than Deffs computed for subpopulation estimates. This is because, to some extent, Deff reflects the degree of intracluster correlation in the sample, and this correlation has less of an effect on an estimate as the sample size within a cluster is reduced. Therefore, applying the population Deff to estimates of standard errors for population subgroups will produce estimates of standard errors that are somewhat overcorrected for the design effect. Nevertheless, these may still be acceptable for many types of statistical inferences. For more information on the estimation for subpopulation domains, see, for example, Cochran (1977).

9.6 CONCLUDING REMARKS

As this chapter demonstrates, survey sampling is based on well-specified assumptions about the sampling process and the measurement process. Under these assumptions, the theory can predict with great accuracy the precision of a sample of any size and can provide very good inferential results. In most practical situations, the assumptions made for statistical inference do not hold exactly and the formulas developed under these assumptions are in reality approximations. For example, the sample may have been drawn by systematic sampling, but simple random sampling is assumed for the purposes of statistical inference. Standard errors may be computed using simple random sampling formulas that are adjusted by approximate design effects. These approximations are made because they save time and survey costs and usually give very good results. A large part of the skill of a good sampling statistician is the ability to choose the simplest and most cost-effective methods for inference that still give acceptable results. In fact, it can be said that sampling applications is more of an approximate science than it is an exact science.

One important assumption made for classical sampling theory that seldom holds in practice is the assumption that the survey data are free of non-sampling errors. Consequently, the results obtained by the application of classical theory are subject to nonsampling error biases. For this reason, postsurvey adjustments are routinely applied to the estimates to compensate for nonsampling errors such as nonresponse and frame undercoverage errors. Although only briefly discussed in this book, postsurvey adjustments such as sample reweighting for unit nonresponse, imputations for item nonresponse, and poststratification for frame undercoverage are an integral part of the estimation process for many large-scale surveys. [See, for example, Cox et al. (1995), Rubin (1987), Little and Rubin (1987), and Lundström and Särndal (2001).]

Survey sampling is a highly specialized and well-developed component of the survey process, yet we were only able to explore a few basic concepts in this chapter. A number of excellent textbooks on sampling are available which are written for elementary, intermediate, and advanced studies of this science. Some of these are provided in the list of references. However, as most experienced sampling statisticians will attest, practical survey design often involves a complex combination of sampling techniques and estimation methods designed to minimize sampling costs and maximize sampling precision. As such, standard textbook formulas are seldom directly applicable in practice, due to the complexities of the sampling process and the necessity to cut costs in the estimation process. Nevertheless, a thorough grounding in classical sampling theory is a prerequisite for practical sampling design. The "art" of sampling, however, is learned only with years of experience.

CHAPTER 10

Practical Survey Design for Minimizing Total Survey Error

In previous chapters we have treated total survey error components, their sources, and their evaluation. The picture we have painted regarding total survey error might appear rather bleak to some survey designers. We have seen that most survey operations are error-prone, the errors across operations often are cumulative so that many small errors can result in an unacceptably large total error, and it is costly to collect information on error sizes and their effects on estimates. Thus, one question that must be addressed in a book on survey quality is: What strategies should be used in the planning and implementation of surveys to minimize total survey error in practical survey work? In this chapter we attempt to address this question. The discussion will integrate many of the concepts and methods treated in earlier chapters and provide guidance on how to apply that information for the design and conduct of surveys.

There are a number of general strategies to minimize MSE which may be expressed as the "rules of the road" for survey work.

1. Whenever possible, use methods known to be reliable, such as the methods that we have described in earlier chapters.
2. As part of the survey design process, develop a plan for allocating resources to each stage of the survey process that explicitly states the share of the total survey resources that should be devoted to each significant operation in the process. Use available information on survey errors from prior studies and from the literature to allocate the resources optimally to survey operations.
3. Incorporate in the survey design a plan for collecting information on survey quality as the survey progresses. This information will be used to monitor the current survey process as well as to provide information for the design of future surveys.

4. Monitor all processes in conformance to quality standards and reallo-
    cate resources and modify the survey design as needed to respond to
    unanticipated problems and error sources.

5. Disseminate information on data quality to data users as well as to data
    producers so that users become aware of any limitations in the data that
    can affect their decisions and producers will have information on how
    the methods used affected data quality. This latter step is very important
    for the planning and implementation of future surveys.

The remainder of this chapter discusses how to convert these strategies into
a plan of action for survey organizations as well as individual researchers. We
begin by reconsidering some of the ideas of Chapter 2, which viewed the
survey design and implementation as a process.

10.1   BALANCE BETWEEN COST, SURVEY ERROR, AND OTHER QUALITY FEATURES

In Figure 2.1 we viewed survey design and implementation as a process
consisting of a number of steps where the goal of the process is to produce a
statistical product that meets or exceeds requirements for a fixed amount of
resources. The discussion in Chapter 2 concerned the interrelationships among
the stages of this process for one survey and how a decision regarding one
component often has an effect on decisions regarding other components. A
key point in that discussion was that it is not possible simply to apply the best
known methods at every stage, since resources and time are always constraints
in practical survey work. The following provides further examples of that
point:

1. The survey topics are sensitive, which would suggest that a self-
    administration mode would be preferred over interviewer-assisted
    modes. However, a critical research objective is to produce the survey
    results very soon after the survey begins. This would suggest the use of
    a telephone survey. Thus, the survey designer must resolve the conflict
    between timeliness and data quality.

2. The survey budget is such that full coverage of the target population
    cannot be afforded even though the sampling frame has nearly 100%
    coverage. For example, the size of the units in the population may be
    markedly skewed toward smaller units, as is the case in many business
    populations, and the smaller units may contribute very little to the totals
    to be estimated in the sample. The survey designer must determine
    whether a *cutoff sampling* strategy should be implemented (i.e., whether
    all units smaller than some cutoff size should be given a zero probabil-
    ity of selection). This decision will reduce costs but will introduce a

coverage bias that will limit statistical inferences and the generalizability of the survey results.

3. Nonresponse is expected to be quite high in the survey and to obtain acceptable response rates while meeting cost constraints, multiple modes are required. However, the survey topics and the type of questions suggest that one particular mode is the preferred one for reducing measurement errors.

These are examples of design conflicts that are a normal part of survey work. In every survey, compromises to some quality dimensions must be made in order that other quality dimensions and costs are held at acceptable levels. Additional compromises may become necessary as the survey gets under way. For example, (1) during the sampling stage, it may be discovered that a complete frame is too expensive to develop and that one or more incomplete frames must be used instead; (2) a question in a CAPI instrument is found to be awkward and potentially confusing to the respondents, but it cannot be fixed within the design stage time frame, so it is implemented anyway; and (3) the organization has won two major surveys, which are to be conducted simultaneously, and this strains the organization's interviewing capability, requiring the hiring of a relatively large number of totally inexperienced interviewers.

In the discussion of the quality dimensions in Chapter 1, we noted that conflicts between relevance, accuracy, accessibility, comparability, coherence, and completeness are inevitable in designing and implementing a survey. They must be balanced not only with respect to their interdependence but with regard to costs. Even though there are methods and techniques to reduce the error and maximize quality, we can seldom use these methods to their full potential because costs or other quality dimensions will suffer. We must compromise and allow errors to enter into the data to get the survey done on time and within budget. Therefore, it is essential to realize that some errors are more costly to reduce than others, and the same allocation of resources can achieve a much larger improvement on data quality when applied appropriately. This knowledge of the error sources and their reduction as a function of costs must be taken into account not only during design of the survey but also throughout the entire survey process.

A common conflict between quality criteria is *accuracy versus timeliness*, since reduction of error usually requires additional time in the schedule. For instance, reducing nonresponse rates may require more follow-up or time for approval to use monetary incentives which can conflict with the demand for up-to-date survey results. One way to resolve this issue where one dimension is quantitative (the resulting reduction of nonresponse error) and the other qualitative (up-to-date or not so up-to-date) is to compromise both dimensions; for example, do less nonresponse follow-up than is desired in order not to overly delay data delivery. Another way to manage the conflict is to retain the highest standards for both accuracy and timeliness but to publish preliminary estimates on the schedule required and then later provide revised estimates that are more accurate. This is a common strategy in some types of business surveys, where it is very important for customers to get a timely estimate. For example, the U.S. Foreign Trade Survey follows this type of procedure.

Another common conflict occurs in balancing *accuracy* and *relevance*. Many users require information on such a detailed level that accuracy must be sacrificed to meet that demand. There are users who want industry coding on a six-digit level rather than a five- or four-digit level, because their decision processes require greater detail. However, industry coding on such a fine level may not be accurate, due to the lack of detail required for responses to the industry question to code at this level. A compromise might be to offer six-digit coding only for industry groups where this detailed coding can be obtained from respondents with acceptable accuracy.

Another common conflict between relevance and accuracy occurs when users want data for small areas although the survey is designed only for national-level estimates. Thus, estimates for most small areas would be based on so few cases that their precision would be unacceptably low. A way out of this dilemma is to provide direct estimates only for the largest metropolitan areas when the sample sizes are such that precision is adequate. For smaller-area data, indirect estimates are provided, which are estimates based on small-area estimation modeling techniques that pool data across many small areas that are similar.

Another conflict that often occurs is between *comparability* and *accuracy*, particularly the conflict between adopting new, improved technology for continuing surveys. The user needs accurate data, but just as important is the need to preserve the integrity of the data series. As new and improved data collection methodology is developed, pressure develops both internally and externally to the survey organizations to adopt these improvements in order to maintain high standards in data accuracy. However, implementation of these new methods will affect the data series, in that estimates of period-to-period changes due to changes in the population cannot be distinguished between changes in the estimates as a result of reduced nonsampling errors. Consequently, the survey organization may be forced to continue with methodologies that produce estimates which are less accurate than they could be. The organization's reputation may begin to suffer as a result of using methods that are viewed by the survey world as outdated and substandard. One solution is to adopt the new methodologies but to do so using a process that allows users to adjust the estimates from the new survey for the effects of changes in the nonsampling error so that an estimate of change that bridges the old and new designs can be computed. This process of *splicing* together the new series and the old series will usually involve essentially conducting two surveys simultaneously for a while: the old survey as it has been conducted and a new survey using an improved design. This overlapping design will enable the organization to compute adjustment factors for the new data series for converting the new estimates to the old estimates for comparison with historical data. This solution is quite costly, however, due to the increase in data collection costs for the overlapping survey.

Of course, a major conflict that occurs for essentially all surveys, as we have seen throughout this book, is the conflict between survey error and survey costs; for example, the choice between providing for a vigorous program of nonresponse reduction or improving frame coverage; larger sample sizes for some population subgroups or a more thorough training program for interviewers; the use of monetary incentives versus more extensive pretesting of the questionnaire; and many, many other design choices that a survey designer must face throughout the entire survey process.

Conflicts between quality dimensions are best solved by interaction between data producers and data users. In most cases the users (or clients) are the most knowledgeable about what the data needs are and how to prioritize conflicting demands on relevance, accessibility, comparability, coherence, and completeness of the data. However, producers are usually most knowledgeable regarding issues of cost, accuracy, and timeliness and how to achieve the quality goals in these areas. Therefore, it is important that producers and users work together to resolve the conflicts between the quality dimensions and the design alternatives (Holt and Jones, 1998). Since the ultimate level of data accuracy is solely a function of the knowledge and expertise of the producer, the reputation of the survey organization for producing quality survey data is key to the client's comfort level and confidence with the process.

The experienced designer can use a number of strategies to arrive at an acceptable allocation of resources provided that there is enough information on the major error sources for the survey and that a reasonable budget to address these error sources is available. Unfortunately, the literature on resource allocation in surveys is rather sparse. A good discussion of the problem is provided by Linacre and Trewin (1993). These authors promote an extensive use of evaluation studies to determine the most cost-effective approaches for dealing with survey problems. They suggest that the mean squared error of survey estimates is a decreasing function of the resources spent on reducing the error. However, the same expenditure of resources can have a smaller or larger effect on total error as shown in Figure 10.1. The figure shows three curves representing the reduction in the MSE by implementing three strategies or plans, denoted by A, B, and C, for reducing the error. All three result in a reduction, but the reduction using plan A is much less than for plan B or C, and plan C produces the greatest reduction. As more resources are spent, plan C becomes increasingly better than the other two MSE reduction plans. Plan C may represent the use of more effective methods, better allocation of resources, and/or a more appropriate schedule than the other two plans. Note that at any level of expenditures, the design can be optimized in the sense of applying the resources in such a way as to minimize the MSE. Even when resources for survey improvement are scarce, there is still an allocation plan that minimizes the MSE for that level of resources.

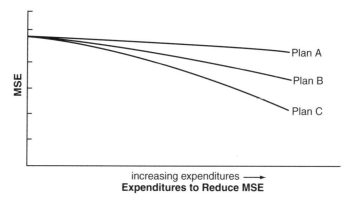

Figure 10.1   Relationship between expenditures to reduce MSE and MSE for three survey plans.

Determining the best strategy (i.e., the one that optimizes survey error subject to costs and timing constraints) usually requires the application of more skill, more information, or both. Dalenius (1971), Linacre and Trewin (1993), Groves (1989), and Fellegi and Sunter (1974) mention a number of problems that impede our ability to optimize surveys:

- A survey organization may not possess the expertise in survey methodology to identify cost-effective ways of minimizing survey error.
- The relationship between cost and error is much more complex than that represented in Figure 10.1. Usually, the way an error source behaves as more resources are directed to it is unknown. For example, we may know that more pretesting of the questionnaire will reduce measurement error, but we do not know how much pretesting is needed to achieve a specified level of accuracy.
- Survey errors are highly interactive across error sources. For example, adding questions to clarify some concept in the questionnaire may reduce measurement error but may also increase respondent burden and reduce the cooperation rate.
- Major surveys are designed to collect information on many items so that any resource allocation model could not possibly be optimal for all survey items. Rather, the focus should be on the most important items in the questionnaire, and even here compromises are likely to be necessary.
- All quality dimensions, not just accuracy and timeliness, and the constraints on them, limit the number of feasible design alternatives.
- It is difficult to determine how much of the survey resources should be devoted to reducing error and how much to the measurement of error and data quality. If we say that all resources should be devoted to reducing error, it may become impossible to optimize future surveys, due to the lack of information on costs, errors, and methods.

- It is not good practice to treat constraints as nonnegotiable boundaries. Quite often, substantial improvements can be realized with a relatively small increase in the overall survey budget. In such cases, a better approach would be to discuss with the client the options and the need for additional resources, explaining the benefits of applying slightly more resources in certain areas of the design. Here it is important for the client to realize that as Deming (1986) warned in his 14 points, always awarding a contract to the lowest bidder is an inferior decision strategy.

Most of the work on modeling error in surveys as a function of costs has been devoted to optimizing sample allocation to reduce sampling error; however, one exception is Groves (1989), who considers cost models with components for sampling and nonsampling error. The gist of his work suggests that the relationships between survey costs and errors are extremely complex, and much of this complexity is unknown. Nevertheless, even very basic models that grossly oversimplify the relationships between cost and error can still offer very good guidance regarding the allocation of resources. Part of the reason for this is that the actual and optimal allocations of resources may be quite different, yet the results can still be quite similar. This is often referred to in the optimization literature as a *flat optimum*.

The quality dimensions always conflict with each other in survey work. *Survey design* is therefore a process of compromise, trading off a loss for one dimension for a gain in another. How to make these design choices is the essence of survey methodology. Design conflicts can usually best be resolved in collaboration between the survey designer and the client or the primary data users.

10.2 PLANNING A SURVEY FOR OPTIMAL QUALITY

Survey planning can take a number of forms, depending on the size of the survey, the type of organization collecting the data, and whether the survey is being planned in response to a request for proposal (RFP) or is a sole-source funded activity. Whatever the situation, planning is an important component of the survey process, just as a road map is an important part of planning for a long trip across a country. Especially for complex survey projects, the survey plan is necessary to make efficient use of the available resources and to allocate resources according to a total survey error minimization strategy. In addition, the plan should describe the objectives of the survey, the technical approach that will be used to achieve the objectives, how the survey will be managed, the deliverables (i.e., interim products during the conduct of the

- Statement of work
  - Target population
  - Survey objectives
- Technical approach
  - Sampling design and procedures
  - Creation of the sampling frame
  - Data collection and prodedures
  - Data processing
  - Database preparation
  - Analysis and reporting
- Management plan
  - Leadership, staff, organizational structure
  - Quality control plan
- Schedule of activities and deliverables
- Budget

Figure 10.2   Example of a typical outline for a survey plan.

survey as well as the final survey data and/or analysis report), and a schedule for the deliverables. Thus, it serves as a means of communicating the survey design to the members of the survey team as well as documenting the design of the survey for future reference. A sample outline of a survey plan appears in Figure 10.2.

The plan should be written in enough detail so that all significant costs for conducting the survey can be estimated. This serves as a check on whether the design of the survey is cost-feasible. If it is not, the plan should be revised by careful redesign of those components that contribute substantially to overall costs while maintaining overall data quality and the survey's ability to achieve the stated objectives. Obviously, this is a critical purpose of the plan since the overall quality of the survey hinges on the survey designers' ability to bring the survey in on time and within budget while maximizing accuracy for the key survey variables.

Despite efforts to plan the survey carefully, it seldom happens that the survey is executed exactly as planned. However, with vigilant quality control systems in place, problems can be discovered early before much damage has been done. For example, in one survey, the CAPI system was programmed in error to an entire section of the questionnaire. This was discovered in the early days of the survey when the frequencies of responses were computed for each question in the questionnaire by the quality control unit. The problem was corrected and new CAPI modules replaced the old, defective version before many cases were affected. Experience has shown repeatedly that early detection and quick response are the keys to successful recovery from unexpected problems during survey implementation.

10.2.1   Use of In-house and External Expertise and Resources

Given the many factors that must be considered in designing a survey, experience and training in the latest techniques for minimizing survey error is a valuable resource in a survey organization. In-house experts in survey methodology should be consulted often during the planning stages of a survey, particularly those with expertise in the type of survey being planned. Obviously, it is much easier to fix a design flaw or inefficiency while the survey is still in the planning stages than after it has been implemented in the field and some data have been collected. It is usually quite difficult to correct major design flaws once the survey is in the field. In addition, a survey that starts off with problems (CAI programs that do not work properly, higher than expected refusal rates, etc.) could lose the confidence and loyalty of the field staff, leading to poor productivity and high nonresponse rates. This problem will add to the existing problems, creating a downward spiral of quality that will be quite difficult to overcome. Meanwhile, irreparable damage will have been done to data quality since the data continue to be collected in turmoil. These problems can be avoided if major problems can be anticipated and dealt with prior to the start of data collection.

One way to expand the knowledge base in survey methods is to participate in professional activities with other survey researchers. For example, statistical conferences are organized on an annual or biannual basis by the American Statistical Association (ASA), the American Association for Public Opinion Research (AAPOR), and the International Statistical Institute (ISI). Parts of these organizations specialize in survey methodology or official statistics. Numerous, one-time conferences on special survey topics are held every year. A list of these can be found in the *AmStat News*, the newsletter for the ASA, as well as the newsletters for other statistical societies.

In addition, international research groups have been organized on specific topics such as nonresponse, editing, database management, and questionnaire design. These are useful for exchanging information on the latest developments in the field. Some survey organizations and individuals have initiated their own informal network of colleagues interested in specific topics. Such networks can be very efficient, leading to information exchange and collaboration. More formal settings include benchmarking activities where organizations get together to study each other's systems and methodologies, resulting in enhanced practices for all involved.

10.2.2   Use of Best Practices Documentation

A number of survey organizations have developed descriptions of the best practices for various survey processes, such as sampling, variance estimation, nonresponse follow-up, editing, and so on. These documents are disseminated throughout the survey organization as standards or guidelines for designing and implementing surveys. When surveys are planned in the organization, these documents are to be consulted so that all surveys are designed consistently and to a high quality standard. These documents are usually referred to as *current best methods* (CBMs), reflecting the intent that the methods change to reflect what is currently regarded as the very best methods for any given survey activity or process. CBMs are sometimes referred to as *best practices* or *standard operating procedures* (SOPs) by some organizations. Even processes such as hiring and training of interviewers, salary adjustments for field staff, documentation and dissemination of survey procedures, and the preparation of budget documents can be addressed with CBMs.

The purpose of CBMs is to ensure that the best practices developed either internally on other surveys within an organization or externally by other organizations, are used by all surveys in the organization. This provides for a consistently high quality output of survey products. If there is considerable variation in the management of survey processes, product quality will not be consistently high in quality. Of course, it is critical that the CBMs represent a consensus within the organization of the current state of the art for any given activity. The process for developing CBMs in an organization is discussed in Section 10.3.2. Once in place, however, CBMs can be an important tool for designing surveys across all parts of the organization for consistent quality.

10.2.3 Applying Findings and Recommendations from the Survey Methods Literature

As described in Section 10.2.2, the survey methods literature is an invaluable resource for identifying optimal methods for survey design. This literature consists of two main bodies of work: textbooks and journal articles, monographs, and other reports. The major textbooks are discussed first. For sampling methodology, key reference texts include: Kish (1965), Cochran (1977), Wolter (1985), and Särndal et al. (1991). These books, like most other books on survey sampling, contain chapters on nonsampling errors, but few books on sampling treat nonsampling errors in any detail.

For discussions of nonsampling error considerations in surveys, notable books include Groves (1989) and Lessler and Kalsbeek (1992). There are also books that treat specific aspects of survey methodology, such as Dillman (1978, 2000) on survey data collection, Groves and Couper (1998) on household survey nonresponse, Wallgren et al. (1999) on graphical presentation of survey results, and Payne (1951), Schuman and Presser (1981), Sudman et al. (1996), and Tourangeau et al. (2000) on cognitive processes and questionnaire design.

Regarding journal articles, monographs, and reports, the literature is abundant. There is a series of edited monographs covering various aspects of survey methodology, all sponsored by the American Statistical Association and other professional societies. The monographs cover panel surveys (Kasprzyk et al., 1989), telephone survey methodology (Groves et al., 1988), measurement errors in surveys (Biemer et al., 1991), business survey methods (Cox et al.,

1995), measurement errors and process quality (Lyberg et al., 1987), computer-assisted survey information collection (Couper et al., 1998), and survey nonresponse (Groves et al., 2002). Other important edited monographs include Madow et al. (1983) on incomplete data in sample surveys, Skinner et al. (1989) on analysis of complex surveys, and Tanur (1992) and Schwarz and Sudman (1992, 1994, 1996) on the cognitive basis of surveys. Many statistical organizations have published monographs or series of methodological publications of general interest. Examples of organizations are the International Statistical Institute, Eurostat, the United Nations, the Food and Agriculture Organization of the United Nations, federal statistical organizations in the United States, OECD, and national statistical offices worldwide. Significant contributions from these organizations have to a large extent been cited in this book. Updated information can always be found on the organizations' Web sites.

Many statistical journals publish articles on survey methodology. Four devoted entirely to the subject are the *Journal of Official Statistics* published by Statistics Sweden, *Survey Methodology* published by Statistics Canada, *Research on Official Statistics* published by Eurostat, and *Proceedings of the American Statistical Association* (Survey Research Methods Section, Social Statistics Section, and Government Statistics Section). Other journals devote some of their space to the subject. Prime examples of these include *Journal of the American Statistical Association, Public Opinion Quarterly, Sociological Methodology, Journal of the Royal Statistical Society* (series A and B), *Sankhyā, Journal of Applied Psychology, International Statistical Review, Bulletin of the International Statistical Institute, Journal of the Market Research Society, Biometrika,* and *American Journal of Public Health.* The U.S. Census Bureau and Statistics Canada publish proceedings from their continuing research conferences and symposia.

A good resource for identifying articles dealing with specific survey topics is the yearly publication *Current Index to Statistics* (CIS), which can be searched by keywords. For example, by searching on the keyword *nonresponse*, it is possible to identify all published work in the CIS database of approximately 100 journals that contain this word in their article titles. It should be noted that work on nonresponse can also be found under headings such as data collection, incentives, response burden, and so on. Also a lot of work is found in unpublished papers written by staff at statistical agencies and other organizations. If the work is of significant importance, it can usually be found in reference lists of work that has been published. Typically, the more publications that cite a particular article, the more important the work is in the field.

Many survey organizations encourage their staff, particularly specialists in survey methodology, to read the literature continuously and to contribute to it frequently. The latter can be difficult, however, since most refereed journals publish less than 25% of all submissions. Nevertheless, the valuable information contained in the published literature cannot be applied in an organization without staff in the organization devoting time and effort in staying current with this literature.

For example, anecdotal evidence might suggest that advance letters increase rather than reduce nonresponse rates. One might argue that an advance letter affords the sample member more time to prepare an excuse not to participate. Although this may be true for some respondents, it has been shown in numerous studies in the literature that this is not the prevailing effect of advance letters. This suggests that intuition and anecdotal experience can be misleading and demonstrates the value of the literature in dispelling incorrect notions about what are good practices. Even if there is no literature that deals exactly with a specific population or survey topic of interest, it is still possible to apply some aspects of the lessons learned from prior research and experimentation to guide intuition about a situation that has not been studied previously.

Still, intuition developed by years of experience with survey work can be very valuable in survey work. This is particularly true for resource allocation decisions where experience is essential for deciding how to allocate the survey budget across the many design, pretesting, data collection, and data processing activities.

Even though it is highly beneficial for the survey researcher to review the literature, this practice can also be quite vexing since there is a great deal of contradictory results in the literature. Some methodological studies are poorly designed and confounding of multiple design factors is not uncommon, especially in nonrefereed material such as conference proceedings and unpublished reports. Misinterpretations of findings occur occasionally as well, since even with well-designed studies, the efficiency or effectiveness of a particular method will depend on the specific survey circumstances. For example, a common finding in survey research is that the same procedures used by different survey organizations for the same survey can produce different results. These differences are sometimes referred to as *house effects* referring to the survey "house" conducting the survey. House effects may be the result of unmeasurable factors such as organizational culture, policies, hiring practices, personnel, and so on. It is up to the survey designer to evaluate how these factors will affect the survey results when using the results from the literature.

> *Survey planning* is a resource-oriented activity that should take advantage of in-house and external expertise, best practices, recommendations from the literature, and well-founded quality guidelines.

10.2.4   Applying Quality Guidelines to the Design of Surveys

Quality guidelines present generally accepted principles for the production of statistics. These are practices that should be followed unless there are very good reasons for not doing so. Guidelines are developed between aspects of design that are considered important and those considered less important regarding the effects on product quality. Quality guidelines have been issued by a number of organizations, including Statistics Canada (1998), the U.K. Office for National Statistics (1997), and the U.S. National Center for Education Statistics (2002). The latter agency combines statistical standards and guidelines into a single document. Following is an example of the subject "achieving acceptable response rates."

Example 10.2.1

Standard 1: The data collection, independent of collection methodology (e.g., whether mailed, over the Internet, or administered by an interviewer either in person or by telephone), must be designed and administered in a manner that encourages respondents to participate.

> *Guideline A.* The method of data collection (e.g., mail, telephone, Internet) should be appropriate for the target population and the objectives of the data collection.
>
> *Guideline B.* The data should be collected at the most appropriate time of the year.
>
> *Guideline C.* The data collection period should be of adequate and reasonable length to achieve good response rates.
>
> *Guideline D.* When appropriate, respondent incentives should be considered.

The American Association for Public Opinion Research (1997) has published a set of 12 guidelines for survey work concerning various aspects of the entire survey process. These are listed in Figure 10.3.

10.2.5 Pretesting and Pilot Surveys

As described in Chapter 8, designing a survey optimally can often require a considerable amount of prior information. By *prior information,* we mean that a survey designer must have a certain degree of knowledge about the population to be studied and even about the characteristics that are the topic of the study. This leads to a paradox since, in principle, the survey designer needs information that will not be available until the survey has been completed. Obviously, this information must be collected in other ways and on a smaller scale than the survey itself. In Chapter 8, a number of methods were discussed for evaluating surveys, including methods for pretesting surveys and choosing between design alternatives. In this section we discuss some of the uses of these methods, particularly the pilot study, for collecting information that can be used for survey planning.

An efficient survey design calls for information on the variability of the population characteristics and data explaining this variability. We have already

---

Quality Guidelines Published by AAPOR

1. Have specific goals for the survey. Objectives should be specific, clear-cut and unambiguous.

2. Consider alternatives to using a survey to collect information. It is not uncommon that certain information needs are best fulfilled by consulting already existing sources, already conducted surveys being one of them.

3. Select samples that well represent the population to be studied. Probability sampling of the right population solves the problem with representativeness.

4. Use designs that balance costs with errors. Consideration must be given to all error sources when budgeting the survey.

5. Take great care in matching question wording to the concepts being measured and the population studied. This is perhaps one of the most important parts of the survey planning process. There are numerous examples of measurement processes that have failed, thereby jeopardizing the entire survey.

6. Pretest questionnaires and procedures to identify problems prior to the survey. It is always better to identify problems ahead of time rather than in the midst of the survey process.

7. Train interviewers carefully on interviewing techniques and the subject matter of the survey. Interviewers who do not have good skills in tracing, motivating, and collecting meaningful data contribute considerably to bad survey quality.

8. Construct quality checks for each step of the survey. Every survey step is a potential contributor to the total survey error and also rework is costly. A quality assurance system in place will guarantee certain quality standards.

9. Maximize cooperation or response rates within the limits of ethical treatment of human subjects. Maximizing cooperation is important but even more important is to minimize the error resulting from nonresponse.

10. Use statistical analytic and reporting techniques appropriate to the data collected. Documentation of all phases of the survey as well as an honest reporting of findings, limitations, and interpretations are crucial to the survey organization's integrity and credibility.

11. Carefully develop and fulfill pledges of confidentiality given to respondents. Any breaches of such pledges can be devastating to the organization insofar that business will suffer and its reputation as well. Usually confidentiality is regulated by statistical acts that are country-specific. The statistician's job is to work out procedures that make disclosure of information on individual sample units virtually impossible.

12. Disclose all methods of the survey to permit evaluation and replication. The documentation should be so detailed that a knowledge research team should be able to replicate a study based on the official documentation.

Figure 10.3 Excerpt from guidelines published by the American Association of Public Opinion Research (1997) for designing surveys.

recognized this need in our discussion of MSE components. We must also have some idea about the errors, costs, and administrative feasibility of the data collection mode and the data processing procedures. For example, the choice between alternative data collection modes and data processing procedures demands extensive knowledge of the advantages and disadvantages of, say, face-to-face versus telephone interviews, optical character recognition versus keying, manual versus automated coding, and dependent versus independent verification. As mentioned earlier, we must be able to make the proper choice of sampling units, sampling system, and estimation system. Without extensive information or knowledge of these methods and procedures, the choices facing the survey researcher become something of a gamble where the researcher hopes for the best possible outcome given his or her expertise, experience, and the general survey conditions. To address this need for information, it is often necessary to conduct pretests and pilot studies specifically designed to generate information that can be used to improve the design of the main survey.

The design and use of pilot studies are sadly neglected in the survey literature. One explanation might be that pilot studies are seen as special cases of ordinary surveys and should be designed as such. However, the problems encountered in the design of pilot studies are different from those encountered in the design of regular sample surveys. The goal of a regular survey or a census is to provide sample estimates or enumerations, whereas the pilot has quite different goals. Design principles for regular surveys are efficient for some pilot survey goals but not for others. For example, pilot studies have the option of using a random or a subjective sample, unlike regular surveys.

The same casual treatment that pilot survey design has received in the literature is also seen in the surveys themselves. Inference is often based on intuition rather than rigorous statistical principles, goals are often loosely defined or vague, and cost-efficiency is seldom an important feature. These are all important aspects of the design of pilot studies that survey designers need to consider. There appears to be no generally accepted terminology in the field of pilot survey design. Such surveys have been referred to as pretests, dress rehearsals, feasibility tests, experiments, embedded experiments, formal tests, informal tests, and methods studies. However, these all have slightly different meanings and emphases (see Chapter 8), yet all these terms have been referred to as *pilot studies*. The following suggested terminology is based on a scan of the literature and personal communications.

Main survey. The survey for which the pilot study activities are performed.

Pretest. Usually, a smaller study using informal qualitative techniques to explore the subject matter and the data collection instrument. Typically, a series of pretests are needed to obtain the information required.

Pilot survey. A survey designed and conducted to obtain information that can improve the main survey. It can be a single survey with multiple goals or a sequence of surveys, each with a limited number of goals. The design depends heavily on the survey's goals but will usually allow for reliable quantitative information and should be conducted at a time when the preliminary design of the main survey can still be adjusted or even changed considerably.

Feasibility study or *feasibility test*. Formal or informal study of methods and procedures conducted when there are doubts or issues related to their practicability. The dividing line between formal and informal studies is not well defined. Basically, formal testing is closely related to experimental design, while informal testing can be very qualitative in nature but still very informative.

Embedded experiments, formal test, and *methodological study.* An embedded experiment, for instance a split–ballot, can be made part of a pilot survey or the main survey to test data collection modes, data or processing systems, and variants of a questionnaire. Such experiments should be strictly designed and usually require large sample sizes.

Dress rehearsal. A miniature of the main survey conducted close to the main survey to reveal weaknesses in the survey design and the survey organization, to provide a base for improving survey methods, to provide survey workers with training and experience, to "prove" the feasibility of the overall operation from start to finish, and to provide realistic data for testing survey operations.

Figure 10.4 lists some design issues that have been evaluated by means of various pilot studies. Of course, it is neither possible nor necessary to have actual data on all these dimensions to produce a good design. As discussed, it is very common to face financial, administrative, methodological, and technical constraints, which automatically reduce the number of design options. Furthermore, the choice of a specific design solution regarding one dimension might affect the number of options for other design dimensions.

The combination of the universe, the test schedule, the survey conditions, and inferential needs determine the test or pilot study that is possible. By and

• Length of recall period	• Feasibility of new equipment
• Choice of mode and mode combinations	• Magnitude of design effects
• Topic sensitivity	• Population variability measures
• Respondent burden	• Interviewer debriefings
• Clarity of concepts and definitions	• Alternative tracing procedures
• Effect of confidentiality pledges	• Extent of editing needed
• Question wording and question context	• Unit and item nonresponse rates
• Questionnaire layout	• Effects of nonresponse rate reduction measures
• Alternative respondent rules	• Expected rates of nonsampling errors due to frames, respondents, interviewers, and data processing
• Time estimates	• Cost components

Figure 10.4 Some topics that have been investigated by pilot studies.

large, accurate estimates require random samples. For instance, the selection of interviewers for formal tests should be a random sample of a pool of interviewers. In formal studies, interviewers should not be assigned on a voluntary basis or assigned because their current workload is light enough to permit extra activities.

One should not put too much faith in the results of pilot studies until the effects of nonrandom sampling, small sample sizes, a limited number of sites or primary sampling units, seasonal variations, and number of alternatives tested are accounted for. The combined effects of pilot study sample size and amount of nonsampling errors can seriously limit the inference. If a pilot survey has many goals, experiments are usually efficient, since different options can be compared simultaneously. For pilot studies, it is usually more sound to draw conclusions from estimation rather than significance tests. Any pilot study should be conducted in a timely fashion so that there is enough time to allow for changes in the final design of the main study. Issues related to pilot studies are also discussed in Brackstone (1976), Jabine (1981), Hunt et al. (1982), Nelson (1985), and Lyberg and Dean (1989).

Pilot studies are necessary when planning information is lacking. They can be conducted in various ways depending on purpose. If a pilot study is deemed necessary by the main survey designer it is important that resources can be set aside for this activity so that the results are timely enough to allow for adjustments of the main survey design.

10.3 DOCUMENTING SURVEY QUALITY

Documenting the survey practice and experience is an important activity for statistical organizations as well as for the field of survey methodology as a whole. The primary purpose of documentation is to communicate the process, procedures, and results from surveys to users of the data as well as other practitioners in the field. It is particularly important for data users since the methods used for collecting data and the limitations of the data will help to prevent misinterpreting the data. In addition, the documentation adds to our knowledge of survey methodology and will help to improve the quality of future surveys. Documentation may take several forms: (1) documentation of survey administrative processes, (2) documentation of recommended or best practices, and (3) quality reporting. These three types of documentation are discussed in this section.

10.3.1 Documentation of Survey Administrative Processes

Earlier in this chapter we discussed the importance of the survey plan for communicating many of the details of the survey process. The survey plan actually defines the quality level of the survey since it describes the activities and levels of effort that will be devoted to each stage of the survey process. In this way, resources are balanced across the various operations involved in the survey in the manner intended by the survey designers. Otherwise, more resources might be consumed during the early stages of the survey, leaving too few resources in the latter stages to achieve the intended quality levels. The plan also delineates the responsibilities of the staff involved and describes how the project team will operate together. The documentation also serves the important role of informing new staff coming onto the project as to the objectives and design of the survey.

As the survey design changes during the implementation stage, which will almost always be the case, the survey plan should be revised accordingly. Thus, the survey plan evolves into the documentation of all the steps in the production process. The survey effort should not be considered complete until the documentation of how all key processes were designed and implemented, including notable problems and successes, is completed. If the survey recurs periodically, this documentation plays an even more crucial role, since improvement work is very difficult to perform without underlying documentation as a basis. This should be written in enough detail so that the process can be understood both by producers and users.

Often, there is a documentation system in place used in all surveys conducted by the organization. Even if there is no such system, the logic of documentation is still relatively simple. For instance, the following would be described for *frame and frame development*:

- Target population and frame population, noting any differences
- Description of the development of the frame and the frame elements
- Information available on the frame
- Process for constructing the frame
- Coverage rates and coverage improvement methods used

For *sampling*, documentation would include:

- Stratification and its purpose
- Sampling design used
- Measures of size used for multistage sampling and how these were constructed
- Sample sizes by stratum and for each sampling stage
- Auxiliary information used
- Anticipated precision in the estimates
- Problems in sampling and any deviations from the sampling plan during implementation

Data collection might include the following:

- Descriptions of modes used
- Procedures to contact sampled units including nonresponse follow-up
- The interview or data collection process
- Interviewer hiring, training, supervision, monitoring, and observation
- Special approaches to reduce nonresponse, including incentives
- Nonresponse rates by key respondent groups, including refusal, non-contact, and other nonresponse rates
- Questionnaires and interview/data collection instructions (possibly in an appendix)

Detailed documentation should be made accessible throughout the organization, and a less detailed version, which emphasizes the major design features, should perhaps be accessible to data users, possibly through the Internet.

10.3.2 Documentation of Recommended and Best Practices

In Section 10.2.2 we discussed the use of CBMs in the survey planning process. As discussed there, a number of survey organizations are creating such documents in an effort to achieve a higher degree of standardization of the statistical production process. The goal, of course, is to adopt standards that are considered "best" in the sense that they represent the most successful, proven methodology. In this section we define some key concepts for defining best practices and for developing CBMs in the organization.

As implied by the name, CBMs should be updated periodically to remain current with new developments in survey methodology. The frequency of these revisions will depend on the rate of progress of research and technological development in the field and the priority given to maintaining the CBMs within the organization.

In Morganstein and Marker (1997) the role of CBMs in the improvement of survey quality is discussed in detail. They state that one of the most frequently identified sources of variation is the difference in performance or even approach among people assigned to do the same task.

Statistics Sweden has developed CBMs for response rate reduction (Japec et al., 1997), editing (Granquist et al., 1997), project work (Statistics Sweden, 1999), disclosure control (Flygare and Block, 2001), questionnaire testing (Lindström et al., 2001) and estimation and nonresponse adjustment (Lundström and Särndal, 2001). Next, as an example, we consider the development of the Swedish nonresponse rate reduction CBM.

The responsibility to control nonresponse in a survey belongs to the survey manager. In recent years, nonresponse rates have not been reduced in most surveys conducted by Statistics Sweden, and for many, they have been increasing. To address this trend, Statistics Sweden decided that a CBM should be developed with the intent of reducing variation in approaches to reducing non-response rates in its surveys.

Since 1986, Statistics Sweden has collected and plotted nonresponse rates for a number of its surveys in a document referred to as the *Nonresponse Barometer*. The number of surveys included in the barometer has increased over the years and it now comprises about 50. The barometer does not, however, contain much information on methodology used for reducing these rates. Thus, it was necessary to collect such data in order to describe how vital processes such as the use of advance letters, data collection strategies, follow-up, questionnaire design, interviewer training, decreasing respondent burden, and the use of incentives are managed for the surveys included in the barometer. Initially, it was assumed that these descriptions could lead to a Pareto analysis, where crucial process steps are identified and best practices for these are described in a CBM.

A study showed that there was a general lack of data on nonresponse and nonrespondents to guide survey managers in their improvement work. The study also showed that procedures and methods varied considerably in similar surveys, even though the general survey conditions were very similar. Finally, the study helped to identify a number of critical and difficult steps in the nonresponse reduction process where guidance was needed.

Armed with these results, work began on developing the CBM. Rather than a "cookbook" approach with step-by-step "recipes" for nonresponse reduction, the CBM took the form of a framework for systematic improvement work by emphasizing the use of known dependable methods and providing guidelines for defining and collecting data on key process variables. This general approach provided some specificity regarding the strategies for nonresponse rate reduction, but also allowed the flexibility needed to accommodate the diverse design parameters across many surveys.

The CBM was developed by a team consisting of six members: three from the research and development department, two statisticians from subject matter departments, and a behavioral scientist specializing in interviewing methodology. The team began by analyzing the study data on the 50 surveys in the barometer mentioned above. Once this was completed, an outline and general contents for the CBM were agreed upon. Much effort was devoted to reviewing the survey methods literature for known, dependable methods and to conducting benchmark studies at other survey organizations. Chapter texts were drafted and reviewed by a group comprising about 15 people from various parts of the organization. A key factor in the success of this work was the very high priority assigned to it by top management.

The resulting CBM is a book consisting of four sections. The first section deals with basic notions such as definitions and calculations of nonresponse rates, reasons and categories of nonresponse, and theories of survey participation. The second section concentrates on what is called the *main processes*, those that are present in virtually all surveys. Processes dealt with include questionnaire design, advance letter design, follow-up procedures, privacy and confidentiality assurance, data collection, and how to combine various measures to achieve low nonresponse rates. The third section concentrates on processes that are not always part of the survey, like handling sensitive questions, respondent burden, interviewer issues, using and administering incentives, using proxy respondents, and administering panel surveys. The fourth and final section provides a framework for identifying and measuring key process variables so that each survey manager can lead his or her own improvement work. Examples of such key process variables are nonresponse rate by sample breakdowns, nonresponse rate by collection mode, tracking hit rate by tracing source, average number of contact attempts, distribution of contact attempts over time, inflow by reminder waves, refusal conversion rate, cost for collecting data on the last 10% of the respondents, and item nonresponse rate per variable.

10.3.3 Quality Reports and Quality Declarations

In various contexts in this chapter, we have discussed the dimensions of survey quality suggested by Eurostat: relevance, accuracy, timeliness, accessibility, comparability, coherence, and completeness. As noted in Chapter 1, this structure for defining quality can be applied to any statistical product, not only surveys. In this section we discuss documents that are intended primarily to provide users of statistical products, surveys in particular, with information on these quality dimensions. Such documents are referred to as quality declarations, quality reports, or quality profiles, although the latter have a broader purpose, as described in Chapter 8.

The primary purposes of the quality declaration are to provide information on the quality characteristics of a product to promote proper use of the product. However, like the quality profile, the quality declaration can also be useful for identifying areas of a survey that are in need of improvement and for improving future surveys. Another framework for quality declarations is the one used for official statistics by Statistics Sweden. It is somewhat simpler than the Eurostat version since, instead of seven dimensions, it includes only five. These are content, accuracy, timeliness, comparability and coherence, and availability and clarity. These are defined as follows:

- *Content* refers to the population parameters estimated by the survey, including target population characteristics, measures, domains, and reference period. This dimension also contains information on comprehensiveness (i.e., how completely the statistical content actually describes the vital aspects of the subject matter field).
- *Accuracy* concerns an overall assessment of total survey error, including various components of the mean squared error.
- *Timeliness* concerns periodicity of the survey, production time, and delivery schedule for key products.

- *Comparability* and *coherence* concern how well different statistics can be used together (i.e., comparability over time, comparability between domains and coherence with other statistics).
- *Availability* and *clarity* concern the physical availability and intellectual clarity of statistics. In this dimension documentation on forms of dissemination, presentation, how to get additional information, access to microdata, and information services are included.

Once these dimensions have been established and defined properly, the quality can be reported. Eurostat has started work on what are called *model quality reports*. A quality report cannot possibly contain information that fully describes the quality of every feature of the statistical product. One reason for this is that it is too expensive to evaluate the total survey error every time a survey is conducted. Further, it may not even be possible to estimate every component of error due to the nature of the phenomenon under study. Thus, a model report aims at producing a declaration that is realistic in terms of methodological and financial resources (see Davies and Smith, 1998). The recommendations regarding the contents of quality reports are the following:

- Produce indicators of survey quality in the absence of actual MSE component estimates. Although they are not direct measures of quality, indicators are by-products of survey processing and are usually strongly correlated with these measures. Examples include weighted and unweighted response rates, frame coverage error rates, and data edit failure rates.
- Quality measures should be produced periodically. Examples are sampling errors, estimates of nonresponse bias, and item reliability.
- Implement a rolling evaluation scheme, where the effect of one or a few error sources are investigated each year. One example could be that in year 1 coverage errors are investigated and reported, in year 2 response errors are evaluated, and in year 3 nonresponse errors are evaluated.
- Document the methods used.

Thus, in practice, the quality report is a mixture of quality estimates and other types of information, such as quality indicators (nonresponse rates, edit failure rates, etc.), pretest results, and metadata (questionnaires, definitions, etc.). For instance, when providing information on the effect of processing errors, it may only be possible to provide estimates for a few, but not all, of the following: variance and bias due to processing errors; rates of processing errors and some methodological notes regarding their estimation; and descriptions of editing, keying, and coding systems along with rates of, say, failed edits for some types of cases.

The accuracy dimension is definitely the most difficult to assess, since there are so many error sources, including sampling, specification, coverage, measurement, nonresponse, and data processing. The other dimensions are easier to handle since they have a metadata character. For instance, it is easy to inform about dates and delays (timeliness), dissemination schemes, publications and databases (accessibility), differences between provisional and final estimates, differences between annual and short-term results, differences between results obtained from different data sources (coherence), and reasons for incompleteness (comprehensiveness).

As discussed in Chapter 8, the *quality profile* brings the quality report one step further. The quality profile is usually much more comprehensive than a quality report and is intended primarily to show where more research is needed to understand total survey error or where changes in the survey design are indicated. As such, a quality profile usually precedes a major redesign of a survey. The reader is referred to Doyle and Clark (2001), Kasprzyk and Kalton (2001), and Chapter 8 for more details.

All documentation efforts have to be planned and implemented over time; otherwise, the task becomes too large to approach. In fact, one common source of reluctance to document quality is that it is usually left until the end of the survey, when resources and time may be the scarcest. Efficient and comprehensive documentation requires a continuing collection of information during the planning and conduct of the survey in such a form that the information can readily be transferred into a document. It is also important that much of the general information and know-how about surveys are stored in a database so that descriptions can be excerpted and reused for multiple documentation efforts in much the same way as proposal writing is accomplished.

Documentation may take several forms, such as documentation of processes, documentation of recommended or best practices, and quality reporting. Documentation is very important for both users and producers.

10.4 ORGANIZATIONAL ISSUES RELATED TO SURVEY QUALITY

10.4.1 Work Environment

The literature on survey quality suggests several characteristics of organizations that appear to be strongly associated with quality of statistical products produced by them. One is the existence of standardized procedures or processes across surveys within the organization. This is not only promoting good quality, as we discussed previously, but is also cost-effective. Standardization ensures that when new findings suggest that major process improvements are possible, the procedures for the entire organization can be revised accordingly in an efficient and uniform manner. Successful organizations almost always encourage and support the uniform implementation of improvements in processes that are deemed to be vital.

A successful organization ensures that important operational deficiencies, large errors, and other failures are analyzed and their root causes understood, so that the true causes of the problems can be addressed. Further, methods of prevention are communicated to the entire organization so that the organization learns from the mistakes. Accordingly, the problems of the system that led to the error should be emphasized, not the involvement in the error of specific persons that may have been responsible for the operation where the error occurred. To do otherwise will tend to discourage open disclosure of problems to the entire organization. If that happens, the likelihood that the error will occur again in other parts of the organization is increased. By the same token, it is also important to share and celebrate successes: proposals that have been won, high response rates that have been achieved, and data that have been delivered successfully to satisfied clients.

It is important to have a work environment that is characterized by collaboration in teams and that utilizes data and lessons learned from previous experiences (see Batcher and Scheuren, 1997). This includes the continuous collection of process data and the use of embedded experiments to advance knowledge in the field. It must be emphasized that collaboration with and knowledge of other statistical organizations both within the country and internationally is extremely useful. The collaboration includes participation in research conferences and network building. The result of these approaches will be a workplace that is engaged in continuous quality improvement (see Morganstein and Hansen, 1990; Lyberg, 2000; Lyberg et al., 2001).

Martin and Straf (1992) have addressed this question of "what constitutes an effective organization" for U.S. federal statistical agencies. With regard to quality and professional standards, they outline six actions which apply not only to government organizations but also to all statistical organizations. According to these authors, an organization should:

- Develop strong staff expertise in the disciplines relevant to its mission as well as in the theory and practice of statistics.
- Develop an understanding of the validity and accuracy of its data and convey the resulting measures of uncertainty to users.
- Undertake ongoing quality assurance programs to improve data validity and reliability and to improve the processes of gathering, compiling, editing, and analyzing data.
- Use modern statistical theory and sound statistical practice in all technical work.
- Develop a strong and continuous relationship with appropriate professional organizations.
- Follow accepted standards in reports and other releases of data on definitions, documentation, descriptions of data collection methodology, measures of uncertainty, and discussions of possible sources of error.

An International Code of Ethics for Survey Workers

1. Statisticians* should guard against predictable misinterpretations or misuse of collected data.

2. Statisticians should use the possibilities to extend the scope of statistical inquiry, and to communicate their findings, for the benefit of the widest possible community.

3. Statisticians should not engage in selecting methods designed to produce misleading results.

4. Statisticians should clarify in advance the respective obligations of employer or sponsor and statistician.

5. Statisticians should assess methodological alternatives impartially.

6. Statisticians should not accept contractual conditions that are contingent upon a particular inquiry outcome.

7. Statistical methods and procedures used should not be kept confidential.

8. Statisticians should permit their methods to be assessed.

9. The advancement of knowledge and the pursuit of information are not themselves sufficient justifications for overriding other social and cultural values.

10. Statistical inquiries involving the active participation of human subjects should be based on their freely given informed consent.

11. On occasions, technical or practical considerations inhibit the achievement of prior informed consent and in these cases the subjects' interests should be safeguarded in other ways.

12. The statisticians should try to minimize disturbance both to subjects themselves and to subjects' relationships with their environment.

13. The identities and records of cooperating or noncooperating·subjects should always be kept confidential.

14. Statisticians should prevent their data from being published in a form that would allow any subject's identity to be disclosed or inferred.

*Note: The term "statistician" is used in the broadest sense to include all survey workers.

Figure 10.5 Excerpt from the International Statistical Institute's Declaration on Professional Ethics—the most general for statistical work. [From the International Statistical Institute (1985).]

10.4.2 Adherence to Ethical Guidelines and Principles

All practical survey work should be guided by the agreed-upon ethical guidelines and principles. Not all survey workers or even survey organizations are aware that there are such guidelines. Many disciplines have had codes of conduct in place for quite some time. The purposes of these have been to list widely held professional values and to guide discussions on how to solve technical and ethical conflicts that might result from sustaining those values.

The Nuremberg Code from 1947 on medical ethics may be, perhaps, the most famous of professional codes, but codes of ethics have been in place in other fields for quite a while. Principles for professional conduct can be found in anthropology, psychology, social research, engineering, business, social work, and market research. Codes of conduct for the statistical and survey professions have been discussed from time to time by the American Statistical Association (1983), the International Statistical Institute (1985), and also some other survey-related organizations. The issues have been discussed extensively in Jowell (1986). There is also a line of code or rule development confined to the production of official statistics. For instance, the United Nations (1994a) has provided a set of fundamental principles of official statistics.

The International Statistical Institute declaration on professional ethics is the most general one for statistical work (Figure 10.5). Here we only list a few principles to suggest the nature of the code and as an example of what is expected from statisticians and survey workers around the world. The reader is referred to the original code for the full text of the code. Note that the term *statistician* is used in the broadest sense to include all persons responsible for planning, conducting, and analyzing surveys.

Note that the survey researcher's or the statistician's ethical responsibility is nontrivial. Especially crucial are the areas of promised use of data and the various ways to convert those who are reluctant to participate to actually participate. Also, it is obvious that proper documentation is very helpful in adhering to many of the guidelines.

The guidelines were adopted by the international statistical community in 1985 as a guide for ethical conduct for all survey researchers. For government employees and others working with official statistics, additional rules apply, and these other thoughts on ethics of official statistics may be found in Gardenier (2000).

References

Ahtiainen, A. (1999), personal communication.

American Association for Public Opinion Research (1997), *Best Practices for Survey and Public Opinion Research and Survey Practices AAPOR Condemns*, AAPOR, Ann Arbor.

American Association for Public Opinion Research (2001), *Standard Definitions: Final Disposition of Case Codes and Outcome Rates for RDD Surveys and In-Person Household Interviews*, AAPOR, Ann Arbor.

American Statistical Association (1983), "Ethical Guidelines for Statistical Practice: Historical Perspective," *The American Statistician*, Vol. 37, No. 1, pp. 1–19.

Anderson, R., Kasper, J., Frankel, M., and associates (1979), *Total Survey Error*, Jossey-Bass, San Francisco.

Bailar, B. A., and Dalenius, T. (1969), "Estimating the Response Variance Components of the U.S. Bureau of the Census' Survey Model," *Sankhyā*, Ser. B, Vol. 31, pp. 341–360.

Bailar, B., Bailey, L., and Stevens, J. (1977), "Measures of Interviewer Bias and Variance," *Journal of Marketing Research*, Vol. 14, pp. 337–343.

Baker, R. P. (1994), "Managing Information Technology in Survey Organizations," *Proceedings of the Annual Research Conference and CASIC Technologies Interchange*, pp. 637–646, U.S. Bureau of the Census, Washington, DC.

Ballou, J., and de Boca, F. K. (1980), "Gender Interaction Effects on Survey Measures in Telephone Interviews," paper presented at the American Association for Public Opinion Research annual conference.

Bassili, J. N. (1996), "The How and Why of Response Latency Measurement in Telephone Surveys," pp. 319–346 in N. Schwarz and S. Sudman (eds.), *Answering Questions: Methodology for Determining Cognitive and Communicative Processes in Survey Research*, Jossey-Bass, San Francisco.

Batcher, M., and Scheuren, F. (1997), "CATI Site Management in a Survey of Service Quality," pp. 573–588 in L. Lyberg, P. Biemer, M. Collins, E. De Leeuw, C. Dippo, N. Schwarz, and D. Trewin (eds.), *Survey Measurement and Process Quality*, Wiley, New York.

Beatty, P., Schechter, S., and Whitaker, K. (1996), "Evaluating Subjective Health Questions: Cognitive and Methodological Investigations," *Proceedings of the Section on Survey Research Methods, American Statistical Association*, pp. 956–961.

377

Bellhouse, D. R. (1998), "London Plague Statistics in 1665," *Journal of Official Statistics*, Vol. 14, No. 2, pp. 207–234.

Berk, M. L., Mathiowetz, N. A., Ward, E. P., and White, A. A. (1987), "The Effect of Prepaid and Promised Incentives: Results of a Controlled Experiment," *Journal of Official Statistics*, Vol. 3, No. 4, pp. 449–457.

Bethlehem, J. (1997), "Integrated Control Systems for Survey Processing," pp. 371–392 in L. Lyberg, P. Biemer, M. Collins, E. De Leeuw, C. Dippo, N. Schwarz, and D. Trewin (eds.), *Survey Measurement and Process Quality*, Wiley, New York.

Bethlehem, J., and van de Pol, F. (1998), "The Future of Data Editing," pp. 201–222 in M. P. Couper, R. P. Baker, J. Bethlehem, C. Z. F. Clark, J. Martin, W. L. Nicholls II, and J. M. O'Reilly (eds.), *Computer Assisted Survey Information Collection*, Wiley, New York.

Bethlehem, J., Keller, W., and Pannekoek, J. (1990), "Disclosure Control of Microdata," *Journal of the American Statistical Association*, Vol. 85, pp. 38–45.

Biemer, P. (1988), "Measuring Data Quality," pp. 273–282 in R. Groves, P. Biemer, L. Lyberg, J. Massey, W. Nicholls II, and J. Waksberg (eds.), *Telephone Survey Methodology*, Wiley, New York.

Biemer, P., and Caspar, R. (1994), "Continuous Quality Improvement for Survey Operations: Some General Principles and Applications," *Journal of Official Statistics*, Vol. 10, pp. 307–326.

Biemer, P., and Fecso, R. (1995), "Evaluating and Controlling Measurement Error in Business Surveys," pp. 257–281 in B. Cox, D. Binder, B. N. Chinnappa, A. Christianson, M. Colledge, and P. Kott (eds.), *Business Survey Methods*, Wiley, New York.

Biemer, P. P., and Forsman, G. (1992), "On the Quality of Reinterview Data with Applications to the Current Population Survey," *Journal of the American Statistical Association*, Vol. 87, No. 420, pp. 915–923.

Biemer, P., and Stokes, L. (1989), "The Optimal Design of Quality Control Samples to Detect Interviewer Cheating," *Journal of Official Statistics*, Vol. 5, No. 1, pp. 23–40.

Biemer, P., and Trewin, D. (1997), "A Review of Measurement Error Effects on the Analysis of Survey Data," pp. 603–632 in L. Lyberg, P. Biemer, M. Collins, E. De Leeuw, C. Dippo, N. Schwarz, and D. Trewin (eds.), *Survey Measurement and Process Quality*, Wiley, New York.

Biemer, P., Groves, R. M., Lyberg, L., Mathiowetz, N., and Sudman, S. (eds.) (1991), *Measurement Errors in Surveys*, Wiley, New York.

Biemer, P., Herget, D., Morton, J., and Willis, G. (2001), "The Feasibility of Monitoring Field Interviewer Performance Using Computer Audio Recorded Interviewing (CARI)," *Proceedings of the Section on Survey Research Methods, American Statistical Association*.

Blom, E., and Lyberg, L. (1998), "Scanning and Optical Character Recognition in Survey Organizations," pp. 499–520 in M. P. Couper, R. P. Baker, J. Bethlehem, C. Z. F. Clark, J. Martin, W. L. Nicholls II, and J. M. O'Reilly (eds.), *Computer Assisted Survey Information Collection*, Wiley, New York.

Bowley, A. L. (1913), "Working Class Households in Reading," *Journal of the Royal Statistical Society*, Vol. 76, pp. 672–691.

Box, G. E. P., Hunter, W. G., and Hunter, J. S. (1978), *Statistics for Experimenters*, Wiley, New York.

Brackstone, G. (1976), *Drawing Inferences from Test Results*, memo, Statistics Canada, Ottawa.

Brackstone, G. (1999), "Managing Data Quality in a Statistical Agency," *Survey Methodology*, Vol. 25, No. 2, pp. 139–149.

Bradburn, N., Sudman, S., and associates (1979), *Improving Interview Method and Questionnaire Design*, Jossey-Bass, San Francisco.

Brewer, M. B., and Lui, L. J. (1996), "Use of Sorting Tasks to Assess Cognitive Structure," pp. 373–387 in N. Schwarz and S. Sudman (eds.), *Answering Questions: Methodology for Determining Cognitive and Communicative Processes in Survey Research*, Jossey-Bass, San Francisco.

Brick, M., and Kalton, G. (1996), "Handling Missing Data in Survey Research," *Statistical Methodology in Medical Research*, Vol. 5, pp. 215–238.

Brooks, C. A., and Bailar, B. A. (1978), *An Error Profile: Employment as Measured by the Current Population Survey*, Statistical Working Paper 3, U.S. Office for Management and Budget, Washington, DC.

Brown, A., Hale, A., and Michaud, S. (1998), "Use of Computer Assisted Interviewing in Longitudinal Surveys," pp. 185–200 in M. P. Couper, R. P. Baker, J. Bethlehem, C. Z. F. Clark, J. Martin, W. L. Nicholls II, and J. M. O'Reilly (eds.), *Computer Assisted Survey Information Collection*, Wiley, New York.

Bushery, J.M. (1981), "Recall Biases for Different Reference Periods in the National Crime Survey," *Proceedings of the Section on Survey Research Methods, American Statistical Association*, pp. 238–273.

Bushnell, D. (1996), "Computer Assisted Occupation Coding," *Proceedings of the Second ASC International Conference*, pp. 165–173, Association for Survey Computing, Chesham, UK.

Campanelli, P., Sturgis, P., and Purdon, S. (1997a), *Can You Hear Me Knocking: An Investigation into the Impact of Interviewers on Survey Response Rates*, Social and Community Planning Research, London.

Campanelli, P., Thomson, K., Moon, N., and Staples, T. (1997b), "The Quality of Occupational Coding in the United Kingdom," pp. 437–453 in L. Lyberg, P. Biemer, M. Collins, E. De Leeuw, C. Dippo, N. Schwarz, and D. Trewin (eds.), *Survey Measurement and Process Quality*, Wiley, New York.

Cannell, C., and Kahn, R. (1968), "Interviewing," pp. 526–595 in G. Lindzey and E. Aronson (eds.), *The Handbook of Social Psychology*, Vol. 2, Addison-Wesley, Reading, MA.

Cannell, C. F., and Oksenberg, L. (1988), "Observation of Behavior in Telephone Interviews," pp. 475–495 in R. Groves, P. Biemer, L. Lyberg, J. Massey, W. Nicholls II, and J. Waksberg (eds.), *Telephone Survey Methodology*, Wiley, New York.

Cannell, C., Groves, R., Magilavy, L., Mathiowetz, N., and Miller, P. (1987), "An Experimental Comparison of Telephone and Personal Health Surveys," *Vital and Health Statistics*, Ser. 2, Vol. 106, pp. 87–138, Public Health Service, Washington, DC.

Carrol, D., Cohen, R., Slider, C., and Thompson, W. (1986), "Use of a Mailed Questionnaire to Augment Response Rates for a Personal Interview Survey," paper presented at the annual meeting of the American Association for Public Opinion Research, May, St. Petersburg, FL.

Carson, C. (2000), "Toward a Framework for Assessing Data Quality," paper presented at Statistical Quality Seminar 2000, December 6–8, Cheju Island, Korea.

Caspar, R., Hubbard, M., Kennedy, J., and Wayne, K. (1993), "Results from Phase I Experimentation of the Alternative NHSDA Questionnaire," RTI Project Report submitted to the Substance Abuse and Mental Health Services Administration, Research Triangle Park, NC.

Chakrabarty, R. P., and Torres, G. (1996), *American Housing Survey: A Quality Profile*, U.S. Department of Housing and Urban Development and U.S. Department of Commerce, Washington, DC.

Chen, B., Creecy, R. H., and Appel, M. V. (1993), "Error Control of Automated Industry and Occupation Coding," *Journal of Official Statistics*, Vol. 9, pp. 729–745.

Christianson, A., and Tortora, R. D. (1995), "Issues in Surveying Businesses: An International Survey," pp. 237–256 in B. Cox, D. Binder, B. N. Chinnappa, A. Christianson, M. Colledge, and P. Kott (eds.), *Business Survey Methods*, Wiley, New York.

Cialdini, R. B. (1984), *Influence: The New Psychology of Modern Persuasion*, Quill (HarperCollins), New York.

Cialdini, R. B. (1990), "Deriving Psychological Concepts Relevant to Survey Participation from the Literatures on Compliance, Helping, and Persuasion," paper presented at the First Workshop on Household Survey Nonresponse, October 15–17, Stockholm, Sweden.

Citteur, C. A. V., and Willenborg, L. C. R. V. (1993), "Public Use Microdata Files: Current Practices at National Statistical Bureaus," *Journal of Official Statistics*, Vol. 9, No. 4, pp. 783–794.

Clayton, R., and Werking, G. S. (1998), "Business Surveys of the Future: The World Wide Web as Data Collection Methodology," pp. 543–562 in M. P. Couper, R. P. Baker, J. Bethlehem, C. Z. F. Clark, J. Martin, W. L. Nicholls II, and J. M. O'Reilly (eds.), *Computer Assisted Survey Information Collection*, Wiley, New York.

Cochran, W. G. (1977), *Sampling Techniques*, 3rd ed., Wiley, New York.

Colledge, M. J. (1995), "Frames and Business Registers: An Overview," pp. 21–47 in B. Cox, D. Binder, B. N. Chinnappa, A. Christianson, M. Colledge, and P. Kott (eds.), *Business Survey Methods*, Wiley, New York.

Colledge, M., and March, M. (1997), "Quality Policies, Standards, Guidelines, and Recommended Practices at National Statistical Agencies," pp. 501–522 in L. Lyberg, P. Biemer, M. Collins, E. De Leeuw, C. Dippo, N. Schwarz, and D. Trewin (eds.), *Survey Measurement and Process Quality*, Wiley, New York.

Collins, M., and Butcher, B. (1982), "Interviewer and Clustering Effects in an Attitude Survey," *Journal of the Market Research Society*, Vol. 25, No. 1, pp. 39–58.

Converse, J. M. (1986), *Survey Research in the United States: Roots and Emergence 1890–1960*, University of California Press, Berkeley, CA.

Converse, J., and Presser, S. (1986), *Survey Questions: Handcrafting the Standardized Questionnaire*, Sage Publications, Thousand Oaks, CA.

Couper, M. P. (1996), "Changes in the Interview Setting Under CAPI," *Journal of Official Statistics*, Vol. 12, No. 3, pp. 301–316.

Couper, M. P., and Nicholls, W. L. (1998), "The History and Development of Computer Assisted Survey Information Collection Methods," pp. 1–21 in M. P. Couper, R. P. Baker, J. Bethlehem, C. Z. F. Clark, J. Martin, W. L. Nicholls II, and J. M. O'Reilly (eds.), *Computer Assisted Survey Information Collection*, Wiley, New York.

Couper, M., Holland, L., and Groves, R. (1992), "Developing Systematic Procedures for Monitoring in a Centralized Telephone Facility," *Journal of Official Statistics*, Vol. 8, No. 1, pp. 63–76.

Couper, M., Hansen, S., and Sadosky, S. A. (1997), "Evaluating Interviewer Use of CAPI Technology," pp. 267–285 in L. Lyberg, P. Biemer, M. Collins, E. De Leeuw, C. Dippo, N. Schwarz, and D. Trewin (eds.), *Survey Measurement and Process Quality*, Wiley, New York.

Couper, M. P., Baker, R. P., Bethlehem, J., Clark, C. Z. F., Martin, J., Nicholls, W. L., II, and O'Reilly, J. M. (eds.) (1998), *Computer Assisted Survey Information Collection*, Wiley, New York.

Couper, M., Traugott, M., and Lamias, M. (2001), "Web Survey Design," *Public Opinion Quarterly*, Vol. 65, No. 2, pp. 230–253.

Cox, B. G., Binder, D. A., Chinnappa, B. N., Christianson, A., Colledge, M. J., and Kott, P. S. (eds.) (1995), *Business Survey Methods*, Wiley, New York.

Creecy, R. H., Masand, B. M., Smith, S. J., and Waltz, D. L. (1992), "Trading MIPS and Memory for Knowledge Engineering," *Communications of the ACM*, Vol. 35, pp. 48–63.

Dalenius, T. (1971), *Principer och metoder för planering av samplingundersökningar* (in Swedish), Intern handbok No. 4, Statistics Sweden, Stockholm, Sweden.

Dalenius, T. (1974), "The Invasion of Privacy in Surveys—An Overview," *Statistisk tidskrift*, Vol. 3.

Dalenius, T. (1985), *Elements of Survey Sampling*, Swedish Agency for Research Cooperation with Developing Countries, Stockholm, Sweden.

Dalenius, T. (1988), *Controlling Invasion of Privacy in Surveys*, Statistics Sweden, Stockholm, Sweden.

Danermark, B., and Swensson, B. (1987), "Measuring Drug Use Among Swedish Adolescents: Randomized Response versus Anonymous Questionnaires," *Journal of Official Statistics*, Vol. 3, pp. 439–448.

Davies, P., and Smith, P. (eds.) (1998), *Model Quality Report in Business Statistics*, final report of SUPCOM project, lot 6, Eurostat, Luxemburg.

De Leeuw, E., and Collins, M. (1997), "Data Collection Methods and Survey Quality: An Overview," pp. 199–220 in L. Lyberg, P. Biemer, M. Collins, E. De Leeuw, C. Dippo, N. Schwarz, and D. Trewin (eds.), *Survey Measurement and Process Quality*, Wiley, New York.

De Leeuw, E., and Hox, J. (1988), "The Effect of Response-Stimulating Factors on Response Rates and Data Quality in Mail Surveys: A Test of Dillman's Total Design Method," *Journal of Official Statistics*, Vol. 4, pp. 241–249.

De Leeuw, E., and van der Zouwen, J. (1988), "Data Quality in Telephone and Face-to-Face Surveys: A Comparative Analysis," pp. 283–299 in R. Groves, P. Biemer, L. Lyberg, J. Massey, W. Nicholls II, and J. Waksberg (eds.), *Telephone Survey Methodology*, Wiley, New York.

DeMaio, T., and Rothgeb, J. (1996), "Cognitive Interviewing Techniques in the Lab and in the Field," pp. 177–196 in N. Schwarz and S. Sudman (eds.), *Answering Questions: Methodology for Determining Cognitive and Communicative Processes in Survey Research*, Jossey-Bass, San Francisco.

Deming, W. E. (1986), *Out of the Crisis*, Cambridge University Press, Cambridge.

De Vries, W. (2001), "Good Practices in Official Statistics," paper presented at the International Conference on Quality in Official Statistics, May, Stockholm, Sweden.

Dielman, L., and Couper, M. (1995), "Data Quality in a CAPI Survey: Keying Errors," *Journal of Official Statistics*, Vol. 11, No. 2, pp. 141–146.

Dijkstra, W. (1987), "Interviewing Style and Respondent Behaviour: An Experimental Study of the Survey-Interview," *Sociological Methods and Research*, Vol. 16, pp. 309–334.

Dijkstra, W., and van der Zouwen, J. (1987), "Styles of Interviewing and the Social Context of the Survey-Interview," pp. 200–211 in H. Hippler, N. Schwarz, and S. Sudman (eds.), *Social Information Processing and Survey Methodology*, Springer-Verlag, New York.

Dijkstra, W., and van der Zouwen, J. (1988), "Types of Inadequate Interviewer Behavior in Survey-Interviews," pp. 24–35 in W. Saris and I. Gallhofer (eds.), *Sociometric Research*, Vol. 1, *Data Collection and Scaling*, St. Martin's Press, New York.

Dillman, D. (1978), *Mail and Telephone Surveys: The Total Design Method*, Wiley-Interscience, New York.

Dillman, D. (1991), "The Design and Administration of Mail Surveys," *Annual Review of Sociology*, Vol. 17, pp. 225–249.

Dillman, D. (2000), *Mail and Internet Surveys: The Tailored Design Method*, Wiley, New York.

Dillman, D. A., Eltinge, J. L., Groves, R. M., and Little, R. J. A. (2002), "Survey Non-response in Design, Data Collection, and Analysis," pp. 3–26 in R. M. Groves, D. A. Dillman, J. L. Eltinge, and R. J. A. Little (eds.), *Survey Nonresponse*, Wiley, New York.

Dodge, H. F., and Romig, H. G. (1944), *Sampling Inspection Tables*, Wiley, New York.

Doyle, P., and Clark, C. Z. F. (2001), "Quality Profiles and Data Users," paper presented at the International Conference on Quality in Official Statistics, May, Stockholm, Sweden.

Economic Commission for Europe (1995), *Glossary of Terms Used in Data Editing*, Work Session on Statistical Data Editing, November 6–9, Working Paper 39, Athens, Greece.

Edwards, S. M., and Cantor, D. (1991), "Toward a Response Model in Establishment Surveys," pp. 211–233 in P. Biemer, R. Groves, L. Lyberg, N. Mathiowetz, and S. Sudman (eds.), *Measurement Errors in Surveys*, Wiley, New York.

Elliott, M. R., and Little, R. J. A. (2000), "Model-Based Alternatives to Trimming Survey Weights," *Journal of Official Statistics*, Vol. 16, No. 3, pp. 191–209.

Eurostat (2000), "Assessment of the Quality in Statistics," Eurostat/A4/Quality/00/General/Standard Report, April 4–5, Luxembourg.

Feather, J. (1973), *A Study of Interviewer Variance*, Department of Social and Preventive Medicine, University of Saskatchewan, Saskatoon, Canada.

Fecso, R. (1991), "A Review of Errors of Direct Observation in Crop Yield Surveys," pp. 327–346 in P. Biemer, R. Groves, L. Lyberg, N. Mathiowetz, and S. Sudman (eds.), *Measurement Errors in Surveys*, Wiley, New York.

Fecso, R., and Pafford, B. (1988), "Response Errors in Establishment Surveys with an Example from an Agribusiness Survey," *Proceedings of the Section on Survey Research Methods, American Statistical Association*, pp. 315–320.

Fellegi, I. (1964), "Response Variance and Its Estimation," *Journal of the American Statistical Association*, Vol. 59, pp. 1016–1041.

Fellegi, I. (1996), "Characteristics of an Effective Statistical System," *International Statistical Review*, Vol. 64, No. 2, pp. 165–187.

Fellegi, I. P., and Brackstone, G. (1999), "Monitoring the Performance of a National Statistical Institute," *Journal of the United Nations Economic Commission for Europe*, Vol. 16, No. 4, pp. 251–266.

Fellegi, I. P., and Holt, D. (1976), "A Systematic Approach to Automatic Edit and Imputation," *Journal of the American Statistical Association*, Vol. 71, pp. 17–35.

Fellegi, I., and Sunter, A. (1974), "Balance Between Different Sources of Survey Errors—Some Canadian Experiences," *Sankhyā*, Ser. C, Vol. 36, pp. 119–142.

Fienberg, S. E., and Tanur, J. M. (1996), "Reconsidering the Fundamental Contributions of Fisher and Neyman on Experimentation and Sampling," *International Statistical Review*, Vol. 64, pp. 237–253.

Fienberg, S. E., and Tanur, J. M. (2001), "History of Sample Surveys," *International Encyclopedia of Social and Behavioral Sciences*, Elsevier Science.

Fienberg, S. E., and Willenborg, L. C. R. J. (1998), "Introduction to the Special Issue: Disclosure Limitation Methods for Protecting the Confidentiality of Statistical Data," *Journal of Official Statistics*, Vol. 14, No. 4, pp. 337–345.

Flygare, A., and Block, H. (2001), *Statistisk röjandekontroll* (in Swedish), Statistics Sweden, Stockholm, Sweden.

Food and Agriculture Organization of the United Nations (1996), "Conducting Agricultural Censuses and Surveys," *FAO Statistical Development Series*, No. 6, FAO, Rome, Italy.

Food and Agriculture Organization of the United Nations (1998), "Multiple Frame Agricultural Surveys," *FAO Statistical Development Series*, No. 10, FAO, Rome, Italy.

Forsman, G., and Schreiner, I. (1991), "The Design and Analysis of Reinterview: An Overview," pp. 279–302 in P. Biemer, R. Groves, L. Lyberg, N. Mathiowetz, and S. Sudman (eds.), *Measurement Errors in Surveys*, Wiley, New York.

Forsyth, B., and Lessler, J. (1991), "Cognitive Laboratory Methods: A Taxonomy," pp. 393–418 in P. Biemer, R. Groves, L. Lyberg, N. Mathiowetz, and S. Sudman (eds.), *Measurement Errors in Surveys*, Wiley, New York.

Fowler, F. J., and Mangione, T. W. (1985), "The Value of Interviewer Training and Supervision," Final Report to the National Center for Health Services Research, Grant 3-R18-HS04189, Hyattsville, MD.

Fowler, F. J., and Mangione, T. W. (1990), *Standardized Survey Interviewing*, Sage Publications, Thousand Oaks, CA.

Franchet, Y. (1999), "Performance Indicators for International Statistical Organisations," *Statistical Journal of the United Nations Economic Commission for Europe*, Vol. 16, No. 4, pp. 241–250.

Freeman, J., and Butler, E. W. (1976), "Some Sources of Interviewer Variance in Surveys," *Public Opinion Quarterly*, Vol. 40, pp. 79–91.

Fuller, W. (1987), *Measurement Error Models*, Wiley, New York.

Gardenier, J. S. (2000), "Ethics of Official Statistics: International Perspectives," U.S. National Center for Health Statistics, Hyattsville, MD.

Geist, J., Wilkinson, R. A., Janet, S., Brother, P. J., Hammond, B., Larsen, N. W., Clear, R. M. S., Matsqui, M. J., Burges, C. J. C., Creecy, R., Hull, J. J., Vogl, T. P., and Wilson, C. L. (1994), *The Second Census Optical Character Recognition Systems Conference: A Report of the National Institute for Standards and Technology*, NIST IR 5452, U.S. Department of Commerce, Washington, DC.

Gillman, D. (2000), "Developing an Industry and Occupation Autocoder for the 2000 Census," paper presented at the annual meetings of the American Statistical Association, Indianapolis, IN.

Granquist, L., and Kovar, J. (1997), "Editing of Survey Data: How Much Is Enough?" pp. 415–435 in L. Lyberg, P. Biemer, M. Collins, E. De Leeuw, C. Dippo, N. Schwarz, and D. Trewin (eds.), *Survey Measurement and Process Quality*, Wiley, New York.

Granquist, L., Andersson, C., Engström, P., Jansson, C., and Ullberg, A. (1997), *Granska effektivt* (in Swedish), Statistics Sweden, Stockholm, Sweden.

Groves, R. (1989), *Survey Errors and Survey Costs*, New York: Wiley.

Groves, R. M., and Couper, M. P. (1993), "Contact-Level Influences on Survey Participation," paper presented at the Fourth International Workshop on Household Survey Nonresponse, September, Bath, U.K.

Groves, R., and Couper, M. (1998), *Nonresponse in Household Interview Surveys*, Wiley-Interscience, New York.

Groves, R. M., and Kahn, R. L. (1979), *Surveys by Telephone: A National Comparison with Personal Interviews*, Academic Press, San Diego, CA.

Groves, R., and McGonagle, K. A. (2001), "A Theory-Guided Interviewer Protocol Regarding Survey Participation," *Journal of Official Statistics*, Vol. 17, No. 2, pp. 249–265.

Groves, R. M., and Tortora, R. D. (1998), "Integrating CASIC into Existing Designs and Organizations: A Survey of the Field," pp. 45–61 in M. P. Couper, R. P. Baker, J. Bethlehem, C. Z. F. Clark, J. Martin, W. L. Nicholls II, and J. M. O'Reilly (eds.), *Computer Assisted Survey Information Collection*, Wiley, New York.

Groves, R., Biemer, P., Lyberg, L., Massey, J., Nicholls, W., II, and Waksberg, J. (1988), *Telephone Survey Methodology*, Wiley, New York.

Groves, R., Fultz, N., and Martin, E. (1991), "Direct Questioning About Comprehension in a Survey Setting," pp. 49–61 in J. Tanur (ed.), *Questions About Questions*, Russell Sage Foundation, New York.

Groves, R. M., Cialdini, R., and Couper, M. P. (1992), "Understanding the Decision to Participate in a Survey," *Public Opinion Quarterly*, Vol. 56, pp. 475–495.

Groves, R. M., Dillman, D. A., Eltinge, J. L., and Little, R. J. A. (eds.) (2002), *Survey Nonresponse*, Wiley, New York.

Hammer, M., and Champy, J. (1995), *Reengineering the Corporation: A Manifesto for Business Revolution*, Nicholas Brealey Publishing, London, U.K.

Hansen, M., Hurwitz, W., and Bershad, M. (1961), "Measurement Errors in Censuses and Surveys," *Bulletin of the International Statistical Institute*, Vol. 38, No. 2, pp. 359–374.

Hansen, M., Hurwitz, W. N., and Pritzker, L. (1964), "The Estimation and Interpretation of Gross Differences and the Simple Response Variance," in C. R. Rao (ed.), *Contributions to Statistics* (presented to P. C. Mahalanobis on the occasion of his 70th birthday), Statistical Publishing Society, Calcutta, India.

Hansen, M., Dalenius, T., and Tepping, B. (1985), "The Development of Sample Surveys of Finite Populations," pp. 327–354 in A. C. Atkinson and S. E. Fienberg (eds.), *A Celebration of Statistics*, The ISI Centenary Volume, Springer-Verlag, New York.

Harris, K. (1974), "Analysis of the Independent Three-Way Verification System in Mortality Medical Coding," memo, U.S. Department of Health, Education, and Welfare, Washington, DC.

Hartley, H. O. (1974), "Multiple Frame Methodology and Selected Applications," *Sankhyā*, Ser. C, Vol. 36, pp. 99–118.

Hatchett, S., and Schuman, H. (1975), "Race of Interviewer Effects Upon White Respondents," *Public Opinion Quarterly*, Vol. 39, No. 4, pp. 523–528.

Hedlin, D. (1993), "A Comparison of Raw and Edited Data of the Manufacturing Survey," unpublished report, Statistics Sweden, Stockholm, Sweden.

Henderson, L., and Allen, D. (1981), "NLS Data Entry Quality Control: The Fourth Followup Survey," National Center for Education Statistics, Office of Educational Research and Improvement, Washington, DC.

Hidiroglou, M. A., Drew, J. D., and Gray, G. B. (1993), "A Framework for Measuring and Reducing Nonresponse in Surveys," *Survey Methodology*, Vol. 19, pp. 81–94.

Hidiroglou, M. A., Särndal, C.-E., and Binder, D. A. (1995a), "Weighting and Estimation in Business Surveys," pp. 477–502 in B. Cox, D. Binder, B. N. Chinnappa, A. Christianson, M. Colledge, and P. Kott (eds.), *Business Survey Methods*, Wiley, New York.

Hidirouglou, M. A., Latouche, M., Armstrong, B., and Gossen, M. (1995b), "Improving Survey Information Using Administrative Records: The Case of the Canadian Employment Survey," *Proceedings of the Bureau of the Census Annual Research Conference*, Suitland, MD, pp. 171–197.

Hogan, H. (1993), "The 1990 Post-Enumeration Survey: Operations and Results," *Journal of the American Statistical Association*, Vol. 88, pp. 1047–1060.

Holt, D., and Jones, T. (1998), "Quality Work and Conflicting Quality Objectives," paper presented at DGINS meeting, May, Eurostat and Statistics Sweden, Stockholm, Sweden.

Horvitz, D. G., and Thompson, D. J. (1952), "A Generalization of Sampling without Replacement from a Finite Universe," *Journal of the American Statistical Association*, Vol. 47, pp. 663–685.

Houston, G., and Bruce, A. G. (1993), "Gred: Interactive Graphical Editing for Business Surveys," *Journal of Official Statistics*, Vol. 9, pp. 81–90.

Hunt, S. D., Sparkman, R. D., and Wilcox, J. B. (1982), "The Pretest in Survey Research: Issues and Preliminary Findings," *Journal of Marketing Research*, Vol. XIX, pp. 269–273.

International Statistical Institute (1985), *Declaration of Professional Ethics*, ISI, Voorburg, The Netherlands.

Ishikawa, K. (1982), *Guide to Quality Control*, 2nd ed., Asian Productivity Organization, Tokyo, Japan.

Jabine, T. B. (1981), "Guidelines and Recommendations for Experimental and Pilot Survey Activities in Connection with the Inter-American Household Survey Program," paper prepared for the Inter-American Statistical Institute, Washington, DC.

Jabine, T. B., and Tepping, B. J. (1973), "Controlling the Quality of Occupation and Industry Data," *Bulletin of the International Statistical Institute*, Vol. 39, pp. 360–392.

Jabine, T., King, K., and Petroni, R. (1990), *Quality Profile for the Survey of Income and Program Participation (SIPP)*, U.S. Bureau of the Census, Washington, DC.

Japec, L., Ahtiainen, A., Hörngren, J., Lindén, H., Lyberg, L., and Nilsson, P. (1997), *Minska bortfallet* (in Swedish), Statistics Sweden, Stockholm, Sweden.

Jenkins, C., and Dillman, D. (1997), "Towards a Theory of Self-Administered Questionnaire Design," pp. 165–196 in L. Lyberg, P. Biemer, M. Collins, E. De Leeuw, C. Dippo, N. Schwarz, and D. Trewin (eds.), *Survey Measurement and Process Quality*, Wiley, New York.

Johnson, A. E. (1995), "Business Surveys as a Network Sample," pp. 219–233 in B. Cox, D. Binder, B. N. Chinnappa, A. Christianson, M. Colledge, and P. Kott (eds.), *Business Survey Methods*, Wiley, New York.

Jones, C., Sebring, P., Crawford, I., Spencer, B., Butz, M., and MacArthur, H. (1986), "High School and Beyond: 1980 Senior Cohort, Second Followup (1984)," Data File User's Manual, CS 85–216, Center for Statistics, Office of Educational Research and Improvement, Washington, DC.

Jowell, R. (1986), "The Codification of Statistical Ethics," *Journal of Official Statistics*, Vol. 2, No. 3, pp. 217–253.

Juran, J. M., and Gryna, Jr, F. M. (1980), *Quality Planning and Analysis*, 2nd ed., McGraw-Hill, New York.

Kahn, R. L., and Cannell, C. F. (1957), *The Dynamics of Interviewing*, Wiley, New York.

Kalton, G., Kasprzyk, D., and McMillen, D. (1989), "Nonsampling Errors in Panel Surveys," in D. Kasprzyk, G. Duncan, G. Kalton, and M. P. Singh (eds.), *Panel Surveys*, Wiley, New York.

Kalton, G., Winglee, M., Krawchuk, S., and Levine, D. (2000), *Quality Profile for SASS: Rounds 1–3: 1987–1995*, (NCES 2000-308), National Center for Education Statistics, U.S. Department of Education, Washington, DC.

Kaplan, R. S., and Norton, D. P. (1996), *The Balanced Scorecard*, Harvard Business School Press, Allston, MA.

Kasprzyk, D., and Kalton, G. (2001), "Quality Profiles in U.S. Statistical Agencies," paper presented at the International Conference on Quality in Official Statistics, May, Stockholm, Sweden.

Kasprzyk, D., Duncan, G., Kalton, G., and Singh, M. P. (eds.) (1989), *Panel Surveys*, Wiley, New York.

Katz, D., and Cantril, H. (1937), "Public Opinion Polls," *Sociometry*, Vol. 1, pp. 155–179.

Keller, W. J. (1994), "Changes in Statistical Technology," in Z. Kennesey (ed.), *The Future of Statistics, An International Perspective*, ISI, Voorburg, The Netherlands.

Keller, W. (1995), "Changes in Statistical Technology," *Journal of Official Statistics*, Vol. 11, No. 1, pp. 115–127.

Kirk, M., Buckles, E., Mims, W., Appel, M., and Johnsen, P. (2001), "Preliminary Results from the Census 2000 Industry and Occupation Coding," U.S. Bureau of the Census, Washington, DC.

Kish, L. (1962), "Studies of Interviewer Variance for Attitudinal Variables," *Journal of the American Statistical Association*, Vol. 57, pp. 92–115.

Kish, L. (1965), *Survey Sampling*, Wiley, New York.

Kish, L. (1995), *The Hundred Years' Wars of Survey Sampling*, Centennial Representative Sampling, Rome, Italy.

Knaus, R. (1987), "Methods and Problems in Coding Natural Language Survey Data," *Journal of Official Statistics*, Vol. 3, pp. 51–60.

Körmendi, E. (1988), "The Quality of Income Information in Telephone and Face to Face Surveys," pp. 341–375 in R. Groves, P. Biemer, L. Lyberg, J. Massey, W. Nicholls, II, and J. Waksberg (eds.), *Telephone Survey Methodology*, Wiley, New York.

Kott, P. S., and Vogel, F. A. (1995), "Multiple-Frame Business Surveys," pp. 185–203 in B. Cox, D. Binder, B. N. Chinnappa, A. Christianson, M. Colledge, and P. Kott (eds.), *Business Survey Methods*, Wiley, New York.

Krosnick, J. A. (1991), "Response Strategies for Coping with the Cognitive Demands of Attitude Measures in Surveys," *Applied Cognitive Psychology*, Vol. 5, pp. 213–236.

Kulka, R. A., and Weeks, M. F. (1988), "Toward the Development of Optimal Calling Protocols for Telephone Surveys: A Conditional Probabilities Approach," *Journal of Official Statistics*, Vol. 4, pp. 319–332.

Lavrakas, P., Stasny, E., and Harpuder, B. (2000), "A Further Investigation of the Last-Birthday Respondent Selection Method and Within Unit Coverage," *Proceedings of the Section on Survey Research Methods, American Statistical Association*.

Lee, L., Brittingham, A., Tourangeau, R., Rasinski, K., Willis, G., Ching, P., Jobe, J., and Black, S. (1999), "Are Reporting Errors Due to Encoding Limitations or Retrieval Failure? Surveys of Child Vaccination as a Case Study," *Journal of Applied Cognitive Psychology*, Vol. 13, pp. 43–63.

Legault, S., and Roumelis, D. (1992), "The Use of Generalized Edit and Imputation Systems (GEIS) for the 1991 Census of Agriculture," Working Paper BSMD-92-010E, Statistics Canada, Ottawa, Canada.

Lepkowski, J., Sadosky, S. A., and Weiss, P. (1998), "Mode, Behavior, and Data Recording Error," pp. 367–388 in M. Couper, R. Baker, J. Bethlehem, C. Clark, J. Martin, W. Nicholls II, and J. O'Reilly (eds.), *Computer Assisted Survey Information Collection*, Wiley, New York.

Lessler, J. T., and Kalsbeek, W. D. (1992), *Nonsampling Errors in Surveys*, Wiley, New York.

Lessler, J., Forsyth, B., and Hubbard, M. (1992), "Cognitive Evaluation of the Questionnaire," pp. 13–52 in C. Turner, J. Lessler, and J. Gfroerer (eds.), *Survey Measurement of Drug Use: Methodological Studies*, National Institute on Drug Use, Rockville, MD.

Linacre, S. J., and Trewin, D. J. (1993), "Total Survey Design—Application to a Collection of the Construction Industry," *Journal of Official Statistics*, Vol. 9, No. 3, pp. 611–621.

Lindström, H., Davidsson, G., Henningsson, B., Björnram, A., and Marklund, H. (2001), *Fråga rätt* (in Swedish), Statistics Sweden, Örebro, Sweden.

Little, R. J. A., and Rubin, D. B. (1987), *Statistical Analysis with Missing Data*, Wiley, New York.

Lohr, S. L. (1999), *Sampling: Design and Analysis*, Duxbury Press, Brooks/Cole Thomson Learning, Pacific Grove, CA.

Lord, F., and Novick, M. R. (1968), *Statistical Theories of Mental Test Scores*, Addison-Wesley, Reading, MA.

Lundström, S., and Särndal, C.-E. (2001), *Estimation in the Presence of Nonresponse and Frame Imperfections*, Statistics Sweden, Örebro, Sweden.

Luppes, M. (1994), "Interpretation and Evaluation of Advance Letters," paper presented at the Fifth International Workshop on Household Survey Nonresponse, Ottawa, Canada.

Luppes. M. (1995), "A Content Analysis of Advance Letters from Expenditure Surveys of Seven Countries," *Journal of Official Statistics*, Vol. 11, No. 4, pp. 461–480.

Lyberg, I., and Lyberg, L. (1991), "Nonresponse Research at Statistics Sweden," *Proceedings of the Section on Survey Research Methods, American Statistical Association*.

Lyberg, L. (1981), "Control of the Coding Operation in Statistical Investigations—Some Contributions," *Urval*, No. 13, Statistics Sweden, Stockholm, Sweden.

Lyberg, L. (1991), "Reducing Nonresponse Rates in Household Expenditure Surveys by Forming Ad Hoc Task Forces," paper presented at the Workshop on Diary Surveys, February 18–20, Stockholm, Sweden.

Lyberg, L. (2000), "Recent Advances in the Management of Quality in Statistical Organizations," paper presented at the Statistical Quality Seminar 2000, December 6–8, Cheju Island, Korea.

Lyberg, L. (2002), "Training of Survey Statisticians in Government Agencies—A Review," invited paper presented at the Joint Statistical Meetings, American Statistical Association, New York.

Lyberg, L., and Cassel, C. (2001), "Survey Sampling: The Field," *International Encyclopedia of Social and Behavioral Sciences*, Elsevier.

Lyberg, L., and Dean, P. (1989), "The Design of Pilot Studies: A Short Review," *R&D Reports*, No. 22, Statistics Sweden, Stockholm, Sweden.

Lyberg, L., and Dean, P. (1992), "Automated Coding of Survey Responses: An International Review," *R&D Reports*, No. 2, Statistics Sweden, Stockholm, Sweden.

Lyberg, L., and Kasprzyk, D. (1991), "Data Collection Methods and Measurement Error," in P. Biemer, R. Groves, L. Lyberg, N. Mathiowetz, and S. Sudman (eds.), *Measurement Errors in Surveys*, pp. 237–257, Wiley, New York.

Lyberg, L., and Kasprzyk, D. (1997), "Some Aspects of Post-Survey Processing," in L. Lyberg, P. Biemer, M. Collins, E. De Leeuw, C. Dippo, N. Schwarz, and D. Trewin (eds.), *Survey Measurement and Process Quality*, pp. 353–370, Wiley, New York.

Lyberg, L., Felme, S., and Olsson, L. (1977), *Kvalitetsskydd av data*, Liber (in Swedish), Stockholm, Sweden.

Lyberg, L., Japec, L., and Biemer, P. (1998), "Quality Improvement in Surveys—A Process Perspective," *Proceedings of the Section on Survey Research Methods, American Statistical Association*, pp. 23–31.

Lyberg, L., Biemer, P., Collins, M., De Leeuw, E., Dippo, C., Schwarz, N., and Trewin, D. (eds.) (1997), *Survey Measurement and Process Quality*, New York: Wiley-Interscience.

Lyberg, L. et al. (2001), "Report from the Leadership Group on Quality," paper presented at the International Conference on Quality in Official Statistics, May, Stockholm, Sweden.

Madansky, A. (1986), "On Biblical Censuses," *Journal of Official Statistics*, Vol. 2, No. 4, pp. 561–569.

Madow, W. G., Nisselson, H., and Olkin,I. (eds.) (1983), *Incomplete Data in Sample Surveys*, Volumes 1–3, Academic Press.

Mahalanobis, P. C. (1946), "Recent Experiments in Statistical Sampling in the Indian Statistical Institute," *Journal of the Royal Statistical Society*, Vol. 109, pp. 325–378.

Maklan, D., and Waksberg, J. (1988), "Within-Household Coverage in RDD Surveys," pp. 51–69 in R. Groves, P. Biemer, L. Lyberg, J. Massey, W. Nicholls II, and J. Waksberg (eds.), *Telephone Survey Methodology*, Wiley, New York.

Martin, M. E., and Straf, M. (1992), *Principles and Practices for Federal Statistical Agencies*, National Academy Press, Washington, DC.

Massey, J. T., and Botman, S. L. (1988), "Weighting Adjustments for Random Digit Dialed Surveys," pp. 143–160 in R. Groves, P. Biemer, L. Lyberg, J. Massey, W. Nicholls II, and J. Waksberg (eds.), *Telephone Survey Methodology*, Wiley, New York.

Mathiowetz, N., and Cannell,C. (1980), "Coding Interviewer Behavior as a Method of Evaluating Performance," *Proceedings of the Section on Survey Research Methods, American Statistical Association*, pp. 525–528.

Moore, J. (1988), "Self/Proxy Response Status and Survey Response Quality—A Review of the Literature," *Journal of Official Statistics*, Vol. 4, pp. 155–172.

Morganstein, D. R., and Hansen, M. (1990), "Survey Operations Processes: The Key to Quality Improvement," pp. 91–104 in G. E. Liepins and V. R. R. Uppuluri (eds.), *Data Quality Control*, Marcel Dekker, New York.

Morganstein, D., and Marker, D. A. (1997), "Continuous Quality Improvement in Statistical Agencies," pp. 475–500 in L. Lyberg, P. Biemer, M. Collins, E. De Leeuw, C. Dippo, N. Schwarz, and D. Trewin (eds.), *Survey Measurement and Process Quality*, Wiley, New York.

Morton-Williams, J. (1993), *Interviewer Approaches*, Dartmouth Publishing Company, Hanover, NH.

Mudryk, W., Burgess, M. J., and Xiao, P. (1996), "Quality Control of CATI Operations in Statistics Canada," *Proceedings of the Section on Survey Research Methods, American Statistical Association*, pp. 150–159.

Mudryk, W., Bougie, B., and Xie, H. (2001a), "Quality Control of ICR Data Capture: 2001 Canadian Census of Agriculture," paper presented at the International Conference on Quality in Official Statistics, May, Stockholm, Sweden.

Mudryk, W., Bougie, B., Xiao, P., and Yeung, A. (2001b), *Statistical Methods in Quality Control at Statistics Canada*, Course Reference Manual, Statistics Canada, Ottawa.

Mudryk, W., Bougie, B., and Xiao, P. (2002), personal communication.

Muscio, B. (1917), "The Influence of the Form of a Question," *British Journal of Psychology*, Vol. 8, pp. 351–389.

Nealon, J. (1983), "The Effects of Male vs. Female Telephone Interviewers," Statistical Research Division, U.S. Department of Agriculture, Washington, DC.

Nelson, D. D. (1985), "Informal Testing as a Means of Questionnaire Development," *Journal of Official Statistics*, Vol. 1, No. 2, pp. 179–188.

Neyman, J. (1934), "On the Two Different Aspects of the Representative Method: The Method of Stratified Sampling and the Method of Purposive Selection," *Journal of the Royal Statistical Society*, Vol. 97, pp. 558–606.

Nicholls, W. L. II, Baker, R. P., and Martin, J. (1997), "The Effect of New Data Collection Technologies on Survey Data Quality," pp. 221–248 in L. Lyberg, P. Biemer, M. Collins, E. De Leeuw, C. Dippo, N. Schwarz, and D. Trewin (eds.), *Survey Measurement and Process Quality*, Wiley, New York.

Ogus, J., Pritzker, L., and Hansen, M. H. (1965), "Computer Editing Methods—Some Applications and Results," *Bulletin of the International Statistical Institute*, Vol. 35, pp. 442–466.

O'Muircheartaigh, C. (1997), "Measurement Errors in Surveys: A Historical Perspective," pp. 1–25 in L. Lyberg, P. Biemer, M. Collins, E. De Leeuw, C. Dippo, N. Schwarz, and D. Trewin (eds.), *Survey Measurement and Process Quality*, Wiley, New York.

O'Muircheartaigh, C., and Marckward, A. M. (1980), "An Assessment of the Reliability of World Fertility Study Data," *Proceedings of the World Fertility Survey Conference*, Vol. 3, pp. 305–379, International Statistical Institute, The Hague, The Netherlands.

O'Reilly, J., Hubbard, M., Lessler, J., Biemer, P., and Turner, C. (1994), "Audio and Video Computer Assisted Self-Interviewing: Preliminary Tests of New Technology for Data Collection," *Journal of Official Statistics*, Vol. 10, pp. 197–214.

Payne, S. (1951), *The Art of Asking Questions*, Princeton University Press, Princeton, NJ.

Pfeffermann, D., Skinner, C., Holmes, D. J., Goldstein, H., and Rasbash, J. (1998), "Weighting for Unequal Selection Probabilities in Multilevel Models" (with discussion), *Journal of the Royal Statistical Society*, Ser. B, Vol. 60, pp. 23–56.

Pierzchala, M. (1990), "A Review of the State of the Art in Automated Data Editing and Imputation," *Journal of Official Statistics*, Vol. 6, No. 4, pp. 355–378.

Pierzchala, M. (1995), "Editing Systems and Software," pp. 425–441 in B. Cox, D. Binder, B. N. Chinnappa, A. Christianson, M. Colledge, and P. Kott (eds.), *Business Survey Methods*, Wiley, New York.

Platek, R., and Särndal, C.-E. (2001), "Can a Statistician Deliver?" *Journal of Official Statistics*, Vol. 17, No. 1, pp. 1–20 and discussion, pp. 21–127.

Ponikowski, C. H., and Meily, S. A. (1989), "Controlling Response Error in Establishment Surveys," *Proceedings of the Section on Survey Research Methods, American Statistical Association*, pp. 258–263.

Ramos, M., Sedivi, B. M., and Sweet, E. M. (1998), "Computerized Self-Administered Questionnaires," pp. 389–408 in M. P. Couper, R. P. Baker, J. Bethlehem, C. Z. F. Clark, J. Martin, W. L. Nicholls II, and J. M. O'Reilly (eds.), *Computer Assisted Survey Information Collection*, Wiley, New York.

Redline, C., and Dillman, D. (2002), "The Influence of Alternative Visual Designs on Respondents' Performance with Branching Instructions in Self-Administered

Questionnaires," pp. 179–193 in R. M. Groves, D. A. Dillman, J. L. Eltinge, and R. J. A. Little (eds.), *Survey Nonresponse*, Wiley, New York.

Rice, S. A. (1929), "Contagious Bias in the Interview," *American Journal of Sociology*, Vol. 35, pp. 420–423.

Riviere, P. (2002), "What Makes Business Statistics Special?" *International Statistical Review*, Vol. 70, No. 1, pp. 145–159.

Roman, A. (1981), "Results from the Methods Development Survey (Phase I)," *Proceedings of the Section on Survey Research Methods, American Statistical Association*, pp. 232–237.

Rosén, B., and Elvers, E. (1999), "Quality Concept for Official Statistics," pp. 621–629 in S. Kotz, C. B. Read, and D. L. Banks (eds.), *Encyclopedia of Statistical Sciences*, Update Vol. 3, Wiley, New York.

Rossi, P., Wright, J., and Anderson, A. (eds.) (1983), *Handbook of Survey Research*, Academic Press, San Diego, CA.

Rubin, D. B. (1987), *Multiple Imputation for Nonresponse in Surveys*, Wiley, New York.

Rugg, D. (1941), "Experiments in Wording Questions II," *Public Opinion Quarterly*, Vol. 5, pp. 91–92.

Rustemeyer, A. (1977), "Measuring Interviewer Performance in Mock Interviews," *Proceedings of the Section on Social Statistics, American Statistical Association*, pp. 341–346.

Ryan, T. P. (2000), *Statistical Methods for Quality Improvement*, Wiley, New York.

Särndal, C.-E., Swensson, B., and Wretman, J. (1991), *Model Assisted Survey Sampling*, Springer–Verlag, New York.

Scheuren, F. (ed.) (1999), "What Is a Survey?" *Pamphlet series*, Section on Survey Research Methods, American Statistical Association.

Scheuren, F. (2001), "How Important Is Accuracy?" *Proceedings of Statistics Canada Symposium*, Statistics Canada, Ottawa, Canada.

Schober, M., and Conrad, F. (1997), "Does Conversational Interviewing Reduce Survey Measurement Error?" *Public Opinion Quarterly*, Vol. 61, pp. 576–602.

Scholtes, P., Joiner, B. L., and Streibel, B. J. (1994), *The Team Handbook*, Joiner Associates Inc., Madison, WI.

Schuman, H., and Converse, J. M. (1971), "The Effect of Black and White Interviewers on Black Responses in 1968," *Public Opinion Quarterly*, Vol. 35, No. 1, pp. 44–68.

Schuman, H., and Presser, S. (1981), *Questions and Answers in Attitude Surveys: Experiments on Question Form, Wording, and Context*, Academic Press, San Diego, CA.

Schwarz, N., and Sudman, S. (eds.) (1992), *Context Effects in Social and Psychological Research*, Springer-Verlag, New York.

Schwarz, N., and Sudman, S. (eds.) (1994), *Autobiographical Memory and the Validity of Retrospective Reports*, Springer-Verlag, New York.

Schwarz, N., and Sudman, S. (eds.) (1996), *Answering Questions: Methodology for Determining Cognitive and Communicative Processes in Survey Research*, Jossey-Bass, San Francisco.

Schwarz, N., Hippler, H., Deutsch, B., and Strack, F. (1985), "Response Order Effects of Category Range on Reported Behavior and Comparative Judgments," *Public Opinion Quarterly*, Vol. 49, pp. 388–395.

Shanks, M. (1989), "Information Technology and Survey Research: Where Do We Go from Here?" *Journal of Official Statistics*, Vol. 5, No. 1, pp. 3–21.

Silberstein, A., and Scott, S. (1991), "Expenditure Diary Surveys and Their Associated Errors," pp. 303–326 in P. Biemer, R. Groves, L. Lyberg, N. Mathiowetz, and S. Sudman (eds.), *Measurement Errors in Surveys*, Wiley, New York.

Sinclair, M., and Gastwirth, J. (1998), "Estimates of the Errors in Classification in the Labour Force Survey and Their Effects on the Reported Unemployment Rate," *Survey Methodology*, Vol. 24, No. 2, pp. 157–169.

Singer, E., and Kohnke-Aguire, L. (1979), "Interviewer Expectation Effects: A Replication and Extension," *Public Opinion Quarterly*, Vol. 43, No. 2, pp. 245–260.

Singer, E., Van Hoewyk, J., and Gebler, N. (1999), "The Effect of Incentives on Response Rates in Interviewer-Mediated Surveys," *Journal of Official Statistics*, Vol. 15, pp. 217–230.

Sirken, M. (1970), "Household Surveys with Multiplicity," *Journal of the American Statistical Association*, Vol. 65, pp. 257–266.

Skinner, C. J., Holt, D., and Smith, T. M. F. (eds.) (1989), *Analysis of Complex Surveys*, Wiley, New York.

Snijkers, G. (2002), *Cognitive Laboratory Experiences on Pretesting Computerized Questionnaires and Data Quality*, Ph. D. thesis, University of Utrecht, The Netherlands.

Speizer, H., and Buckley, P. (1998), "Automated Coding of Survey Data," pp. 223–243 in M. P. Couper, R. P. Baker, J. Bethlehem, C. Z. F. Clark, J. Martin, W. L. Nicholls II, and J. M. O'Reilly (eds.), *Computer Assisted Survey Information Collection*, Wiley, New York.

Statistics Canada (1998), *Statistics Canada Quality Guidelines*, Ottawa, Canada.

Statistics Denmark (2000), *Strategy 2005*, Statistics Denmark, Copenhagen, Denmark.

Statistics New Zealand (not dated), *Protocols for Official Statistics*, Statistics New Zealand, Wellington, New Zealand.

Statistics Sweden (1999), *Projektarbete vid SCB* (in Swedish), Statistics Sweden, Stockholm, Sweden.

Stephan, F. F. (1948), "History of the Uses of Modern Sampling Procedures," *Journal of the American Statistical Association*, Vol. 43, pp. 12–39.

Stevens, J., and Bailar, B. (1976), "The Relationship Between Various Interviewer Characteristics and the Collection of Income Data," *Proceedings of the Section on Social Statistics*, American Statistical Association, pp. 785–789.

Suchman, S., and Jordan, B. (1990), "Interactional Troubles in Face to Face Survey Interviews," *Journal of the American Statistical Association*, Vol. 85, pp. 232–241.

Sudman, S., and Bradburn, N. (1982), *Asking Questions: A Practical Guide to Questionnaire Design*, Jossey-Bass, San Francisco.

Sudman, S., Bradburn, N., and associates (1974), *Response Effects in Surveys*, Aldine, Chicago.

Sudman, S., Bradburn, N. M., and Schwarz, N. (1996), *Thinking About Answers: The Applications of Cognitive Processes to Survey Methodology*, Jossey-Bass, San Francisco.

Sudman, S., Willimack, D., Nichols, E., and Mesenbourg, T. L. (2000), "Exploratory Research at the U.S. Census Bureau on the Survey Response Process in Large

Companies," *Proceedings of the Second International Conference on Establishment Surveys*, American Statistical Association, pp. 327–337.

Sweeney, L. (2000), "Uniqueness of Simple Demographics in the U.S. Population," LIDAP-WP4, Laboratory for International Data Privacy, Carnegie Mellon University, Pittsburgh, PA.

Swires-Hennessy, E., and Drake, M. (1992), "The Optimum Time at Which to Conduct Survey Interviews," *Journal of the Market Research Society*, Vol. 34, pp. 61–72.

Sykes, W., and Collins, M. (1988), "Effects of Mode of Interview: Experiments in the U. K." pp. 301–320 in R. Groves, P. Biemer, L. Lyberg, J. Massey, W. Nicholls II, and J. Waksberg (eds.), *Telephone Survey Methodology*, Wiley, New York.

Taguchi, G. (1986), *Introduction to Quality Engineering*, Asian Productivity Center, Tokyo, Japan.

Tanur, J. (ed.) (1992), *Questions About Questions: Inquiries into the Cognitive Bases of Surveys*, Russel Sage Foundation, New York.

Thomas, P. (1994), "Optical Character Recognition and Document Image Processing," U. K. Employment Department, 8th International Round Table on Business Survey Frames, Heerlen, The Netherlands.

Tourangeau, R., and Smith, T. W. (1998), "Collecting Sensitive Information with Different Modes of Data Collection," pp. 431–453 in M. P. Couper, R. P. Baker, J. Bethlehem, C. Z. F. Clark, J. Martin, W. L. Nicholls, II, and J. M. O'Reilly (eds.), *Computer Assisted Survey Information Collection*, Wiley, New York.

Tourangeau, R., Jobe, J. B., Pratt, W. F., Smith, T., and Rasinski, K. (1997), "Design and Results of the Women's Health Study," pp. 344–365 in L. Harrison and A. Hughes (eds.), *The Validity of Self-Reported Drug Use: Improving the Accuracy of Survey Estimates*, National Institute on Drug Abuse, Rockville, MD.

Tourangeau, R., Rips, L., and Rasinski, K. (2000), *The Psychology of Survey Response*, Cambridge University Press, Cambridge, U.K.

Tozer, C., and Jaensch, B. (1994), "CASIC Technologies Interchange, Use of OCR Technology in the Capture of Business Survey Data," *Proceedings of the U.S. Bureau of the Census Annual Research Conference*, Arlington, VA.

Tucker, C. (1992), "The Estimation of Instrument Effects on Data Quality in the Consumer Expenditure Diary Survey," *Journal of Official Statistics*, Vol. 8, No. 1, pp. 41–61.

Tucker, C. (1997), "Methodological Issues Surrounding the Application of Cognitive Psychology in Survey Research," *Bulletin de Methodologie Sociologique*, Vol. 55, pp. 67–92.

Turner, C. F., and Martin, E. (eds.) (1984), *Surveying Subjective Phenomena*, Volumes I and II, Russell Sage Foundation, New York.

Turner, C. F., Ku, L., Sonenstein, F. L., and Platek, J. H. (1996), "Impact of Audio-CASI on Bias in Reporting of Male–Male Sexual Contacts," in R. Warnecke (ed.), *Health Survey Research Methods*, Hyattsville, MD: National Center for Health Statistics.

Turner, C. F., Forsyth, B. H., O'Reilly, J. M., Cooley, P. C., Smith, T. K., Rogers, S. M., and Miller, H. G. (1998), "Automated Self-Interviewing and the Survey Measurement of Sensitive Behaviors," pp. 455–473 in M. P. Couper, R. P. Baker, J. Bethlehem, C. Z. F. Clark, J. Martin, W. L. Nicholls II, and J. M. O'Reilly (eds.), *Computer Assisted Survey Information Collection*, Wiley, New York.

Turner, M. J. (1994), "General Survey Processing Software: An Architectural Model," *Proceedings of the Annual Research Conference*, pp. 596–606, U.S. Bureau of the Census, Washington, DC.

U.K. Office for National Statistics (1997), *Statistical Quality Checklist*, U.K. Government Statistical Service, London, U.K.

United Nations (1994a), Report of the Special Session of the Statistical Commission, April 11–15, E/1994/20, New York.

United Nations (1994b), "Statistical Data Editing: Methods and Techniques," Vol. 1, *Statistical Standards and Studies*, No. 44, United Nations Statistical Commission for Europe, Geneva, Switzerland.

U.S. Bureau of the Census (1974), *Coding Performance in the 1970 Census, Evaluation and Research Program PHC(E)-8*, U.S. Government Printing Office, Washington, DC.

U.S. Bureau of the Census (1977), "1976 Census of Camden, New Jersey," memo, U.S. Department of Commerce, Washington, DC.

U.S. Bureau of the Census (1992), "Marriage, Divorce, and Remarriage in the 1990's," *Current Population Reports*, P-23–120, U.S. Department of Commerce, Washington, DC.

U.S. Bureau of the Census (1993), "Memorandum for Thomas C. Walsh from John H. Thompson, Subject: 1990 Quality Assurance Evaluation," M. Roberts, author, October 18.

U.S. Bureau of the Census (1996), "Census Bureau Strategic Plan 1997–2001," *www.census.gov*.

U.S. Bureau of the Census (2002), "Census Bureau Best Practices for Coding Verification," memo, February 8.

U.S. Department of Labor and U.S. Bureau of the Census (2000), "Current Population Survey: Design and Methodology," *Technical Paper 63*, U.S. Department of Commerce, Washington, DC.

U.S. Energy Information Administration (1996), "Residential Energy Consumption Survey Quality Profile," U.S. Department of Energy, Washington, DC.

U.S. Federal Committee on Statistical Methodology (1980), *Report on Statistical Uses of Administrative Records*, Statistical Policy Working Paper 6, U.S. Office of Management and Budget, Washington, DC.

U.S. Federal Committee on Statistical Methodology (1988), *Quality in Establishment Surveys*, Statistical Policy Working Paper 15, U.S. Office of Management and Budget, Washington, DC.

U.S. Federal Committee on Statistical Methodology (1990), *Data Editing in Federal Statistical Agencies*, Statistical Policy Working Paper 18, U.S. Office of Management and Budget, Washington, DC.

U.S. Federal Committee on Statistical Methodology (2001), *Measuring and Reporting the Quality of Survey Data*, Statistical Policy Working Paper 31, U.S. Office of Management and Budget, Washington, DC.

U.S. National Center for Education Statistics (2002), *NCES Statistical Standards*, *http://nces.ed.gov/statprog/pdf/2002standards.pdf*.

Van den Brakel, J. A. (2001), *Design and Analysis of Experiments Embedded in Complex Sample Surveys*, Erasmus University, Rotterdam, The Netherlands.

Vezina, S. M. (1994), "Imaging and Intelligent Character Recognition," pilot project prepared by Operations and Integration Division, Statistics Canada, Ottawa, Canada.

Wallgren, A., Wallgren, B., Persson, U., Jorner, U., and Haaland, J. (1999), *Graphing Statistics and Data*, Sage Publications, Thousand Oaks, CA.

Warner, S. L. (1965), "Randomized Response: A Survey Technique for Eliminating Evasive Answer Bias," *Journal of the American Statistical Association*, Vol. 60, pp. 63–69.

Weeks, M. F. (1992), "Computer-Assisted Survey Methods Information Collection: A Review of CASIC Methods and Their Implications for Survey Operations," *Journal of Official Statistics*, Vol. 8, No. 4, pp. 445–465.

Wenzowski, M. J. (1996), "Advances in Automated and Computer Assisted Coding Software at Statistics Canada," *Proceedings of the Annual Research Conference and Technology Interchange*, pp. 1117–1126, U.S. Bureau of the Census, Washington, DC.

Willenborg, L., and De Waal, T. (1996), *Statistical Disclosure Control in Practice*, Lecture Notes in Statistics, Vol. 111, Springer-Verlag, New York.

Williams, D. (1942), "Basic Instructions for Interviewers," *Public Opinion Quarterly*, Vol. 6, pp. 634–641.

Williams, E. (1964), "Experimental Comparison of Face to Face and Mediated Communication: A Review," *Psychological Bulletin*, Vol. 84, pp. 963–976.

Wilson, D., and Olesen, E. (2002), "Perceived Race of Interviewer Effects in Telephone Interviews," paper presented at the Annual Meetings of the American Association for Public Opinion Research, St Petersburg, FL.

Wolter, K. M. (1985), *Introduction to Variance Estimation*, Springer-Verlag, New York.

Wolter, K. (1986), "Some Coverage Error Models for Census Data," *Journal of the American Statistical Association*, Vol. 81, pp. 338–346.

Woltman, H. F., Turner, A. G., and Bushery, J. M. (1980), "A Comparison of Three Mixed-Mode Interviewing Procedures in the National Crime Survey," *Journal of the American Statistical Association*, Vol. 75, No. 371, pp. 534–543.

Zanutto, E., and Zaslavsky, A. (2002), "Using Administrative Records to Impute for Nonresponse," pp. 403–415 in R. M. Groves, D. A. Dillman, J. L. Eltinge, and R. J. A. Little (eds.), *Survey Nonresponse*, Wiley, New York.

Zarkovich, S. (1966), *Quality of Statistical Data*, Food and Agricultural Organization of the United Nations, Rome, Italy.

Index

WILEY SERIES IN SURVEY METHODOLOGY
Established in Part by WALTER A. SHEWHART AND SAMUEL S. WILKS

Editors: *Robert M. Groves, Graham Kalton, J. N. K. Rao, Norbert Schwarz, Christopher Skinner*

The *Wiley Series in Survey Methodology* covers topics of current research and practical interests in survey methodology and sampling. While the emphasis is on application, theoretical discussion is encouraged when it supports a broader understanding of the subject matter.

The authors are leading academics and researchers in survey methodology and sampling. The readership includes professionals in, and students of, the fields of applied statistics, biostatistics, public policy, and government and corporate enterprises.

*Now available in a lower priced paperback edition in the Wiley Classics Library.